美 国 数 学 会 经 典 影 印 系 列

出版者的话

近年来，我国的科学技术取得了长足进步，特别是在数学等自然科学基础领域不断涌现出一流的研究成果。与此同时，国内的科研队伍与国外的交流合作也越来越密切，越来越多的科研工作者可以熟练地阅读英文文献，并在国际顶级期刊发表英文学术文章，在国外出版社出版英文学术著作。

然而，在国内阅读海外原版英文图书仍不是非常便捷。一方面，这些原版图书主要集中在科技、教育比较发达的大中城市的大型综合图书馆以及科研院所的资料室中，普通读者借阅不甚容易；另一方面，原版书价格昂贵，动辄上百美元，购买也很不方便。这极大地限制了科技工作者对于国外先进科学技术知识的获取，间接阻碍了我国科技的发展。

高等教育出版社本着植根教育、弘扬学术的宗旨服务我国广大科技和教育工作者，同美国数学会（American Mathematical Society）合作，在征求海内外众多专家学者意见的基础上，精选该学会近年出版的数十种专业著作，组织出版了"美国数学会经典影印系列"丛书。美国数学会创建于 1888 年，是国际上极具影响力的专业学术组织，目前拥有近30000 会员和 580 余个机构成员，出版图书 3500 多种，冯·诺依曼、莱夫谢茨、陶哲轩等世界级数学大家都是其作者。本影印系列涵盖了代数、几何、分析、方程、拓扑、概率、动力系统等所有主要数学分支以及新近发展的数学主题。

我们希望这套书的出版，能够对国内的科研工作者、教育工作者以及青年学生起到重要的学术引领作用，也希望今后能有更多的海外优秀英文著作被介绍到中国。

高等教育出版社

2016 年 12 月

AMS
AMERICAN
MATHEMATICAL
SOCIETY

美国数学会经典影印系列

Partial Differential Equations:

An Accessible Route through Theory and Applications

偏微分方程：理论和应用

András Vasy

高等教育出版社·北京

Contents

Preface

This book is intended as an introduction to partial differential equations (PDE) for advanced undergraduate mathematics students or beginning graduate students in applied mathematics, the natural sciences and engineering. The assumption is that the students either have some background in basic real analysis, such as norms, metric spaces, ODE existence and uniqueness, or they are willing to learn the required material as the course goes on, with this material provided either in the text of the chapters or in the notes at the end of the chapters. The goal is to teach the students PDE in a mathematically complete manner, without using more advanced mathematics, but with an eye toward the larger PDE world that requires more background. For instance, distributions are introduced early because, although conceptually challenging, they are, nowadays, the basic language of PDE and they do not require a sophisticated setup (and they prevent one from worrying too much about differentiation!). Another example is that L^2-spaces are introduced as completions, their elements are shown to be distributions, and the L^2-theory of the Fourier series is developed based on this. This avoids the necessity of having the students learn measure theory and functional analysis, which are usually prerequisites of more advanced PDE texts, but which might be beyond the time constraints of students in these fields.

As for the aspects of PDE theory covered, the goal is to cover a wide range of PDE and emphasize phenomena that are general, beyond the cases which can be studied within the limitations of this book. While first order scalar PDE can be covered in great generality, beyond this the basic tools give more limited results, typically restricted to constant coefficient PDE. Nonetheless, when plausible, more general tools and results, such as energy estimates, are discussed even in the variable coefficient setting. At the end of the book these are used to show solvability of elliptic non-constant coefficient PDE via duality based arguments with the text also providing the basic Hilbert space tools required (Riesz representation).

In terms of mathematical outlook, this book is more advanced than Strauss's classic text [6]—but does not cover every topic Strauss covers— though it shares its general outlook on the field. It assumes much less background than Evans' [1] or Folland's [2] text; Folland's book covers many similar topics but with more assumption on the preparation of the students. For an even more advanced text see Taylor's book [7] (which has some overlaps with this book) which, however, in a sense has a similar outlook on the field: this would be a good potential continuation for students for a second PDE course. This text thus aims for a middle ground; it is hoped that this will bring at least aspects of modern PDE theory to those who cannot afford to go through a number of advanced mathematics courses to reach the latter.

Since PDE theory necessarily relies on basic real analysis as we recall, more advanced topics develop as we progress. Good references for further real analysis background are Simon's book [4] for multivariable calculus and basic real analysis topics, and Johnsonbaugh and Pfaffenberger [3] for the metric space material.

The chapters have many concrete PDE problems, but some of them also have some more abstract real analysis problems. The latter are not necessary for a good understanding of the main material, but give a more advanced overview.

The last two chapters of the text are more advanced than the rest of the book. They cover solvability by duality arguments and variational problems. While no additional background is required since the basic Hilbert space arguments are provided, the reader will probably find these chapters more difficult. However, these chapters do show that even sophisticated PDE theory is within reach after working through the previous chapters!

In practice, in a 10-week quarter at Stanford most of the (main chapter) material in Chapters 1–14 is covered in a very fast-paced manner. In a semester it should be possible to cover the whole book at a fast pace, or most of the book at a more moderate pace.

Chapter 1

Introduction

Partial differential equations (PDE) are ubiquitous in applications of mathematics, and indeed they play an important role in 'pure mathematics' as well. In this course we study these equations, and we start by describing their general types. However, before the outset it is good to keep in mind that while PDE are generalizations of ordinary differential equations (ODE), and thus some ODE notions often have analogues, one should not have expectations such as 'explicit solutions' that are common in ODE theory. Not having such explicit solutions also means that there is a greater emphasis on qualitative features of the equations. Typically this analysis requires some sophisticated tools, of which we can only develop a very small part. Thus often our approach is to study special cases, in which more direct techniques can be applied (e.g. one has explicit or semi-explicit solutions), and then emphasize the features that are still present in more general circumstances. These qualitative tools must also be emphasized even in numerical solutions of PDE, since without this qualitative understanding one may use numerical methods that result in extremely inaccurate (or completely wrong) solutions even if one decreases the step size/mesh size.

1. Preliminaries and notation

First, we recall some notation. Readers for whom this is familiar may want to skip to (1.1). It is useful to start by recalling a bit of linear algebra: the derivative is a linear operator, and linear algebra provides the background for thinking about it. Recall first that a vector space V over a field $\mathbb{F} = \mathbb{R}$ or $\mathbb{F} = \mathbb{C}$ (called the field of scalars) is a set with two operations, the addition of vectors and multiplication of vectors by scalars,

$$+ : V \times V \to V, \qquad \cdot : \mathbb{F} \times V \to V,$$

with the usual properties: there is $0 \in V$ and for $x \in V$ there is $-x \in V$ such that

$$x + y = y + x, \ x + (y + z) = (x + y) + z, \ x + 0 = x, \ x + (-x) = 0$$

for all $x, y, z \in V$, and

$$(\lambda + \mu)x = (\lambda x) + (\mu x), \ \lambda(x + y) = \lambda x + \lambda y, \ (\lambda\mu)x = \lambda(\mu x), \ 1 \cdot x = x$$

for all $\lambda, \mu \in \mathbb{F}$ and for all $x, y \in V$. One says that V is *finite dimensional* if there is a finite basis of V, i.e. a finite set, say $\{e_j\}_{j=1}^n \subset V$, such that elements x of V can uniquely be written as $x = \sum_{j=1}^n x_j e_j$; one says $n = \dim V$. Such a basis *identifies* V with $\mathbb{R}^n$ or $\mathbb{C}^n$; namely, one has a bijective map $V \to \mathbb{R}^n$ (let's take $\mathbb{F} = \mathbb{R}$), $x \mapsto (x_1, \ldots, x_n)$.

If V, W are vector spaces over the field $\mathbb{F} = \mathbb{R}$ or $\mathbb{F} = \mathbb{C}$, a *linear map* $L : V \to W$ is a map satisfying

$$L(x + y) = Lx + Ly, \ L(cx) = cL(x), \ x, y \in V, \ c \in \mathbb{F}.$$

For instance, if $V = \mathbb{R}^n$, $W = \mathbb{R}^k$, and one writes $x = (x_1, \ldots, x_n) \in V$, then linear maps $V \to W$ are given by $k \times n$ matrices $(a_{ij})_{i=1,\ldots,k,j=1,\ldots,n}$ via

$$(Lx)_i = \sum_{j=1}^n a_{ij} x_j.$$

In general, if V and W are finite dimensional, one always has such a matrix representation of a linear map *after one chooses a basis of V and W*, say $\{e_j\}_{j=1}^{\dim V}$ and $\{f_i\}_{i=1}^{\dim W}$. Indeed, elements of V are then written as $x = \sum_j x_j e_j$, and then $Lx = \sum_j x_j Le_j = \sum_j x_j a_{ij} f_i$, where we write $Le_j = \sum a_{ij} f_i$; identifying V with $\mathbb{R}^n$ via the map $x \mapsto (x_1, \ldots, x_n)$ and W with $\mathbb{R}^m$ via $y \mapsto (y_1, \ldots, y_m)$ means that the map $x \mapsto Lx = y$ is given by $(x_1, \ldots, x_n) \mapsto (y_1, \ldots, y_m)$ with $y_i = \sum a_{ij} x_j$.

Also notice that linear maps $\mathbb{R}^n \to \mathbb{R}$ are of the form $Lx = \sum_{j=1}^n a_j x_j$, so the set of these linear maps can be identified with $\mathbb{R}^n$ via identifying L with $a = (a_1, \ldots, a_n)$. The same holds in a vector space V if one chooses a basis of V. However, in general one needs to make a choice for such an identification, and it is best to regard linear maps $V \to \mathbb{F}$ as a vector space V^* distinct from V.

If V or W are infinite dimensional, then one also wants some topology on V and W, and a corresponding notion of continuity for L. A typical setting is if V and W have norms, $\|.\|_V$, resp. $\|.\|_W$. Recall that a *norm* on a vector space V is a map

$$\|.\|_V : V \to [0, \infty)$$

such that
$$\|x\|_V = 0 \text{ if and only if } x = 0$$
and
$$\|\lambda x\|_V = |\lambda| \|x\|_V, \quad \|x + y\|_V \le \|x\|_V + \|y\|_V$$
for all scalars λ and vectors $x, y \in V$, with the last inequality being the *triangle inequality*. In a normed vector space V one says that a sequence x_j *converges* to x if $\|x_j - x\|_V \to 0$ as $j \to \infty$, i.e. if for all $\epsilon > 0$ there exists $N \in \mathbb{N}$ (natural number, i.e. non-negative integer) such that whenever $j \ge N$, one has $\|x_j - x\|_V < \epsilon$.

Notice that $\mathbb{R}^n$ and $\mathbb{C}^n$ are normed spaces with the *Euclidean norms*,
$$\|(x_1, \ldots, x_n)\|_{\mathbb{R}^n} = \left(\sum_{j=1}^n x_j^2 \right)^{1/2}$$
and
$$\|(x_1, \ldots, x_n)\|_{\mathbb{C}^n} = \left(\sum_{j=1}^n |x_j|^2 \right)^{1/2}.$$

In general, one is usually interested in norms *up to equivalence*. That is, two norms $\|.\|_V$ and $\|.\|_V'$ on V are equivalent if there exists $C > 0$ such that for all $x \in V$,
$$C^{-1}\|x\|_V \le \|x\|_V' \le C\|x\|_V.$$
For instance, an equivalent norm on $\mathbb{R}^n$ is
$$\|(x_1, \ldots, x_n)\|_{\mathbb{R}^n}' = \sum_{j=1}^n |x_j|;$$

indeed, on finite dimensional vector spaces all norms are equivalent (but this is false on infinite dimensional spaces!). A main reason for this is that if two norms are equivalent, the notion of convergence is the same for them, i.e. if a sequence converges with respect to one of them, it also converges with respect to the other, directly from the definition.

Turning to linear maps between normed vector spaces V, W, one says that a linear map $L : V \to W$ is *bounded* if there is a constant $C \ge 0$ such that for all $x \in V$,
$$\|Lx\|_W \le C\|x\|_V.$$
One writes $\mathcal{L}(V, W)$ for the set of bounded linear maps from V to W. If V is finite dimensional, then any linear map from V is automatically bounded. Moreover, if V, W are normed vector spaces, $L : V \to W$ linear is bounded if and only if it is continuous, i.e. if and only if for any sequence x_j in V converging to some $x \in V$ (i.e. $\|x_j - x\|_V \to 0$ as $j \to \infty$), one also has $Lx_j \to Lx$ in W (i.e. $\|Lx_j - Lx\|_W \to 0$ as $j \to \infty$); see the appendix to this chapter for more details.

We mention here that bounded linear maps between normed vector spaces themselves form a normed vector space. One adds linear maps point-wise, and similarly for multiplication by scalars: for $A, B \in \mathcal{L}(V, W)$, $\lambda \in \mathbb{F}$, the linear maps $A + B, \lambda A \in \mathcal{L}(V, W)$ are given by

$$(A + B)(x) = Ax + Bx, \quad (\lambda A)(x) = \lambda Ax,$$

and the norm on these is

$$\|A\|_{\mathcal{L}(V,W)} = \sup_{\|x\|_V \leq 1} \|Ax\|_W.$$

Examples of infinite dimensional vector spaces are given by function spaces, such as $C([0, 1]) = C^0([0, 1])$ of continuous functions on the interval $[0, 1]$, with pointwise addition and multiplication by scalars, i.e. for $f, g \in C([0, 1])$, $\lambda \in \mathbb{F}$, $f + g, \lambda f \in C([0, 1])$ are given by

$$(f + g)(x) = f(x) + g(x), \quad (\lambda f)(x) = \lambda f(x),$$

and with the norm

$$\|f\|_{C([0,1])} = \sup_{x \in [0,1]} |f(x)|$$

or bounded continuous functions $C_\infty(\mathbb{R}^n) = C_\infty^0(\mathbb{R}^n)$ on $\mathbb{R}^n$:

$$\|f\|_{C_\infty(\mathbb{R}^n)} = \sup_{x \in \mathbb{R}^n} |f(x)|.$$

(Note that continuous functions on an interval are automatically bounded, i.e. $\sup_{x \in [0,1]} |f(x)|$ is finite. The same is true for any *compact*, i.e. closed and bounded, subset of $\mathbb{R}^n$. See Problem 1.6.) In particular $f_j \to f$ in $C([0, 1])$ as $j \to \infty$ means that

$$\sup_{x \in [0,1]} |f_j(x) - f(x)| \to 0 \text{ as } j \to \infty,$$

i.e. given any $\epsilon > 0$ there is $N > 0$ such that for $j \geq N$, $|f_j(x) - f(x)| < \epsilon$. In other words, convergence in $C([0, 1])$ to f means that for sufficiently large j, the function f_j is within an ϵ-interval around the corresponding values of f, i.e. the graph of f_j is contained in an 'ϵ-strip' around the graph of f.

For a function u on a domain Ω in (i.e. open subset of) $\mathbb{R}^n$ which takes values in $\mathbb{R}$ or $\mathbb{C}$, we say that u is *differentiable* if for each $x \in \Omega$ there exists $L_x : \mathbb{R}^n \to \mathbb{R}$ or $\mathbb{R}^n \to \mathbb{C}$ (real)-linear such that if $E(x, h)$ is given by

(1.1) $$u(x + h) = u(x) + L_x h + E(x, h),$$

i.e.

$$E(x, h) = u(x + h) - u(x) - L_x h,$$

then $\lim_{h \to 0} \frac{|E(x,h)|}{\|h\|} = 0$. The function L_x is called the derivative of u at x and is often written as $(\partial u)(x)$ or $(Du)(x)$. We mostly refrain from the latter notation due to the standard notation used in the Fourier transform, which will be studied later in this book. Also in this book we usually write $|.|$ for the norm on $\mathbb{R}^n$ or $\mathbb{C}^n$ and reserve the double bar $\|.\|$ for norms on function spaces, i.e. we would typically write $\lim_{h\to 0} \frac{|E(x,h)|}{|h|} = 0$.

Writing $x = (x_1, \ldots, x_n)$, the components

$$(\partial_j u)(x) = \frac{\partial u}{\partial x_j}(x) = (\partial u)(x)e_j,$$

where e_j is the jth unit vector, are the *partial derivatives* of u. One says that a function u is *continuously differentiable*, or C^1, if the map $\partial u : \Omega \to \mathcal{L}(\mathbb{R}^n, \mathbb{R})$ is continuous or equivalently if $\partial_j u : \Omega \to \mathbb{R}$ are continuous. Notice that

$$\partial_j u(x) = \lim_{t \to 0} \frac{u(x + te_j) - u(x)}{t}.$$

Conversely, if these limits exist and are continuous functions of x, then indeed u is differentiable (and automatically continuously so), but this requires a bit of work to show (in calculus; see [**4**, Chapter 2.4]).

We remark that the definition of differentiability works equally well if $u : \Omega \to \mathbb{R}^k$ or $u : \Omega \to \mathbb{C}^k$, or indeed $u : \Omega \to V$, where V is either finite dimensional or normed. One requires in the definition that for each $x \in \Omega$ there exists $L_x : \mathbb{R}^n \to V$ (real-)linear such that

$$u(x + h) = u(x) + L_x h + E(x, h),$$

where $\lim_{h\to 0} \frac{\|E(x,h)\|_V}{|h|} = 0$.

If n is small, one often uses labels such as x, y, z, t for the components of the domain (or independent) variable, and correspondingly writes partial derivatives as, say,

$$\partial_z u = \frac{\partial u}{\partial z} = u_z.$$

Higher derivatives are defined similarly, e.g. since ∂u is a $\mathcal{L}(\mathbb{R}^n, \mathbb{R})$- (or $\mathcal{L}(\mathbb{R}^n, \mathbb{C})$-) valued function, which can be identified with an $\mathbb{R}^n$- (or $\mathbb{C}^n$-) valued function, its derivative, $\partial^2 u$, can be defined as above (if it exists). In general, for $\ell \geq 2$, we say that u is C^ℓ if it is $C^{\ell-1}$, and its $(\ell - 1)$st derivative, $\partial^{\ell-1} u$, is C^1. Recall that if u is C^2, then the mixed derivatives satisfy $\partial_i \partial_j u = \partial_j \partial_i u$.

In PDE theory one usually uses the so-called *multiindex* notation. A multiindex is an n-tuple of natural numbers (non-negative integers), $\alpha = (\alpha_1, \ldots, \alpha_n) \in \mathbb{N}^n$. One then writes

$$\partial^\alpha u = \partial_1^{\alpha_1} \ldots \partial_n^{\alpha_n} u$$

for the αth derivative. One writes

$$|\alpha| = \alpha_1 + \ldots + \alpha_n$$

for the total number of derivatives. Thus, $\partial^m u$ can be considered as the collection $\{\partial^\alpha u : |\alpha| = m\}$.

2. Partial differential equations

Suppose $m \in \mathbb{N}$. An mth order PDE is an equation of the form

$$F(x, u, \partial u, \ldots, \partial^m u) = 0,$$

where F is a given function of its arguments. For instance, on $\mathbb{R}^2 = \mathbb{R}_x \times \mathbb{R}_y$, $u_x^2 + u_y^2 = 1$ is a first order PDE; in this case we rewrite it as $u_x^2 + u_y^2 - 1 = 0$, which is of the desired form if

$$F(x, y, z, p, q) = p^2 + q^2 - 1,$$

and we substitute $z = u$, $p = u_x$, $q = u_y$. (Here F happens to be independent of z, as well as of x and y.) This is called an eikonal equation. One has to be somewhat careful, of course, since F identically 0 still gives a (trivial) PDE (of any order) $0 = 0$, so the structure of F is rather important.

The best behaved general class of PDE (within which there are many classes) is *linear* PDE. These are equations of the form

$$\sum_{|\alpha| \leq m} a_\alpha(x)(\partial^\alpha u)(x) = f(x),$$

where a_α and f are given functions. If $f = 0$ (the identically zero function), the PDE is called *homogeneous*; otherwise it is *inhomogeneous*. If the a_α are independent of x (i.e. are constants) the equation is called a *constant coefficient equation*. A typical example is

$$\Delta u = f, \quad \Delta = \sum_{j=1}^{n} \partial_j^2,$$

which is *Laplace's equation*. Here u is for instance the steady-state distribution of heat in the presence of heat sources given by f or the electrostatic potential in the presence of charges f (in appropriate units). Another example is the *wave equation* on $\mathbb{R}^{n+1} = \mathbb{R}_x^n \times \mathbb{R}_t$,

$$\partial_t^2 u - \sum_{j=1}^{n} \partial_j^2 u = f,$$

which describes approximately small vibrations of an n-dimensional membrane, or, from Maxwell's equations, the time-dependent electromagnetic field in the presence of forcing (corresponding to charges and currents). Yet another example is the *heat equation*

$$\partial_t u = k\Delta u.$$

These three equations are examples of the three most common types of linear equations: elliptic, hyperbolic and parabolic. we study these in more detail later.

A general feature of PDEs is that the terms with the highest number of derivatives, called *principal terms*, matter most typically. Thus, when moving to non-linear PDE, it makes a big difference in which terms the non-linearity shows up. An mth order PDE which is of the form

$$\sum_{|\alpha|=m} a_\alpha(x)(\partial^\alpha u)(x) = F(x, u, \partial u, \ldots, \partial^{m-1}u),$$

i.e. where the non-linearity at most enters in $(m-1)$st order terms, is called *semilinear*. Notice that the left-hand side contains the principal terms! For instance,

$$\Delta u = u^2 + \sum_{j=1}^{n}(\partial_j u)^2 + x^2$$

is a semilinear equation. If one allows the coefficients a_α on the left-hand side to depend on u as well as its derivatives up to order $m-1$,

$$\sum_{|\alpha|=m} a_\alpha(x, u, \partial u, \ldots, \partial^{m-1}u)(\partial^\alpha u)(x) = F(x, u, \partial u, \ldots, \partial^{m-1}u),$$

the equation is called *quasilinear*. An example is

$$(1 + u_x^2)u_{xx} + (1 + u_y^2)u_{yy} = f(x, y).$$

Equations which are not quasilinear are called *fully non-linear*.

The general expectation one may have is that semilinear equations can be handled as perturbations of linear equations. An example of this is the Picard iteration in ODE theory, in which one writes an ODE as

$$u_t = F(t, u),$$

and one integrates to

$$u(t) = u(0) + \int_0^t F(s, u(s))\, ds.$$

Here integration is the inverse of differentiation (given $u(0)$), which is the linear term on the left-hand side of the ODE. Given this step (i.e. that one could solve the linear equation by integration) one can have an iterative argument to solve the ODE. (Picard iteration is discussed in the additional material in Chapter 3.) Quasilinear and general non-linear equations are much more delicate.

This example already shows one more important feature of PDEs: typically there are additional conditions one imposes, such as u on a given hypersurface being specified. Typical examples are *Dirichlet boundary conditions* for Laplace's equation in a domain Ω,

$$\Delta u = f, \ u|_{\partial\Omega} = g,$$

where f, g are given functions on Ω, resp. the boundary, $\partial\Omega$; *Neumann boundary boundary conditions*

$$\Delta u = f, \ \left.\frac{\partial u}{\partial n}\right|_{\partial\Omega} = g,$$

where $\frac{\partial u}{\partial n}$ is the normal derivative, $\partial u \cdot \hat{n}$, where $\hat{n}$ is the inward pointing normal vector; and *initial conditions* for the wave or the heat equations on $\mathbb{R}_x^n \times \mathbb{R}_t$ or $\mathbb{R}_x^n \times (0, \infty)_t$,

$$\partial_t^2 u - \Delta u = f, \ u|_{t=0} = \phi, \ u_t|_{t=0} = \psi,$$

resp.

$$\partial_t u - k\Delta u = f, \ u|_{t=0} = \phi,$$

where ϕ and ψ are given functions on the initial hypersurface $\{t = 0\} = \mathbb{R}^n \times \{0\}$ and f is a given function on $\mathbb{R}_x^n \times \mathbb{R}_t$ or $\mathbb{R}_x^n \times (0, \infty)_t$. Again, the additional conditions are called *homogeneous* if the specified data vanish identically; otherwise they are called *inhomogeneous* (as in 'homogeneous Dirichlet boundary conditions').

It is important to discuss what 'solving the PDE' means. Ideally one would like that given a PDE and extra conditions (such as boundary and initial conditions), the solution should

(i) exist (there is a solution),

(ii) be unique (only one solution exists), and

(iii) the solution (which is unique) should depend continuously on various parameters, such as forcing f, and initial and boundary conditions as above.

To be precise, one should specify in what space the solution exists, is unique and depends continuously on the data. For instance, the ODE

$$u_t = u, \ u(0) = a$$

is well-posed in $C^1([0, 1])$, i.e. given $a \in \mathbb{C}$ there is a unique solution in $C^1([0, 1])$, namely $u(t) = ae^t$, which depends continuously on a. Here we put the norm

$$\|u\|_{C^1([0,1])} = \sup_{t\in[0,1]} |u| + \sup_{t\in[0,1]} |\partial u|$$

on $C^1([0, 1])$, so the continuous dependence means that if $a_j \to a$ in $\mathbb{R}$ as $j \to \infty$, then the solutions $u_j(t) = a_je^t$, $u(t) = ae^t$ satisfy

$$\lim_{j\to\infty} \|u_j - u\|_{C^1([0,1])} = 0.$$

Equivalently, this means that there is a constant $C \geq 0$ such that

$$\|u\|_{C^1([0,1])} \leq C|a|.$$

It is also well-posed in C^2, etc.

As another example, consider the ODE

$$u_t - u = f, \ u(0) = a,$$

so

$$u(t) = ae^t + \int_0^t e^{t-s} f(s) \, ds,$$

which is an example of Duhamel's principle (to be discussed later) and which one obtains by multiplying the ODE by e^{-t}, in which case it becomes $\partial_t(e^{-t}u) = e^{-t}f$, and integrating. Now one needs to specify where both f and u lie. For $f \in C^0([0,1])$ and $a \in \mathbb{C}$, the ODE is well-posed in $C^1([0,1])$; thus the unique solution, described above, depends continuously on both a and f. Again, this means either that $a_j \to a$ in $\mathbb{R}$ and $f_j \to f$ in $C^0([0,1])$ implies that the solutions satisfy $u_j \to u$ in $C^1([0,1])$, or equivalently that one has an estimate

$$\|u\|_{C^1([0,1])} \leq C(\|f\|_{C^0([0,1])} + |a|).$$

Note that this estimate follows easily from the explicit solution formula since on $[0,1]$, the exponential functions is $\leq e$. So for $t \in [0,1]$,

$$|u(t)| \leq |a|e + \int_0^t e|f(s)| \, ds \leq |a|e + te \sup_{s \in [0,t]} |f(s)| \leq e(|a| + \sup_{s \in [0,1]} |f(s)|),$$

and thus

$$\|u\|_{C^0([0,1])} = \sup_{t \in [0,1]} |u(t)| \leq e(|a| + \|f\|_{C^0([0,1])}),$$

while the ODE gives

$$|u'(t)| = |u(t) + f(t)| \leq |u(t)| + |f(t)|,$$

so

$$|u(t)| + |u'(t)| \leq 2|u(t)| + |f(t)| \leq 2e|a| + (2e+1)\|f\|_{C^0([0,1])}$$
$$\leq (2e+1)(|a| + \|f\|_{C^0([0,1])}),$$

so

$$\|u\|_{C^1([0,1])} \leq (2e+1)(|a| + \|f\|_{C^0([0,1])}),$$

showing the claimed well-posedness estimate. The equation is also well-posed if one takes $f \in C^1$ and $u \in C^2$, but not if one takes $f \in C^0$ and $u \in C^2$ (since the typical solution u for such f is merely C^1).

With these preliminaries, we are ready to embark on our study of PDEs. We shall first look at some places where PDEs arise, then continue to first order PDE, followed by distribution theory, a qualitative study of second order PDE, the Fourier transform, a study of boundaries and forcing (Duhamel's principle), separation of variables, inner product spaces and Fourier series, Bessel functions and finishing with some more advanced considerations such as solvability via duality and variational problems. While there is much more to PDE theory than what we can cover, this should provide a reasonable first introduction; the reader is of course encouraged to take further courses for an in-depth understanding of this important field.

Additional material: More on normed vector spaces and metric spaces

We start by recalling the notion of continuity:

Definition 1.1. A map $F : V \to W$ between normed vector spaces V, W is *continuous* if $x_j \to x$ in V implies that $F(x_j) \to F(x)$ in W.

If V_1, V_2 are vector spaces, then the direct sum, $V_1 \oplus V_2$, is the set $V_1 \times V_2$ (i.e. ordered pairs (v_1, v_2) with $v_j \in V_j$) with componentwise addition and multiplication by scalars, i.e.

$$(v_1, v_2) + (w_1, w_2) = (v_1 + w_1, v_2 + w_2), \ \lambda(v_1, v_2) = (\lambda v_1, \lambda v_2),$$

$v_j, w_j \in V_j, \lambda \in \mathbb{F}$. If V_1, V_2 are normed, then $V_1 \oplus V_2$ is normed with, for example, the norm

$$\|(x_1, x_2)\|_{V_1 \oplus V_2} = \|x_1\|_{V_1} + \|x_2\|_{V_2}.$$

It thus also makes sense to talk about maps $F : V_1 \times V_2 \to W$ being continuous when V_1, V_2, W are normed. Notice that sequences $x^{(j)} = (x_1^{(j)}, x_2^{(j)})$ in $V_1 \oplus V_2$ converge to $x = (x_1, x_2)$ if and only if $x_1^{(j)} \to x_1$ in V_1 and $x_2^{(j)} \to x_2$ in V_2.

The following lemma states that in normed vector spaces the vector space operations are (jointly) continuous; this is a generalization of the corresponding facts in $\mathbb{R}$ or $\mathbb{R}^n$ (with the same proofs!).

Lemma 1.2. *Suppose V is a normed vector space. If $x_j \to x$, $y_j \to y$ in V, then $x_j + y_j \to x + y$. Further, if $x_j \to x$ in V and $\lambda_j \to \lambda$ in $\mathbb{F}$, then $x_j \lambda_j \to x\lambda$ in V.*

Proof. Starting with addition, if $x_j \to x$ and $y_j \to y$, then $x_j + y_j \to x + y$ since

$$\|(x_j + y_j) - (x + y)\|_V = \|(x_j - x) + (y_j - y)\|_V \leq \|x_j - x\|_V + \|y_j - y\|_V$$

by the triangle inequality, and the right-hand side is going to 0 as $j \to \infty$. Similarly, for multiplication by scalars, if $\lambda_j \to \lambda$ in $\mathbb{F}$ and $x_j \to x \in V$, then

$\lambda_j x_j \to \lambda x$ in V since

$$\|\lambda_j x_j - \lambda x\|_V = \|\lambda_j(x_j - x) + (\lambda_j - \lambda)x\|_V$$
$$\leq \|\lambda_j(x_j - x)\|_V + \|(\lambda_j - \lambda)x\|_V$$
$$= |\lambda_j|\|x_j - x\|_V + |\lambda_j - \lambda|\|x\|_V.$$

Now $|\lambda_j - \lambda| \to 0$ means that the last term on the right-hand side is tending to 0, while $\|x_j - x\|_V \to 0$ and $|\lambda_j|$ bounded (as follows from $\lambda_j \to \lambda$: $|\lambda_j| \leq |\lambda| + |\lambda_j - \lambda|$ by the triangle inequality, and where the second term is < 1 for large j) shows that the first term on the right-hand side is tending to 0, so indeed $\lambda_j x_j \to \lambda x$ in V.   □

In the proof we used the fact that convergent sequences in the scalars are bounded; this is also true for convergent sequences in a normed vector space V. Namely, if $v_j \to v$ in V, then for j sufficiently large, say $j \geq N$, one has $\|v_j - v\| \leq 1$; hence

$$\|v_j\| = \|(v_j - v) + v\| \leq \|v_j - v\| + \|v\| \leq \|v\| + 1.$$

Thus, for all but finitely many j, $\|v_j\| \leq \|v\| + 1$; hence the sequence $\{\|v_j\|\}_{j=1}^{\infty}$ is bounded:

$$\|v_j\| \leq \max\{\|v_1\|, \ldots, \|v_{N-1}\|, \|v\| + 1\}$$

for all j.

Next, we note that for linear maps between normed vector spaces continuity is the same as boundedness:

Lemma 1.3. *Suppose V and W are normed vector spaces and $L : V \to W$ is linear. Then L is continuous if and only if it is bounded.*

Proof. First, suppose that L is bounded, i.e. there is $C \geq 0$ such that $\|Lx\|_W \leq C\|x\|_V$ for all $x \in V$. Suppose $x_j \to x$ in V. We claim that $Lx_j \to Lx$ in W. Indeed,

$$\|Lx_j - Lx\|_W = \|L(x_j - x)\|_W \leq C\|x_j - x\|_V,$$

and the right-hand side goes to 0 since $x_j \to x$ in V, proving that $Lx_j \to Lx$ in W, and thus that L is continuous.

Conversely, suppose that L is *not* bounded. Thus, no $C > 0$ gives a bound as in the definition of boundedness, so taking $C = j$ as a positive integer, there is $\tilde{x}_j \in V$ such that $\|L\tilde{x}_j\|_W > j\|\tilde{x}_j\|_V$; notice that this means $\tilde{x}_j \neq 0$. Now let $x_j = \frac{\tilde{x}_j}{j\|\tilde{x}_j\|_V}$. Then $\|x_j\|_V = 1/j$, so $x_j \to 0 = x$ as $j \to \infty$. On the other hand, $\|Lx_j\|_W = \frac{1}{j\|\tilde{x}_j\|_V}\|L\tilde{x}_j\|_W > 1$, so in particular, Lx_j does *not* converge to $0 = Lx$ as $j \to \infty$. Thus, L is not continuous.   □

A typical continuous (but not linear) function on a normed vector space is the norm. To see its continuity, note that if $x_j \to x$ in a normed vector space V, then

(1.2) $$\left| \|x_j\|_V - \|x\|_V \right| \le \|x_j - x\|_V$$

would show that $\|x_j\|_V \to \|x\|_V$ in $\mathbb{R}$. But (1.2) follows from the triangle inequalities,

$$\|x_j\|_V \le \|x\|_V + \|x_j - x\|_V, \quad \|x\|_V \le \|x_j\|_V + \|x - x_j\|_V,$$

so

$$\pm(\|x_j\|_V - \|x\|_V) \le \|x_j - x\|_V,$$

giving (1.2), and thus the continuity of the norm.

We now turn to metric spaces, which are a generalization of normed vector spaces.

Definition 1.4. A *metric space* (M, d) is a set M together with a *distance function* or *metric* $d : M \times M \to [0, \infty)$ such that

(i) $d(x, y) = 0$ if and only if $x = y$,

(ii) $d(x, y) = d(y, x)$ for all $x, y \in M$,

(iii) (triangle inequality) $d(x, z) \le d(x, y) + d(y, z)$ for all $x, y, z \in M$.

Any normed vector space V is a metric space with distance function $d : V \times V \to [0, \infty)$ being given by $d(x, y) = \|x - y\|_V$ (called the *induced metric*), with the triangle inequality following from that for the norm.

The notion of convergence of a sequence x_j in a metric space is:

Definition 1.5. A sequence x_j in a metric space (M, d) converges to $x \in M$ if $d(x_j, x) \to 0$ as $j \to \infty$, i.e. for all $\epsilon > 0$ there exists N such that $j \ge N$ implies $d(x_j, x) < \epsilon$.

Writing out the definition, we see that this agrees with our definition of convergence in a normed vector space with the induced metric.

The notion of sequential continuity is:

Definition 1.6. Suppose (M, d) and (M', ρ) are metric spaces. A map $F : M \to M'$ is *sequentially continuous* if $x_j \to x$ in M implies $F(x_j) \to F(x)$ in M'.

This agrees with the notion of continuity discussed above for normed vector spaces.

The notion of continuity is:

Definition 1.7. Suppose (M, d) and (M', ρ) are metric spaces. A map $F : M \to M'$ is *continuous* if for all $x \in M$ and for all $\epsilon > 0$ there exists $\delta > 0$ such that $x' \in M$ with $d(x', x) < \delta$ implies $\rho(F(x'), F(x)) < \epsilon$.

Lemma 1.8. *Suppose (M, d) and (M', ρ) are metric spaces, and $F : M \to M'$. Then F is continuous if and only if it is sequentially continuous.*

Proof. Suppose first that F is continuous. Suppose $x_j \to x$ in M. Let $\epsilon > 0$ be given. By the continuity of F there is $\delta > 0$ such that $d(x', x) < \delta$ implies $\rho(F(x'), F(x)) < \epsilon$. Since $x_j \to x$, there is N such that $j \geq N$ implies $d(x_j, x) < \delta$. Thus, for $j \geq N$, $\rho(F(x_j), F(x)) < \epsilon$, so $F(x_j) \to F(x)$ indeed.

Conversely, suppose that F is *not* continuous. This means that there is an $\epsilon > 0$ such that *no* δ works in the definition of continuity, i.e. for all $\delta > 0$ there is $x' \in M$ such that $d(x', x) < \delta$ but $\rho(F(x'), F(x)) \geq \epsilon$. In particular, $\delta = 1/j$ does not work, so there is $x_j \in M$ such that $d(x_j, x) < 1/j$ but $\rho(F(x_j), F(x)) \geq \epsilon$. Thus, $x_j \to x$, but $F(x_j)$ does not converge to $F(x)$, so F is *not* sequentially continuous. This completes the proof. $\square$

We also recall the definition of open and closed sets:

Definition 1.9. Suppose (M, d) is a metric space.

A set $A \subset M$ is *open* if for all $x \in A$ there is $\epsilon > 0$ such that the ϵ-ball around x, $B_\epsilon(x) = \{y \in M : d(y, x) < \epsilon\}$, satisfies $B_\epsilon(x) \subset A$.

A set $A \subset M$ is *closed* if for all sequences $x_j \in A$ convergence to some $x \in M$ implies $x \in A$.

A set is open if and only if its complement is closed. Further, any union of open sets is open, while the intersection of a finite number of open sets is also open. Similarly, any intersection of closed sets is closed, while the union of a finite number of closed sets is closed. In addition, both M and the empty set $\emptyset$ are both open and closed.

A typical set $A \subset M$ is neither open nor closed. However, it is possible to manufacture a closed set out of A, namely its *closure* $\overline{A}$. This consists of points $x \in M$ for which there is a sequence $x_j \in A$ such that $x_j \to x$ as $j \to \infty$; such points are called *limit points* of A. Note that $A \subset \overline{A}$. One can take the constant sequence, $x_j = x$ for all j, to show that if $x \in A$, then $x \in \overline{A}$. Further, $\overline{A}$ is closed, and indeed it is the smallest closed set containing A in the sense that if C is closed and $A \subset C$, then $\overline{A} \subset C$.

There is an analogous notion for open sets, giving the *interior* A° of A. Here $A^\circ \subset A$ is open, and it is the largest open set contained in A. It is defined by $x \in M$ in A° if there is $\epsilon > 0$ such that $B_\epsilon(x) \subset A$ (which necessitates $x \in A$, and hence $A^\circ \subset A$ follows).

The final concept we recall is that of a compact set, which in turn requires that of the *subsequence* of a sequence. If $\{x_j\}_{j=1}^{\infty}$ is a sequence, a subsequence is a sequence of the form $\{x_{j_k}\}_{k=1}^{\infty}$, where $\{j_k\}_{k=1}^{\infty}$ is strictly increasing, i.e. $j_{k+1} > j_k$ for all k. Concretely this means that a subsequence of a sequence arises by skipping some of the elements of the sequence but keeping the order, so $x_1, x_3, x_7, \ldots$ is a subsequence of $x_1, x_2, x_3, \ldots$ but $x_1, x_7, x_3, \ldots$ and $x_1, x_3, x_3, x_7, \ldots$ are not.

Definition 1.10. Suppose (M, d) is a metric space. A subset A of M is *(sequentially) compact* if for all sequences x_j in A there is a subsequence x_{j_k}, $k = 1, 2, \ldots$, such that x_{j_k} converges to some $x \in A$ as $k \to \infty$.

The standard definition of compactness is in terms of open covers. In a metric space the notion of compactness and sequential compactness, as defined above, coincide, and we only use the latter here.

In $\mathbb{R}^n$, compact subsets are exactly subsets which are closed and *bounded*, i.e. closed subsets A for which there is $R > 0$ such that $A \subset B_R(0) = \{x \in \mathbb{R}^n : \|x\| < R\}$. This is false in general metric spaces, indeed in general normed vector spaces. However, in $\mathbb{R}^n$ (or any finite dimensional normed real or complex vector space), if $A \subset \mathbb{R}^n$ is bounded, then $\overline{A}$ is compact. Indeed, it is automatically closed (being a closure), and it is also bounded, for if $A \subset B_R(0)$, i.e. $x' \in A$ implies $\|x'\| < R$, then for $x \in \overline{A}$ there is a sequence $x_j \in A$ such that $x_j \to x$, so $\|x_j\| < R$, and thus $\|x\| = \lim_{j \to \infty} \|x_j\| \leq R$, so $\overline{A}$ is also bounded and thus compact.

One of the main properties of compact sets is that continuous real-valued functions are bounded on (non-empty) compact sets and indeed attain their maximum and minimum. That is, if $A \subset M$ is compact, $A \neq \emptyset$, and $f : M \to \mathbb{R}$ is continuous, then $\sup_{x \in A} |f(x)|$ is finite (i.e. there is $C > 0$ such that $|f(x)| \leq C$ for all $x \in A$), and indeed there are $x_1, x_2 \in A$ such that $f(x_1) \leq f(x) \leq f(x_2)$ for all $x \in A$.

Another important property of compact sets is that continuous maps from a compact set are *uniformly continuous*. Namely, a map $f : M \to M'$, where (M, d), (M', ρ) are metric spaces, is called uniformly continuous if for all $\epsilon > 0$ there exists $\delta > 0$ such that for all $x, x' \in M$, $d(x, x') < \delta$ implies $\rho(f(x), f(x')) < \epsilon$. Thus, every uniformly continuous map is continuous, but not vice versa. The point is that in the definition of continuity we are given $\epsilon > 0$ and $x \in M$ before we need to come up with $\delta > 0$, while in the definition of uniform continuity we are only given $\epsilon > 0$, i.e. δ is *not allowed to depend on* $x \in M$. Now, the important result for continuous maps on compact metric spaces is that the converse is also true: continuous maps *are* uniformly continuous on compact spaces, i.e. if for M compact, $f : M \to M'$ is continuous, then in fact it is uniformly continuous; see Problem 1.7.

Problems

Problem 1.1. Classify the following PDEs by degree of non-linearity (linear, semilinear, quasilinear, fully nonlinear):

    (i) $(\cos x)u_x + u_y = u^2$.

    (ii) $uu_{tt} = u_{xx}$.

    (iii) $u_x - e^x u_y = \cos x$.

    (iv) $u_{tt} - u_{xx} + e^u u_x = 0$.

Problem 1.2. Show *Taylor's theorem with an integral remainder* on $\mathbb{R}$: if f is a C^{k+1} function, then

$$f(x) = \sum_{j \le k} \frac{f^{(j)}(x_0)}{j!}(x-x_0)^j + (x-x_0)^{k+1} \int_0^1 \frac{(1-s)^k}{k!} f^{(k+1)}(x_0+s(x-x_0))\, ds.$$

Use this to show that if f is C^{k+1} and $f(x_0) = 0$ then

$$g(x) = \frac{f(x)}{x - x_0}$$

is a C^k function.

    Note: Notice that

$$\int_0^1 \frac{(1-s)^k}{k!} f^{(k+1)}(x_0 + s(x - x_0))\, ds$$

is a continuous function of x; if f is C^∞, it is in fact a C^∞ function of x as can be seen by differentiation under the integral sign. This is the main advantage of the integral remainder version vs. the mean value theorem based version.

    Hint: If $k = 0$, by the fundamental theorem of calculus,

$$f(x) = f(x_0) + \int_{x_0}^x f'(t)\, dt = f(x_0) + (x - x_0)\int_0^1 f'(x_0 + s(x - x_0))\, ds,$$

where we wrote $t = x_0 + s(x-x_0)$ and changed variables, so $dt = (x-x_0)\, ds$. To continue one writes the integrand as $1 \cdot f'(x_0 + s(x - x_0))$ and integrates by parts, making the indefinite integral of 1 to be $s - 1$ (and after k steps, starting from the above expression, one gets $\frac{1}{k!}(s - 1)^k$ for this).

Problem 1.3. Show Taylor's theorem with an integral remainder on $\mathbb{R}^n$: if f is a C^{k+1} function on $\mathbb{R}^n$, then

$$f(x) = \sum_{|\alpha| \le k} \frac{f^{(\alpha)}(x_0)}{j!}(x - x_0)^\alpha$$

$$+ \sum_{|\alpha| = k+1} (x - x_0)^\alpha \int_0^1 \frac{k+1}{\alpha!}(1-s)^k f^{(\alpha)}(x_0 + s(x - x_0))\, ds.$$

Here $y^\alpha = y_1^{\alpha_1} \ldots y_n^{\alpha_n}$, for $\alpha \in \mathbb{N}^n$ a multiindex.

Use this to show that if f is C^{k+1} with $f(x_0) = 0$, then there exist C^k functions f_j, $j = 1, \ldots, n$, such that $f(x) = \sum_{j=1}^{n} (x_j - (x_0)_j) f_j(x)$.

Hint: Consider the function $\phi(t) = f(x_0 + t(x - x_0))$, $t \in [0, 1]$, and use the argument (or the result) of the previous problem to compute $\phi(1)$.

Problem 1.4. The following problems are on the additional material.

Show that if (M, d) is a metric space, $A \subset M$ is open if and only if its complement $A^c = M \setminus A = \{x \in M : x \notin A\}$ is closed.

Hint: To show that A open implies A^c closed, suppose for the sake of contradiction that $x_n \in A^c$, $x_n \to x$ as $n \to \infty$, $x \notin A^c$, so $x \in A$. Now use the definition of A being open. To show that A^c closed implies A open, show instead that if A is not open, then A^c is not closed; to do so, use that A not open means that there is $x \in A$ such that for all $\epsilon > 0$, $B_\epsilon(x)$ is *not* a subset of A, i.e. there exists $x' \in A^c \cap B_\epsilon(x)$. Use this with $\epsilon = 1/n$, n a positive integer, to define a sequence x_n in A^c; now use the definition of convergence and of A^c being closed.

Problem 1.5. If (M, d) is a metric space, $A \subset M$, let $\overline{A}$ be the set of limit points of A, i.e. points $x \in M$ for which there exist $x_n \in A$ such that $x_n \to x$ as $n \to \infty$. Show that $\overline{A}$ is closed.

Hint: Suppose $\{x_n\}_{n=1}^{\infty}$ is a sequence in $\overline{A}$ converging to $x \in M$. That means that for each n there is a sequence in A converging to x_n, so in particular there is y_n such that $d(y_n, x_n) < 1/n$. What can you say about the convergence of $\{y_n\}_{n=1}^{\infty}$?

Problem 1.6.

(i) Show that if (M, d) is a metric space, $A \subset M$ is compact, non-empty, and $f : A \to \mathbb{R}$ is continuous, then f is bounded on A.

   Hint: Suppose f is not bounded on A. Then for each $C > 0$ there exists $x \in A$ such that $|f(x)| > C$ (since C is not an upper bound for $|f|$ on A). Thus, for each positive integer n, there is $x_n \in A$ such that $|f(x_n)| > n$. Now use compactness and continuity.

(ii) Under the same assumptions as in (i) show that f attains its maximum on A, i.e. there exists $y \in A$ such that for $x \in A$, $f(x) \leq f(y)$. (Thus, the maximum is attained at y.)

   Hint: Use that f is bounded to let $C = \sup_{x \in A} f(x)$, and let $x_n \in A$ be such that $f(x_n) > C - 1/n$; this exists by the least upper bound property of C. Now use compactness and continuity to get $y \in A$ such that $f(y) = C$.

Problem 1.7. Show that if (M, d), (M', ρ) are metric spaces, M is compact, and $f : M \to M'$ is continuous, then f is uniformly continuous.

Hint: Suppose on the contrary that f is *not* uniformly continuous; i.e. there is $\epsilon > 0$ for which no $\delta > 0$ works in the definition of uniform continuity, in particular $\delta_j = 1/j$ does not (here j is a positive integer). That is, for all j, there are points $x_j, x'_j \in M$ such that $d(x_j, x'_j) < 1/j$ but $\rho(f(x_j), f(x'_j)) \geq \epsilon$. Now use compactness to get a subsequence of $\{x_j\}_{j=1}^{\infty}$ that converges to some $x \in M$, and try to get a contradiction with the continuity of f at x.

Chapter 2

Where do PDE come from?

1. An example: Maxwell's equations

PDE often arise naturally from physical problems, where they may be formulated based on experiments, typically taking into account expected properties (like rotational invariance of space), sometimes in equivalent integral forms. A good example is *Maxwell's equations.* If $\Omega \subset \mathbb{R}^3$ is a region in free space, so $\Omega \times \mathbb{R}_t$ is the corresponding space-time region, Maxwell's equations for the electric field E and the magnetic field B (which are both maps $\Omega \times \mathbb{R} \to \mathbb{R}^3$) corresponding to a charge density ρ and current density J (which are maps $\Omega \times \mathbb{R} \to \mathbb{R}$ and $\Omega \times \mathbb{R} \to \mathbb{R}^3$ respectively) are:

$$\nabla \cdot E = \frac{\rho}{\epsilon_0},$$
$$\nabla \cdot B = 0,$$
$$\nabla \times E = -\frac{\partial B}{\partial t},$$
$$\nabla \times B = \mu_0 \Big(J + \epsilon_0 \frac{\partial E}{\partial t} \Big).$$

Here ϵ_0 and μ_0 are (positive) physical constants, reflecting the choice of physical units, called the permittivity and permeability of free space, respectively, and ∇ is the gradient, i.e. equivalently the differential ∂; see Problem 2.1. Here, for now, we assume that E and B are C^1 and ρ, J are continuous, so all the expressions make sense as continuous functions.

Due to the Gauss-Stokes theorems, these are equivalent to integrated forms. In the case of the first two, one integrates over a region D and uses the divergence theorem to express the left-hand side as a surface integral:

$$\int_{\partial D} E \cdot \hat{n} \, dS = \int_D \frac{\rho}{\epsilon_0} \, dx,$$

$$\int_{\partial D} B \cdot \hat{n} \, dS = 0,$$

where $\hat{n}$ is the outward unit normal. The second two equations are similar, but one integrates over a surface S with boundary given by an oriented curve C and uses Stokes' theorem to obtain

$$\int_C E \cdot ds = -\int_S \frac{\partial B}{\partial t} \, dS,$$

$$\int_C B \cdot ds = \mu_0 \int_S \left(J + \epsilon_0 \frac{\partial E}{\partial t} \right) dS.$$

One can also go back to the differential version of Maxwell's equation, using the Gauss-Stokes theorems to rewrite the left-hand side and noting that if one has for instance $\int_D f \, dx = \int_D g \, dx$ for two continuous functions for every domain D, then in fact $f = g$. Indeed for suitably small domains D around a point x_0 the integrals divided by the volume of D will be close to $f(x_0)$, resp. $g(x_0)$, so a limiting argument gives the conclusion. (See also the argument below for removing the perturbation v in deriving the Euler-Lagrange equation!)

Note here that if there are no charges and currents, one obtains

$$\nabla \cdot E = 0 = \nabla \cdot B$$

and

$$\nabla \times E = -\frac{\partial B}{\partial t}, \ \nabla \times B = c^{-2} \frac{\partial E}{\partial t}, \ c = \frac{1}{\sqrt{\mu_0 \epsilon_0}}.$$

In particular, as (see Problem 2.1)

$$\nabla \times (\nabla \times u) = \nabla(\nabla \cdot u) - \nabla^2 u$$

for any C^2 function $u : \Omega \to \mathbb{R}^3$, one gets

$$-\nabla^2 E = \nabla(\nabla \cdot E) - \nabla^2 E = \nabla \times (\nabla \times E)$$

$$= -\nabla \times \frac{\partial B}{\partial t} = -\frac{\partial}{\partial t} \nabla \times B = -\frac{\partial}{\partial t} c^{-2} \frac{\partial E}{\partial t},$$

so

$$\nabla^2 E = c^{-2} \frac{\partial^2 E}{\partial t^2},$$

or with $\Delta = \nabla^2$,

$$c^2 \Delta E = \frac{\partial^2 E}{\partial t^2};$$

i.e. E satisfies the wave equation with speed of waves c. Similarly, B satisfies the same wave equation. The waves are the electromagnetic waves, i.e. light.

Another special case of the Maxwell system is electrostatics, i.e. when there is no current and the fields are independent of time t. Then we have $\nabla \times E = 0$, which implies that E is the gradient of a function (in a simply connected region Ω); conventionally one writes $E = -\nabla V$ and calls V the electrostatic potential. Then $\nabla \cdot E = -\nabla^2 V$, so we obtain

$$\Delta V = -\frac{\rho}{\epsilon_0}.$$

Thus, V solves an inhomogeneous Laplace equation, called Poisson's equation. Note that V is only defined up to adding a constant, which does not affect the negative gradient E.

These examples also explain the role of boundary conditions. For instance, suppose Ω is a region in free space whose boundary is a perfect conductor. Then the electric field must be normal to $\partial\Omega$ or $\partial\Omega \times \mathbb{R}_t$ in the time dependent setting, for otherwise it would instantenously move the free charges in the perfect conductor $\partial\Omega$ to arrange this, and the magnetic field tangent to $\partial\Omega$ (or $\partial\Omega \times \mathbb{R}_t$) elsewhere (otherwise it would instantenously generate currents to arrange this). Thus, if $E = -\nabla V$, as in the electrostatic case, $\partial\Omega$ is a level surface of V (since we have that ∇V is normal to $\partial\Omega$), i.e. V is constant on the boundary. This is the case of a Dirichlet boundary condition: one specifies the potential the boundary is at (i.e. $V|_{\partial\Omega}$ is given), and one attempts to find the potential inside by solving Poisson's equation.

2. Euler-Lagrange equations

Often PDE arise from somewhat different considerations, such as critical points of a functional, called an *Euler-Lagrange functional*. This is an integral of the unknown function $u : \Omega \to \mathbb{R}^k$ and its derivatives over the region Ω (we assume u is C^1 here):

$$(2.1) \qquad I(u) = \int_\Omega F(x, u, \partial u)\, dx,$$

where F is a given real-valued function $F : \Omega \times \mathbb{R}^k \times \mathbb{R}^{nk} \to \mathbb{R}$. For instance, we may have, with u scalar (so $k = 1$) $F(x, z, p) = \frac{1}{2}\sum_{j=1}^n p_j^2$, and then

$$I(u) = \frac{1}{2}\int_\Omega \sum_{j=1}^n (\partial_j u)^2\, dx.$$

One then wants to determine minima and maxima, or more generally critical points, of I. Here a critical point means that if one perturbs u by a function v, which is assumed to be compactly supported in the open set Ω (so 0 near

$\partial\Omega$), i.e. considers $u + sv$ where s is a small real parameter, then

$$\frac{d}{ds}I(u + sv)|_{s=0} = 0.$$

Thus the Euler-Lagrange functional evaluated on the perturbed function has a critical point when the perturbation vanishes. Notice that maxima and minima of the Euler-Lagrange functional are certainly critical points, for the function $s \mapsto I(u + sv)$ has a maximum or minimum in this case at $s = 0$, thus its derivative vanishes there.

Now, if u is a critical point of I of the form (2.1), then one has

$$0 = \frac{d}{ds}I(u + sv)|_{s=0} = \int_\Omega \frac{d}{ds}F(x, u + sv, \partial u + s\partial v)|_{s=0} \, dx$$

(2.2)

$$= \int_\Omega \left(\partial_z F(x, u, \partial u)v + \sum_{j=1}^n \partial_{p_j} F(x, u, \partial u)\partial_j v \right) dx,$$

where we used the chain rule

$$\frac{d}{ds}F(x, u + sv, \partial u + s\partial v) = \partial_z F(x, u + sv, \partial u + s\partial v)v$$

$$+ \sum_{j=1}^n \partial_{p_j} F(x, u + sv, \partial u + s\partial v)\partial_j v$$

and evaluated the result at $s = 0$. In order to obtain an equation for u, i.e. to eliminate v, it is useful to rewrite (2.2) so that the derivatives $\partial_j v$ are eliminated from it. This can be done via integration by parts, assuming u is C^2, putting the derivatives on $\partial_{p_j} F(x, u, \partial u)$. Thus, we have

$$0 = \int_\Omega \left(\partial_z F(x, u, \partial u)v - \sum_{j=1}^n \partial_j \partial_{p_j} F(x, u, \partial u)v \right) dx$$

$$= \int_\Omega \left(\partial_z F(x, u, \partial u) - \sum_{j=1}^n \partial_j \partial_{p_j} F(x, u, \partial u) \right) v \, dx$$

and there are no boundary terms from the integration by parts since v vanishes at $\partial\Omega$. Since this expression vanishes for all C^1 functions v which vanish near $\partial\Omega$, we conclude that

(2.3) $$\partial_z F(x, u, \partial u) - \sum_{j=1}^n \partial_j \partial_{p_j} F(x, u, \partial u) = 0$$

in Ω. (To see that $\int fv \, dx = 0$ for all such v implies $f = 0$, take $x_0 \in \Omega$, use the continuity of f, namely that it takes values close to $f(x_0)$ near x_0, and use v which is 0 outside a small open set containing x_0; see the argument after Lemma 5.7 to conclude $f(x_0) = 0$.) Now recall that in (2.3), ∂_j simply stands for the jth derivative of the function $x \mapsto \partial_{p_j} F(x, u(x), \partial u(x))$. Thus, we should rewrite it using the chain rule:

$$\partial_j \partial_{p_j} F(x, u, \partial u) = \partial_{x_j} \partial_{p_j} F(x, u, \partial u) + \partial_z \partial_{p_j} F(x, u, \partial u) \partial_j u$$

$$+ \sum_{k=1}^{n} \partial_{p_k} \partial_{p_j} F(x, u, \partial u) \partial_k \partial_j u.$$

We then finally obtain the *Euler-Lagrange equation* for (2.1):
(2.4)
$$\partial_z F(x, u, \partial u) - \sum_{j=1}^{n} \Big(\partial_{x_j} \partial_{p_j} F(x, u, \partial u) + \partial_z \partial_{p_j} F(x, u, \partial u) \partial_j u$$

$$+ \sum_{k=1}^{n} \partial_{p_k} \partial_{p_j} F(x, u, \partial u) \partial_k \partial_j u \Big) = 0.$$

In the special case $F(x, z, p) = \frac{1}{2} \sum_{l=1}^{n} p_l^2$, we have $\partial_z F = 0$, $\partial_{x_j} F = 0$, while $\partial_{p_j} \partial_{p_k} F = 0$ unless $j = k$, and if $j = k$, $\partial_{p_j} \partial_{p_k} F = 1$. Thus, the Euler-Lagrange equation becomes

$$\sum_{j=1}^{n} \partial_j^2 u = 0,$$

which is exactly Laplace's equation. On the other hand, if we are given a function f and consider the Euler-Lagrange functional given by $F(x, z, p) = \frac{1}{2} \sum_{j=1}^{n} p_j^2 + fz$, i.e.

$$I(u) = \int \Big(\frac{1}{2} \sum_{j=1}^{n} (\partial_j u)^2 + fu \Big) \, dx,$$

then the Euler-Lagrange equation becomes (as $\partial_z F = f$ now)

$$f - \sum_{j=1}^{n} \partial_j^2 u = 0,$$

i.e. $\Delta u = f$, which is Poisson's equation. Note that for the electrostatic equation arising from Maxwell's equations, f is essentially the charge (up to an overall constant) that causes, or forces, the electric field, so correspondingly f is often called a *forcing*. These Euler-Lagrange functionals will play a role in Chapter 7, where they will be called the *energy* corresponding to the PDE, and in Chapter 18, where they will be used to give a variational characterization of the eigenvalues of Laplace-like operators and to show the completeness of the corresponding collection of eigenfunctions as an orthonormal basis for $L^2(\Omega)$.

It is important to keep in mind that what we have shown is that any minima or maxima (or more generally critical points) for the Euler-Lagrange functional *must* satisfy the Euler-Lagrange equation (2.4). However, it is by no means clear at first that the Euler-Lagrange equation (subject to whatever constraints one has on u) has any solutions. Indeed, even if one has, say, a non-negative functional, such as that given by $F(x, z, p) = \frac{1}{2} \sum_{l=1}^{n} p_l^2$, so one knows that $I(u)$ is bounded below (where Ω is a bounded domain), all one can conclude is that there is a sequence u_j such that $I(u_j)$ converges to $\inf_u I(u)$. Whether u_j even has any convergent subsequences relies on compactness properties based on more advanced analysis of PDEs, starting from energy estimates, which is a topic we take up in Chapter 18 using the energy methods of Chapter 7.

These Euler-Lagrange functionals can arise from physical considerations, such as the energy above. They can also arise from geometric problems. For instance, given a curve C in $\mathbb{R}^3$ (say a circle) which can be spanned by a surface S (so C is the boundary of S), one may want to find S with boundary C which has minimal area. A physical interpretation is soap films on a wire C: surface tension causes the area minimization, but this is of direct geometric interest as well. Such surfaces are called minimal surfaces. Let's assume for simplicity that with $\mathbb{R}^3 = \mathbb{R}^2 \times \mathbb{R}$ we can write S as the graph of a function u on $\Omega \subset \mathbb{R}^2$, i.e. it is of the form $(x, u(x)) \in \mathbb{R}^2 \times \mathbb{R}$, so C is the graph of u over $\partial\Omega$. The area of S is then

$$\int_\Omega \sqrt{1 + |\nabla u(x)|^2}\, dx,$$

which is an Euler-Lagrange functional with $F(x, z, p) = \sqrt{1 + \sum_{l=1}^{2} p_l^2}$. The Euler-Lagrange equation then becomes

$$\sum_{j,k=1}^{2} (\partial_{p_k}\partial_{p_j} F)(x, u(x), \partial u(x))(\partial_k \partial_j u)(x) = 0,$$

i.e.

$$0 = \sum_{j,k=1}^{2} \partial_{p_k}\left(\frac{p_j}{\sqrt{1 + \sum_{l=1}^{2} p_l^2}}\right)(x, u(x), \partial u(x))(\partial_k \partial_j u)(x)$$

$$= \frac{\partial_1^2 u + \partial_2^2 u}{\sqrt{1 + \sum_{l=1}^{2}(\partial_l u)^2}} - \sum_{j,k=1}^{2} \frac{\partial_j u\, \partial_k u}{(1 + \sum_{l=1}^{2}(\partial_l u)^2)^{3/2}}(\partial_k \partial_j u)(x).$$

This can be rewritten, after multiplying by $\sqrt{1 + \sum_{l=1}^{2}(\partial_l u)^2}$, as

$$\Delta u - \sum_{j,k=1}^{2} \frac{\partial_j u\, \partial_k u}{1 + \sum_{l=1}^{2}(\partial_l u)^2}(\partial_k \partial_j u)(x) = 0,$$

which is called the minimal surface equation. Notice that this is a non-linear equation, due to the second term, but it is linear in the highest, second, derivatives, with coefficients however depending on the lower order (at most first order) derivatives, so it is a quasilinear equation.

Problems

Problem 2.1. Recall that if $\Omega \subset \mathbb{R}^3$ is a domain and $u = (u_1, u_2, u_3) : \Omega \to \mathbb{R}^3$ is C^1, then the divergence of u is the function $\nabla \cdot u : \Omega \to \mathbb{R}$,

$$\nabla \cdot u = \partial_1 u_1 + \partial_2 u_2 + \partial_3 u_3,$$

while the curl of u is the vector field $\nabla \times u : \Omega \to \mathbb{R}^3$,

$$\nabla \times u = (\partial_2 u_3 - \partial_3 u_2, \partial_3 u_1 - \partial_1 u_3, \partial_1 u_2 - \partial_2 u_1).$$

Further, for C^2 functions u, $\nabla^2 u : \Omega \to \mathbb{R}^3$ is the vector field

$$\nabla^2 u = (\nabla^2 u_1, \nabla^2 u_2, \nabla^2 u_3)$$
$$= ((\partial_1^2 + \partial_2^2 + \partial_3^2)u_1, (\partial_1^2 + \partial_2^2 + \partial_3^2)u_2, (\partial_1^2 + \partial_2^2 + \partial_3^2)u_3).$$

Show that

$$\nabla \times (\nabla \times u) = \nabla(\nabla \cdot u) - \nabla^2 u.$$

Problem 2.2. As shown in the text, every solution of Maxwell's equations in free space solves the wave equation (both E and B do). The converse is not true.

(i) Show that the functions $E = (f(x - ct), 0, 0)$, $B = (0, 0, 0)$, where f is any given real-valued C^2 function, solve the wave equation, $c^2 \Delta E = \partial_t^2 E$, $c^2 \Delta B = \partial_t^2 B$, on $\mathbb{R}_{x,y,z}^3 \times \mathbb{R}_t$, but do not satisfy Maxwell's equations unless f is constant.

(ii) Show that if E, B solve the wave equation (as $\mathbb{R}^3$-valued functions), then so do the derivatives $\partial_x E, \partial_y E, \partial_z E, \partial_t E$, and similarly for B (assume that E, B are C^3). Use this to conclude that $\nabla \cdot E, \nabla \cdot B$ (which are real-valued functions), as well as $\partial_t E - c^2 \nabla \times B$ and $\partial_t B + \nabla \times E$, solve the wave equation.

(iii) Show that, on the other hand, if E and B is such that

$$\nabla \cdot E|_{t=0} = 0 = \nabla \cdot B|_{t=0},$$

which are just the first two Maxwell equations enforced at the initial time $t = 0$, and

$$\frac{\partial E}{\partial t}\Big|_{t=0} = c^2 \nabla \times B|_{t=0}, \quad \frac{\partial B}{\partial t}\Big|_{t=0} = -\nabla \times E|_{t=0},$$

so the second two Maxwell's equations are also enforced at the initial time, and E, B solve the wave equation, then in fact $\nabla \cdot E$, $\nabla \cdot B$, as well as $\partial_t E - c^2 \nabla \times B$ and $\partial_t B + \nabla \times E$, have the property that they vanish at $t = 0$ and so do their derivatives with respect to t.

     Hint: Show and use that $\nabla \cdot (\nabla \times u) = 0$ for any $\mathbb{R}^3$-valued C^2 function u.

Note: In Chapter 7 we show that the solution of the wave equation with given initial data (for the wave equation for u, $u|_{t=0}$ and $(\partial_t u)|_{t=0}$ being given) is unique. Since the identically 0 function solves the wave equation, this means that in case (iii), $\nabla \cdot E$, $\nabla \cdot B$, as well as $\partial_t E - c^2 \nabla \times B$ and $\partial_t B + \nabla \times E$, vanish for all t, i.e. Maxwell's equations are satisfied. Thus, one can solve Maxwell's equations by solving the wave equation; one just needs to have the initial data satisfy the Maxwell equations at the initial time.

Problem 2.3. Consider the Euler-Lagrange functional

$$I(u) = \int_\Omega F(x, u, \partial u) \, dx$$

given by

$$F(x, z, p) = \frac{1}{2} c(x)^2 \sum_{j=1}^n p_j^2 + \frac{1}{2} q(x) z^2 + fz,$$

where c, q, f are given functions (speed of waves, potential and forcing, respectively), and show that the corresponding Euler-Lagrange equation is

$$\nabla \cdot (c^2 \nabla u) - qu = f,$$

which in the special case of constant c reduces to

$$c^2 \Delta u - qu = f.$$

Problem 2.4. Find the Euler-Lagrange equation for the functional

$$I(u) = \int_\Omega F(x, u, \partial u) \, dx$$

given by

$$F(x, z, p) = \frac{1}{2\sqrt{\det a}} \sum_{l,m=1}^n a_{lm}(x) p_l p_m,$$

where $a = (a_{lm})_{l,m=1}^n$ is a positive definite n-by-n matrix and $\det a$ is its determinant. Show that the resulting equation is a second order linear PDE, of the form $Lu = 0$, where

$$L = -\sum_{j,k=1}^n a_{jk}(x) \partial_j \partial_k + L_1$$

for some first order operator L_1. L is called the *Laplace-Beltrami operator* corresponding to a_{lm}, whose matrix is called the (dual) Riemannian metric.

(Note that one can think of $F(x, z, p)$ as giving the squared length of a vector (the inner product with itself; see Chapter 13), p, at x.)

Chapter 3

First order scalar semilinear equations

First order scalar semilinear equations have the form

(3.1) $$a(x, y)u_x + b(x, y)u_y = c(x, y, u);$$

here we assume that a, b, c are at least C^1, given real-valued functions. Let V be the vector field on $\mathbb{R}^2$ given by

$$V(x, y) = (a(x, y), b(x, y)),$$

so $a(x, y)u_x + b(x, y)u_y$ is the *directional derivative* of u along V. Let $\gamma = \gamma(s)$ be an *integral curve* of V. That is, $\gamma(s) = (x(s), y(s))$ has tangent vector $V = V(x(s), y(s))$ for each s. Explicitly, this says that

(3.2) $$x'(s) = a(x(s), y(s)), \ \ y'(s) = b(x(s), y(s)).$$

Now let $v(s) = u(\gamma(s)) = u(x(s), y(s))$. Thus, by the chain rule,

$$
\begin{aligned}
v'(s) &= x'(s)\, u_x(x(s), y(s)) + y'(s)\, u_y(x(s), y(s)) \\
&= a(x(s), y(s))\, u_x(x(s), y(s)) + b(x(s), y(s))\, u_y(x(s), y(s)) \\
&= c(x(s), y(s), u(x(s), y(s))),
\end{aligned}
$$

where in the last step we used the PDE. Thus,

$$v'(s) = c(x(s), y(s), v(s)),$$

i.e. v satisfies an ODE along each integral curve of V.

To solve the PDE, we parameterize the integral curves by an additional parameter r, i.e. the integral curves are $\gamma_r = \gamma_r(s) = (x_r(s), y_r(s))$, where r

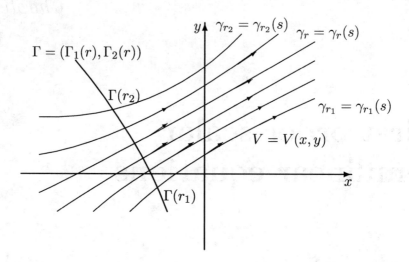

Figure 3.1. (Projected) characteristic curves $\gamma = \gamma_r(s)$, shown by thin lines, for the vector field $V = V(x,y)$, shown by arrows, going through the initial curve $\Gamma = \Gamma(r)$.

is in an interval (or the whole real line), and each γ_r is an integral curve for V, i.e.

$$(3.3) \qquad x'_r(s) = a(x_r(s), y_r(s)), \;\; y'_r(s) = b(x_r(s), y_r(s)),$$

so $v_r(s) = u(\gamma_r(s))$ solves

$$(3.4) \qquad v'_r(s) = c(x_r(s), y_r(s), v_r(s)).$$

Note that here the subscript r denotes a parameter, not a derivative! We may equally well write $x_r(s) = x(r,s)$, $y_r(s) = y(r,s)$, and thus we will do so; we adopted the subscript notation to emphasize that along each integral curve r is fixed, i.e. is a constant.

Which parameterization should we use? We are normally also given initial conditions along a curve $\Gamma = \Gamma(r)$ with $\Gamma(r) = (\Gamma_1(r), \Gamma_2(r))$, namely

$$u(\Gamma(r)) = \phi(r),$$

where ϕ is a given function. For example, we are given an initial condition on the x-axis: $u(x,0) = \phi(x)$, in which case we may choose $r = x$, $\Gamma(r) = (r,0)$. Then we want the integral curve with parameter r to go through $\Gamma(r)$ at 'time' 0, i.e. we want

$$\gamma_r(0) = \Gamma(r),$$

and we want

$$v_r(0) = u(\gamma_r(0)) = \phi(r).$$

Combining these, we have two groups of ODEs: a system for the integral curves of V, also called *(projected) characteristic curves*, with initial conditions given by Γ (see Figure 3.1), and a scalar ODE along the integral curves with initial condition given by ϕ:

$$(3.5) \qquad \begin{aligned} x_r'(s) &= a(x_r(s), y_r(s)), \ x_r(0) = \Gamma_1(r), \\ y_r'(s) &= b(x_r(s), y_r(s)), \ y_r(0) = \Gamma_2(r), \end{aligned}$$

and

$$(3.6) \qquad v_r'(s) = c(x_r(s), y_r(s), v_r(s)), \ v_r(0) = \phi(r).$$

We solve these by first solving the ODE in (3.5) with the initial conditions, and then solving the ODE in (3.6) with the initial conditions. Then finally we express (r, s) in terms of x, y, i.e. we invert the map

$$(r, s) \mapsto (x(r, s), y(r, s))$$

to get $r = R(x, y)$, $s = S(x, y)$, and then our solution is

$$u(x, y) = v_{R(x,y)}(S(x, y)).$$

Let's do concrete examples, starting with a simple (constant coefficient) homogeneous linear PDE.

Example 3.1. Solve $au_x + bu_y = 0$, where a, b are constants, $a \neq 0$, with the initial condition $u(0, y) = e^y$.

Solution. Our initial curve will be $\Gamma(r) = (0, r)$, and as $y = r$ along Γ, the initial condition is $u(\Gamma(r)) = e^r$. The equations for the (projected) characteristic curves are

$$x_r'(s) = a, \ x_r(0) = 0,$$
$$y_r'(s) = b, \ y_r(0) = r.$$

The solution is

$$x_r(s) = as, \ y_r(s) = r + bs.$$

The ODEs along the characteristic curves are

$$v_r'(s) = 0, \ v_r(0) = e^r.$$

The solution is $v_r(s) = e^r$. Now we need to express r, s in terms of x, y. First, $s = a^{-1}x$, and next $r = y - bs = y - (b/a)x$. Thus, the solution of the PDE is

$$u(x, y) = e^{y-(b/a)x} = e^y e^{-(b/a)x}.$$

Note that the ODE along the characteristic in this case (vanishing c) stated that u is constant along characteristics. In general, the ODE along the characteristic propagates, or transports, the values of u on the initial curve, so this ODE may be called a transport equation.

Note also that we used $a \neq 0$ in the solution. In general, while we can always solve the ODEs for small s at least (see the appendix to the chapter), there is no guarantee that the map $(r,s) \mapsto (x,y)$ is invertible or that the inverse map is differentiable. The problem is (locally) caused by integral curves $\gamma_r(s)$ which at $s=0$ are *tangent* to $\Gamma(r)$. For our problem, Γ was a parameterization of the y-axis, and thus we had to make sure that V is not tangent to the y-axis, i.e. has non-zero x-component, which is precisely the statement $a \neq 0$. In general, we call the initial value problem *non-characteristic* if V is not tangent to Γ. Note that, now writing $x=x(r,s)$, $y=y(r,s)$,

$$\frac{\partial x}{\partial r}(r,0) = \Gamma_1'(r), \quad \frac{\partial y}{\partial r}(r,0) = \Gamma_2'(r),$$

since $x(r,0)=\gamma_1(r)$, $y(r,0)=\gamma_2(r)$, while from the equation of the characteristic curves,

$$\frac{\partial x}{\partial s}(r,0) = a(\Gamma(r)), \quad \frac{\partial y}{\partial s}(r,0) = b(\Gamma(r)).$$

The inverse function theorem, recalled in the appendix to the chapter, tells us that the map $(r,s) \mapsto (x(r,s),y(r,s))$ is invertible near a point $(r_0,0)$ if the Jacobian matrix

$$\begin{bmatrix} \partial_r x & \partial_s x \\ \partial_r y & \partial_s y \end{bmatrix}$$

is invertible, i.e. if its columns are linearly independent, which is *exactly* the statement that V is not tangent to the initial curve Γ.

Also, when V is tangent to Γ, the difficulty is not an artifact of our method: the PDE tells us the derivative of u along V, while the initial condition tells us the derivative of u along Γ, so if V is tangent to Γ, typically these two conditions will contradict each other.

A more interesting example is the following:

Example 3.2. Solve $u_x + yu_y = y^2$, with the initial condition $u(0,y)=\sin y$.

Solution. Now the initial curve is $\Gamma(r)=(0,r)$, the y-axis, so $\Gamma'(r)=(0,1)$ and $V=(1,y)$, so indeed this is a non-characteristic initial value problem. The equations for the characteristic curves are

$$x_r'(s)=1, \quad x_r(0)=0,$$
$$y_r'(s)=y_r(s), \quad y_r(0)=r.$$

The solution is $x_r(s)=s$, $y_r(s)=re^s$. The ODE along the characteristic curves is

$$v_r'(s)=y_r^2(s), \quad v_r(0)=\sin r.$$

Rewriting y in terms of r,s, and dropping the subscript r to simplify notation,

$$v'(s)=r^2 e^{2s}, \quad v(0)=\sin r.$$

The solution of the ODE is

$$v(s) = \sin r + (r^2/2)(e^{2s} - 1).$$

Now expressing r, s in terms of x, y, $s = x$ while $r = ye^{-s} = ye^{-x}$. Thus,

$$u(x, y) = \sin(ye^{-x}) + (1/2)y^2 e^{-2x}(e^{2x} - 1) = \sin(ye^{-x}) + (1/2)y^2 - (1/2)y^2 e^{-2x}.$$

We can make this into a semilinear PDE by changing the right-hand side:

Example 3.3. Solve $u_x + yu_y = u^2$ with the initial condition $u(0, y) = \sin y$.

Solution. The characteristic curves are unchanged, as only the right-hand side of the PDE was altered. Thus, $x_r(s) = s$, $y_r(s) = re^s$, and conversely $s = x$, while $r = ye^{-x}$. The ODE along the characteristic curve now is

$$v'(s) = v(s)^2, \quad v(0) = \sin r.$$

The solution is, from $\int_0^{s_0} \frac{v'}{v^2} ds = \int_0^{s_0} 1\, ds = s_0$, $\frac{-1}{v}\big|_0^s = s$, i.e. $\frac{1}{\sin r} - \frac{1}{v} = s$,

$$v = \left(\frac{1}{\sin r} - s\right)^{-1} = \frac{\sin r}{1 - s(\sin r)}.$$

(Note that some of the algebraic manipulations above don't work when $v = 0$, so one should check the result at the end of the calculations!) Thus,

$$u(x, y) = \frac{\sin(ye^{-x})}{1 - x\sin(ye^{-x})}.$$

We remark that in this case the solution of the ODE along the characteristic curve blows up when the denominator vanishes (though it is non-zero for sufficiently small x, namely $|x| < 1$!), which is fairly typical of non-linear ODE, and thus all one can expect is local solutions near the initial curve Γ in general.

There is a slightly different way of looking at our characteristic curves. Namely, there is some arbitrariness in the way they are parameterized, since the PDE can be rewritten by multiplying through by any (non-zero) factor. In particular, if $a \neq 0$ near a point on Γ, we can divide by a, so the new PDE is

$$(3.7) \qquad u_x + \frac{b(x, y)}{a(x, y)} u_y = \frac{c(x, y, u)}{a(x, y)}.$$

In this version the vector field $V = (a, b)$ is replaced by $W = (1, b/a)$, which is in the same direction as (a, b), so the integral curves have the same image but are parameterized differently. The equations for the integral curves of W are

$$x'(s) = 1, \quad y'(s) = b(x(s), y(s))/a(x(s), y(s)), \quad x(0) = \Gamma_1(r), \quad y(0) = \Gamma_2(r).$$

Thus, $x(s) = s + \Gamma_1(r)$, i.e. x is simply a shifted version of the variable s. We might as well use x as the variable along the integral curve instead, i.e. let $y = y(x)$, so

$$\frac{dy}{dx} = \frac{y'(s)}{x'(s)} = b(x, y)/a(x, y).$$

We can thus completely eliminate the variable s, and we only need to keep the parameter r, i.e. in full notation,

$$\frac{dy_r}{dx} = \frac{y_r'(s)}{x_r'(s)} = b(x, y_r(x))/a(x, y_r(x)).$$

If our initial curve is $\Gamma(r) = (\Gamma_1(r), \Gamma_2(r))$, then we need

$$y_r(\Gamma_1(r)) = \Gamma_2(r)$$

so that $\Gamma(r) = (\Gamma_1(r), y_r(\Gamma_1(r)))$. In addition, the initial condition for the ODE along the characteristic curves is

$$v_r(\Gamma_1(r)) = u(\Gamma_1(r), \Gamma_2(r)) = \phi(r).$$

Let us redo our Example 3.2:

Example 3.4. (Example 3.2 restated.) Solve $u_x + y u_y = y^2$, with the initial condition $u(0, y) = \sin y$.

Solution. The initial curve is $\Gamma(r) = (0, r)$ as before, so the initial condition for characteristic curves is

$$y_r(0) = r.$$

Thus, the characteristic curves solve

$$\frac{dy_r}{dx} = y_r, \ y_r(0) = r.$$

The solution is

$$y_r(x) = r\,e^x,$$

so $r = ye^{-x}$. The ODE along the characteristic curves is now

$$\frac{dv_r}{dx} = y_r^2, \ v_r(0) = \sin r.$$

Substituting $y_r = re^x$, we get

$$\frac{dv_r}{dx} = r^2\,e^{2x}, \ v_r(0) = \sin r,$$

so

$$v_r(x) = \sin r + (r^2/2)(e^{2x} - 1),$$

and thus

$$u(x, y) = \sin(ye^{-x}) + (1/2)y^2 e^{-2x}(e^{2x} - 1) = \sin(ye^{-x}) + (1/2)y^2 - (1/2)y^2 e^{-2x}.$$

Of course, there are numerous variations one can do — for instance, one can ignore the initial conditions initially and instead find a 'general solution', and then try to match the initial conditions. This may be convenient if the initial condition is on a curve that is hard to handle — instead one can pretend that one a has (unknown) initial conditions on a simple curve nearby, e.g. a coordinate axis, find the general solution in terms of these unknown initial data on the simple curve, and then match this general solution with the initial condition.

It is also important to realize that while the solution of the ODEs and the inversion of the coordinate change $(r, s) \mapsto (x(r, s), y(r, s))$ may not be easy to do 'explicitly', the analysis results giving the existence of solutions and invertibility (under the assumption of an invertible Jacobian matrix) are actually rather explicit: they are given by a contraction mapping argument, and one could in principle implement these on a computer, with a stable result. (In other words, numerically these are always locally solvable.)

Another important point is that while here we discussed solving the scalar PDE in 2 independent variables, x and y, the method works for *any number* of independent variables $x_1, \ldots, x_n$; of course the ODEs for the characteristic curves may then be harder to solve. Thus, writing $x = (x_1, \ldots, x_n)$ for the PDE

$$\sum_{j=1}^{n} a_j(x)u_{x_j}(x) = c(x, u(x)), \ \ u(\Gamma(r)) = \phi(r),$$

where $r = (r_1, \ldots, r_{n-1})$ is an $n - 1$-dimensional variable parameterizing the initial *hypersurface* $\Gamma(r) = (\Gamma_1(r), \ldots, \Gamma_n(r))$, the a_j are C^1 and real along with the Γ_j, the vector field V becomes

$$V = (a_1, \ldots, a_n),$$

and the equation for the characteristic curves is

$$\frac{\partial x_j}{\partial s}(r, s) = a_j(x(r, s)), \ j = 1, \ldots, n,$$
$$x_j(r, 0) = \Gamma_j(r).$$

As above, $v(r, s) = u(x(r, s))$ satisfies

$$\frac{\partial v}{\partial s}(r, s) = c(x(r, s), v(r, s)), \ \ v(r, 0) = \phi(r).$$

Thus, the same methods as above work for the non-characteristic initial value problem, i.e. if V is not tangent to the hypersurface Γ.

Example 3.5. Solve $u_x + yu_y + zu_z = u$, with the initial condition $u(0, y, z) = y^2 + z^2$.

Solution. To avoid confusion, we'll label the variables as (x_1, x_2, x_3). Now the initial surface is $\Gamma(r_1, r_2) = (0, r_1, r_2)$, the $x_2 x_3$-plane, so the tangent plane of Γ is the span of $(0, 1, 0)$ and $(0, 0, 1)$, and $V = (1, x_2, x_3)$, so indeed this is a non-characteristic initial value problem. The equations for the characteristic curves are

$$x_1'(r, s) = 1, \quad x_1(r_1, r_2, 0) = 0,$$
$$x_2'(r, s) = x_2(r, s), \quad x_2(r_1, r_2, 0) = r_1,$$
$$x_3'(r, s) = x_3(r, s), \quad x_3(r_1, r_2, 0) = r_2.$$

The solution is $x_1(r, s) = s$, $x_2(r, s) = r_1 e^s$, $x_3(r, s) = r_2 e^s$. The ODE along the characteristic curves is

$$v_r'(s) = v_r(s), \quad v_r(0) = r_1^2 + r_2^2.$$

The solution of the ODE is $v_r(s) = (r_1^2 + r_2^2)e^s$. Now expressing r, s in terms of x, $s = x_1$, while $r_1 = e^{-s}x_2 = e^{-x_1}x_2$, $r_2 = e^{-s}x_3 = e^{-x_1}x_3$. Thus,

$$u(x) = (e^{-2x_1}x_2^2 + e^{-2x_1}x_3^2)e^{x_1} = (x_2^2 + x_3^2)e^{-x_1}.$$

Returning to the original notation,

$$u(x, y, z) = (y^2 + z^2)e^{-x}.$$

Although the title of this chapter is scalar equations, the theory is unchanged for systems of equations, provided the principal (i.e. first order) terms involve the same vector field. (If they do not, the situation is very different; see for example the treatment of the system (8.12) in Chapter 8 by the Fourier transform.) For instance, consider the system

$$V(x, y) \begin{pmatrix} u_1(x, y) \\ \cdots \\ u_N(x, y) \end{pmatrix} = \begin{pmatrix} F_1(x, y, u_1(x, y), \ldots, u_N(x, y)) \\ \cdots \\ F_N(x, y, u_1(x, y), \ldots, u_N(x, y)) \end{pmatrix},$$

with initial conditions

$$\begin{pmatrix} u_1(\Gamma(r)) \\ \cdots \\ u_N(\Gamma(r)) \end{pmatrix} = \begin{pmatrix} \phi_1(r) \\ \cdots \\ \phi_N(r) \end{pmatrix},$$

where the left-hand side of the PDE means simply that we apply V componentwise, i.e. we have

$$\begin{pmatrix} V(x, y)u_1(x, y) \\ \cdots \\ V(x, y)u_N(x, y) \end{pmatrix}$$

there. As before, we consider integral curves $\gamma(r, s) = \gamma_r(s) = (x(r, s), y(r, s))$ of V through an initial curve $\Gamma = \Gamma(r)$. We find these solving the same ODEs as before, and the criterion for inverting the change of coordinates $(r, s) \mapsto$

$(x(r,s), y(r,s))$ is still the non-tangency of $V(\Gamma(r))$ to Γ. Substituting in $v_j(r,s) = u_j(\gamma_r(s))$ yields the system

$$\frac{\partial}{\partial s}\begin{pmatrix} v_1(r,s) \\ \cdots \\ v_N(r,s) \end{pmatrix} = \begin{pmatrix} F_1(x(r,s), y(r,s), v_1(r,s), \ldots, v_N(r,s)) \\ \cdots \\ F_N(x(r,s), y(r,s), v_1(r,s), \ldots, v_N(r,s)) \end{pmatrix},$$

which is a system of ODEs (just like the system of ODEs for (x,y) was already a system even in the scalar case!), and which is thus solvable at least for small s.

Example 3.6. Solve the system of ODE $u_x + yu_y = w$, $w_x + yw_y = u$, with initial conditions $u(0,y) = y$, $w(0,y) = 1$.

Solution. The characteristic curves are the same as in Example 3.3, so $x(r,s) = s$, $y(r,s) = re^s$, and the inverse map is $s = x$, $r = e^{-x}y$. The ODE for (u,w), written as (v_1, v_2), is now

$$\frac{\partial v_1}{\partial s} = v_2(r,s), \ v_1(r,0) = r,$$

$$\frac{\partial v_2}{\partial s} = v_1(r,s), \ v_2(r,0) = 1.$$

Recall from ODE system theory that one should rewrite the right-hand side as a linear system

$$\frac{\partial}{\partial s}\begin{pmatrix} v_1 \\ v_2 \end{pmatrix} = \begin{pmatrix} 0 & 1 \\ 1 & 0 \end{pmatrix}\begin{pmatrix} v_1 \\ v_2 \end{pmatrix}, \qquad \begin{pmatrix} v_1 \\ v_2 \end{pmatrix}(r,0) = \begin{pmatrix} r \\ 1 \end{pmatrix}.$$

The solution of this ODE system then is

$$\begin{pmatrix} v_1 \\ v_2 \end{pmatrix} = \exp\left(s\begin{pmatrix} 0 & 1 \\ 1 & 0 \end{pmatrix}\right)\begin{pmatrix} r \\ 1 \end{pmatrix}.$$

Here the exponential can be calculated either by summing a Taylor series, namely it is

$$\sum_{n=0}^{\infty} \frac{s^n}{n!}\begin{pmatrix} 0 & 1 \\ 1 & 0 \end{pmatrix}^n \begin{pmatrix} r \\ 1 \end{pmatrix},$$

or by diagonalizing the matrix by conjugation by another matrix, and exponentiating the diagonal entries (i.e. the eigenvalues) of this matrix. In this case the eigenvectors are $\begin{pmatrix} 1 \\ 1 \end{pmatrix}$ of eigenvalue 1 and $\begin{pmatrix} 1 \\ -1 \end{pmatrix}$ with eigenvalue -1, so the diagonalization is (using the eigenvectors as the columns in the first factor on the right-hand side, and the eigenvalues as the diagonal entries of the second factor)

$$\begin{pmatrix} 0 & 1 \\ 1 & 0 \end{pmatrix} = \begin{pmatrix} 1 & 1 \\ 1 & -1 \end{pmatrix}\begin{pmatrix} 1 & 0 \\ 0 & -1 \end{pmatrix}\begin{pmatrix} 1 & 1 \\ 1 & -1 \end{pmatrix}^{-1},$$

and the solution is thus (since $\begin{pmatrix} 1 & 1 \\ 1 & -1 \end{pmatrix}^{-1} = \frac{1}{2} \begin{pmatrix} 1 & 1 \\ 1 & -1 \end{pmatrix}$)

$$\begin{pmatrix} v_1 \\ v_2 \end{pmatrix} = \frac{1}{2} \begin{pmatrix} 1 & 1 \\ 1 & -1 \end{pmatrix} \begin{pmatrix} e^s & 0 \\ 0 & e^{-s} \end{pmatrix} \begin{pmatrix} 1 & 1 \\ 1 & -1 \end{pmatrix} \begin{pmatrix} r \\ 1 \end{pmatrix}$$

$$= \frac{1}{2} \begin{pmatrix} e^s & e^{-s} \\ e^s & -e^{-s} \end{pmatrix} \begin{pmatrix} 1 & 1 \\ 1 & -1 \end{pmatrix} \begin{pmatrix} r \\ 1 \end{pmatrix}$$

$$= \begin{pmatrix} \cosh s & \sinh s \\ \sinh s & \cosh s \end{pmatrix} \begin{pmatrix} r \\ 1 \end{pmatrix} = \begin{pmatrix} r \cosh s + \sinh s \\ r \sinh s + \cosh s \end{pmatrix}.$$

Substituting in r and s we finally obtain

$$\begin{pmatrix} u(x,y) \\ w(x,y) \end{pmatrix} = \begin{pmatrix} ye^{-x} \cosh x + \sinh x \\ ye^{-x} \sinh x + \cosh x \end{pmatrix} = \begin{pmatrix} \frac{y}{2}(1 + e^{-2x}) + \sinh x \\ \frac{y}{2}(1 - e^{-2x}) + \cosh x \end{pmatrix}$$

for the solution.

Here we assumed we were working in $\mathbb{R}^2$; changing to $\mathbb{R}^n$ is purely a matter of notational complexity.

Additional material: More on ODE and the inverse function theorem

The main tool for proving general theorems about ODEs and the inverse function theorem is the *contraction mapping theorem.* The contraction mapping theorem concerns maps $f : X \to X$, (X, d) a metric space, and their fixed points. A point x is a *fixed point* of f if $f(x) = x$, i.e. f fixes x. A *contraction mapping* is a map $f : X \to X$ such that there is $\theta \in (0, 1)$ such that $d(f(x), f(y)) \leq \theta d(x, y)$ for all $x, y \in X$. There is one more important hypothesis in the theorem: *completeness* of the metric space. This is a notion that will play a key role in the setting of inner product spaces, but here we give a preview:

Definition 3.1. Suppose that (X, d) is a metric space.

We say that a sequence $\{x_n\}_{n=1}^{\infty}$ is a *Cauchy sequence* if $d(x_n, x_m) \to 0$ as $n, m \to \infty$. Explicitly, this means that given any $\epsilon > 0$ there is $N > 0$ such that for $n, m \geq N$, $d(x_n, x_m) < \epsilon$, i.e. beyond a certain point, all elements of the sequence are within ϵ of each other.

We say that a metric space X is *complete* if every Cauchy sequence converges.

One important example of a complete metric space is $\mathbb{R}$ and another one is $\mathbb{R}^n$. Further examples will be discussed after the theorem.

Theorem 3.2. *Suppose X is a complete metric space and $f : X \to X$ is a contraction mapping. Then f has a unique fixed point x.*

Remark 3.3. Note that if (Z, d) is a complete metric space and X is a closed subset, then X is complete (with the relative metric). Indeed, if $\{x_n\}_{n=1}^{\infty}$ is a Cauchy sequence in X, then it is such in Z and thus it converges to some $x \in Z$. But then x is a limit point of X, and X is closed, so $x \in X$, proving the desired completeness (every Cauchy sequence in X converges to a point in X).

Remark 3.4. Note that any contraction mapping is continuous; one can take $\delta = \epsilon$ in the definition of continuity.

Proof. First suppose x, x' are fixed points of f. Then $f(x) = x$, $f(x') = x'$, so
$$d(x, x') = d(f(x), f(x')) \leq \theta d(x, x'),$$
so $(1 - \theta)d(x, x') \leq 0$, so $d(x, x') \leq 0$, so $d(x, x') = 0$, and so $x = x'$, showing uniqueness.

We now turn to existence. To do so, take an arbitrary $x_0 \in X$, and define x_n inductively: $x_{n+1} = f(x_n)$, $n \geq 0$. We claim that $\{x_n\}_{n=0}^{\infty}$ is Cauchy. To see this, notice first that for all $n \geq 1$,
$$d(x_{n+1}, x_n) = d(f(x_n), f(x_{n-1})) \leq \theta d(x_n, x_{n-1}),$$
so by induction,
$$d(x_{n+1}, x_n) \leq \theta^n d(x_1, x_0).$$
Thus, for $n > m$, using the triangle inequality,
$$d(x_n, x_m) \leq d(x_n, x_{n-1}) + \ldots + d(x_{m+1}, x_m) \leq (\theta^{n-1} + \theta^{n-2} + \ldots + \theta^m)d(x_1, x_0).$$
Summing the finite geometric series,
$$d(x_n, x_m) \leq \theta^m(1 + \theta + \ldots + \theta^{n-m-1})d(x_1, x_0)$$
$$= \theta^m d(x_1, x_0)\frac{1 - \theta^{n-m}}{1 - \theta} \leq \theta^m \frac{d(x_1, x_0)}{1 - \theta}.$$
Since $\lim_{m \to \infty} \theta^m = 0$, we have that given $\epsilon > 0$ there exists N such that $m \geq N$ implies $\theta^m < \epsilon \frac{1-\theta}{1+d(x_1, x_0)}$, and thus for $n, m \geq N$, $d(x_n, x_m) < \epsilon$, proving the Cauchy claim.

But X is complete, so $x = \lim_{n \to \infty} x_n$ exists. Since f is continuous and $\lim x_n = x$, sequential continuity shows that $\lim f(x_n) = f(x)$. But $f(x_n) = x_{n+1}$, so $\lim_{n \to \infty} x_{n+1} = f(x)$. Since $\lim_{n \to \infty} x_n = \lim_{n \to \infty} x_{n+1}$, we deduce that $f(x) = x$, so x is a fixed point of f as claimed. $\square$

Note that the proof of the contraction mapping theorem is *constructive*: starting with any $x \in X$, one forms the sequence $x, f(x), f(f(x))$, $f(f(f(x))), \ldots$, and the theorem guarantees that this sequence converges to the fixed point. Thus, by taking sufficiently many iterations, one can approximate the fixed point arbitrarily well.

We now use this in the simplest ODE setting. An ODE is an equation of the form

$$(3.8) \qquad x'(t) = F(t, x(t)), \ x(t_0) = x_0,$$

where $F : \mathbb{R} \times \mathbb{R}^n \to \mathbb{R}^n$ is a given function, $t_0 \in \mathbb{R}$, $x_0 \in \mathbb{R}^n$, and one is looking for a solution x at least on an interval near t_0, say $[t_0 - \delta, t_0 + \delta]$, $\delta > 0$, so $x \in C^1([t_0 - \delta, t_0 + \delta])$. We impose the minimal requirement that F be continuous for the following discussion.

Rather than considering the ODE directly, we rewrite it as an integral equation using the fundamental theorem of calculus:

$$x(t) = x(t_0) + \int_{t_0}^t x' = x_0 + \int_{t_0}^t F(\tau, x(\tau)) \, d\tau.$$

Note that any solution of the original ODE solves the integral equation

$$(3.9) \qquad x(t) = x_0 + \int_{t_0}^t F(\tau, x(\tau)) \, d\tau,$$

and conversely, if $x \in C^0([t_0 - \delta, t_0 + \delta])$ merely solving (3.9), then in fact x is C^1 by the fundamental theorem of calculus (since F is continuous, being the composite of continuous functions!), and therefore solves the ODE.

Thus, from now on we consider (3.9). We consider the equation as a fixed point claim for a map $T : X \to X$, $X = C^0([t_0 - \delta, t_0 + \delta]; \mathbb{R}^n)$, i.e. $\mathbb{R}^n$-valued continuous functions on $[t_0 - \delta, t_0 + \delta]$, which alternatively can be thought of as n-tuples $(x_1, \ldots, x_n)$ of real-valued continuous functions. As discussed earlier, X is a normed vector space, thus a metric space, with metric

$$d(x, y) = \sup_{t \in [t_0 - \delta, t_0 + \delta]} \|x(t) - y(t)\|,$$

with the norm on the right-hand side being that in $\mathbb{R}^n$. (So if $n = 1$, it is simply an absolute value.) Concretely, let

$$Tx(t) = x_0 + \int_{t_0}^t F(\tau, x(\tau)) \, d\tau;$$

so certainly $T : X \to X$. Now, as $d(x, y) = \sup\{\|x(t) - y(t)\| : t \in [t_0 - \delta, t_0 + \delta]\}$, we have

$d(Tx, Ty)$

$$= \sup \left\{ \left\| (x_0 + \int_{t_0}^t F(\tau, x(\tau)) \, d\tau) - (x_0 + \int_{t_0}^t F(\tau, y(\tau)) \, d\tau) \right\| \right.$$

$$\left. : t \in [t_0 - \delta, t_0 + \delta] \right\}$$

$$= \sup \left\{ \left\| \int_{t_0}^t (F(\tau, x(\tau)) - F(\tau, y(\tau)) \, d\tau \right\| : t \in [t_0 - \delta, t_0 + \delta] \right\}.$$

So let us suppose that F is *globally Lipschitz* in the second variable, i.e. for some $\delta_0 > 0$,

(3.10) $\|F(t,x) - F(t,y)\| \le M\|x-y\|, \ x,y \in \mathbb{R}^n, t \in [t_0 - \delta_0, t_0 + \delta_0].$

Then, if $\delta \le \delta_0$, $t \ge t_0$ (with a similar calculation if $t < t_0$),

$$\left\| \int_{t_0}^{t} (F(\tau, x(\tau)) - F(\tau, y(\tau)) \, d\tau) \right\|$$

(3.11)
$$\le \int_{t_0}^{t} \|(F(\tau, x(\tau)) - F(\tau, y(\tau))\| \, d\tau \le \int_{t_0}^{t} M\|x(\tau) - y(\tau)\| \, d\tau$$

$$\le \int_{t_0}^{t} M \sup\{\|x(s) - y(s)\| : \ s \in [t_0 - \delta, t_0 + \delta]\} \, d\tau$$

$$= M(t - t_0) \sup\{\|x(s) - y(s)\| : \ s \in [t_0 - \delta, t_0 + \delta]\}$$

$$\le M\delta \sup\{\|x(s) - y(s)\| : \ s \in [t_0 - \delta, t_0 + \delta]\} = M\delta d(x,y).$$

Thus,

$$d(Tx, Ty) \le M\delta d(x,y),$$

so it is a contraction mapping provided $\delta < \frac{1}{M}$.

So take $\delta < 1/M$, $\delta \le \delta_0$. Then T is a contraction mapping on the metric space $C^0([t_0 - \delta, t_0 + \delta]; \mathbb{R}^n)$ (continuous maps from $[t_0 - \delta, t_0 + \delta]$ to $\mathbb{R}^n$). Further, X is *complete*, as sketched in Problem 3.6 below. Thus, T has a unique fixed point $x \in C^0([t_0 - \delta, t_0 + \delta]; \mathbb{R}^n)$, which as explained means that the ODE has a unique solution on $[t_0 - \delta, t_0 + \delta]$:

Theorem 3.5. *Suppose F is globally Lipschitz on $[t_0 - \delta_0, t_0 + \delta_0] \times \mathbb{R}^n$ in the sense of (3.10). Then there is $\delta > 0$, $\delta \le \delta_0$ such that the ODE (3.8) has a unique C^1 solution on $[t_0 - \delta, t_0 + \delta]$.*

This is the simplest local existence and uniqueness theorem for ODE. The only unsatisfactory assumption in it is (3.10). Notice that the estimate of (3.10) is a very reasonable assumption *locally*, namely if it is replaced by

(3.12) $\|F(t,x) - F(t,y)\| \le M\|x-y\|, \ x,y \in \overline{B}_R(0), t \in [t_0 - \delta_0, t_0 + \delta_0],$

where $\overline{B}_R(0)$ is the closed ball of radius R in $\mathbb{R}^n$: $\overline{B}_R(0) = \{x \in \mathbb{R}^n : \|x\| \le R\}$. Indeed, in this case it follows from F being C^1, since by the fundamental theorem of calculus (see also Problem 1.2),

$$F(t,x) - F(t,y) = \sum_{j=1}^{n} (x_j - y_j) \int_0^1 \partial_j F(t, y + s(x-y)) \, ds$$

$$= \int_0^1 \partial F(t, y + s(x-y))(x-y) \, ds,$$

where ∂_j, ∂ denote derivatives in the second slot. Thus,

$$\|F(t,x) - F(t,y)\| \leq \int_0^1 \|\partial F(t, y + s(x-y))(x-y)\| \, ds$$

$$\leq \int_0^1 \|\partial F(t, y + s(x-y))\| \|x - y\| \, ds$$

$$= \|x - y\| \int_0^1 \|\partial F(t, y + s(x-y))\| \, ds,$$

so if F is C^1, so $\partial_j F$ are bounded on the compact set $[t_0 - \delta_0, t_0 + \delta_0] \times \overline{B}_R(0)$,

$$\|F(t,x) - F(t,y)\| \leq C\|x - y\|,$$

$$C = \sup\{\|\partial F(z)\| : z \in [t_0 - \delta_0, t_0 + \delta_0] \times \overline{B}_R(0)\}.$$

The more natural ODE theorem is then:

Theorem 3.6 (Local existence and uniqueness theorem for ODE). *Suppose F is locally Lipschitz on $[t_0 - \delta_0, t_0 + \delta_0] \times \overline{B}_R(0)$, $R > 0$, in the sense of (3.12). Then there is $\delta > 0$, $\delta \leq \delta_0$ such that for $x_0 \in \overline{B}_{R/2}(0)$ the ODE (3.8) has a unique C^1 solution on $[t_0 - \delta, t_0 + \delta]$ with $\|x(t)\| \leq R$ on $[t_0 - \delta, t_0 + \delta]$.*

Proof. Let $C_0 = \sup\{\|F(z)\| : \in [t_0 - \delta_0, t_0 + \delta_0] \times \overline{B}_R(0)\}$. Then for $\|x_0\| \leq R/2$, $\delta \leq \delta_0$, $x \in C^0([t_0 - \delta, t_0 + \delta]; \mathbb{R}^n)$ with $\|x(t)\| \leq R$ for $t \in [t_0 - \delta, t_0 + \delta]$, and for $t \geq t_0$ (with a similar formula for $t < t_0$),

$$\|(Tx)(t)\| \leq \|x_0\| + \int_{t_0}^t \|F(\tau, x(\tau))\| \, d\tau \leq R/2 + C_0(t - t_0) \leq R/2 + C_0\delta.$$

Thus, if $\delta < R/(2C_0 + 1)$, then

$$\|(Tx)(t)\| \leq R, \ t \in [t_0 - \delta, t_0 + \delta].$$

Thus, if we let $X = \{x \in C^0([t_0 - \delta, t_0 + \delta]; \mathbb{R}^n) : \sup\|x\| \leq R\}$, then $T : X \to X$ provided $\delta < R/(2C_0 + 1)$ (and $\delta \leq \delta_0$).

Further, for $x \in X$, we have by (3.12)

$$\|F(t, x(t)) - F(t, y(t))\| \leq M\|x(t) - y(t)\|, \ t \in [t_0 - \delta, t_0 + \delta],$$

so the calculation of (3.11) applies. This gives that if in addition $\delta < M^{-1}$, then T is a contraction mapping, and thus the contraction mapping theorem gives a unique fixed point. This gives a unique solution to the ODE with the property that $\|x(t)\| \leq R$ for $t \in [t_0 - \delta, t_0 + \delta]$, and proves the theorem. $\square$

In fact, it is not hard to strengthen this theorem, showing that the solution is C^{k+1} if F is in addition C^k, and also that the solution depends smoothly on the initial condition x_0.

Finally let us make the contraction mapping argument explicit in the case of a concrete ODE,

$$x'(t) = x(t), \ x(0) = 1.$$

Notice that for this ODE $F(t,x) = x$ simply, which is in fact globally Lipschitz with Lipschitz constant $M = 1$. To obtain the solution, let us start with any continuous function on an interval including 0; it is reasonable to start with the constant function satisfying the initial condition, i.e. we have $x_1(t) = 1$. Then $x_2(t) = (Tx_1)(t) = 1 + \int_0^t 1 \, dt = 1 + t$; $x_3(t) = (Tx_2)(t) = 1 + \int_0^t (1+t) \, dt = 1 + t + \frac{t^2}{2}$; in general $(Tx_n)(t) = \sum_{j=0}^{n-1} \frac{t^j}{j!}$, as shown by induction. This is the Taylor series at 0 of the solution $x(t) = e^t$ of the ODE; this illustrates the convergence to x. Note that we could have started with any other (continuous) x_1; in this case the particular approximating sequence we get would be different, but it would still converge to e^t (on suitable intervals).

The inverse function theorem can also be proved by the contraction mapping theorem. Here we simply state it and refer to [**4**] for its proof.

Theorem 3.7. *Suppose $\Omega \subset \mathbb{R}^n$ is open, $F : \Omega \to \mathbb{R}^n$ is C^1, $x_0 \in \Omega$, and $(\partial F)(x_0)$ is invertible. Then there exist open sets $O, U \subset \mathbb{R}^n$ such that $x_0 \in O$, $F(O) = U$, such that $F|_O$ is invertible with a C^1 inverse $F^{-1} : U \to O$ and with $(\partial F^{-1})(y) = [(\partial F)(F^{-1}(y))]^{-1}$ for $y \in U$.*

Further, if F is C^k, $k \geq 1$, then F^{-1} is also C^k.

Problems

Problem 3.1.

(i) Solve
$$u_x + (\sin x)u_y = y, \ u(0,y) = 0.$$
(ii) Sketch the projected characteristic curves for this PDE.

Problem 3.2. Solve
$$u_x + yu_y = u^2, \ u(0,y) = \cos y,$$
for $|x|$ small.

Problem 3.3.

(i) Solve
$$yu_x + xu_y = 0, \ u(0,y) = e^{-y^2}.$$
(ii) In which region is u uniquely determined?

Problem 3.4.

(i) Solve $u_x + u_t = u^2$, $u(x,0) = e^{-x^2}$.

(ii) Show that there is $T > 0$ such that u blows up at time T, i.e. u is continuously differentiable for $t \in [0, T)$, x is arbitrary, but for some x_0, $|u(x_0, t)| \to \infty$ as $t \to T-$. What is T?

Problem 3.5. This problem and the following one serve to solidify the understanding of completeness, which is a key tool in the additional material section. They are, however, not required for a good understanding of the main chapter.

(i) Show that every convergent sequence in a metric space (X, d) is Cauchy.

(ii) Show that every Cauchy sequence in a metric space (X, d) is bounded.

(iii) Show that if K is a compact subset of X, then K (with the relative metric, i.e. the metric restricted to $K \times K$) is complete. (In particular, all compact metric spaces are complete.)

(iv) Show that $\mathbb{R}^n$ is complete.

Problem 3.6.

(i) Suppose A is a non-empty set, and let $\ell^\infty(A)$ consist of bounded real-valued functions on A, i.e. functions $f : A \to \mathbb{R}$ for which $f(A) = \{f(x) : x \in A\}$ is bounded. Let $\|f\|_{\ell^\infty} = \sup\{|f(x)| : x \in A\}$. Show that $\ell^\infty(A)$ is a normed vector space and is complete.

Note: By a simple rewriting of the definition of completeness, completeness of $\ell^\infty(A)$ is equivalent to the following statement. If $f_n \in \ell^\infty(A)$ is such that for all $\epsilon > 0$ there is N such that $n, m \geq N$ implies that for all $x \in A$, $|f_n(x) - f_m(x)| \leq \epsilon$ (i.e. '$\{f_n\}_{n=1}^\infty$ is uniformly Cauchy'), then there exists $f \in \ell^\infty(A)$ such that for all $\epsilon > 0$ there is N such that $n \geq N$ implies that for all $x \in A$, $|f(x) - f_n(x)| \leq \epsilon$ (i.e. '$\{f_n\}_{n=1}^\infty$ converges uniformly to f'). Also, if $A = \{1, \ldots, n\}$, then $\ell^\infty(A) = \mathbb{R}^n$ simply.

(ii) Suppose now (M, d) is a compact non-empty metric space (if you wish, you may assume $M \subset \mathbb{R}^d$), and let $C(M)$ be the set of continuous real-valued functions on M. Show that $C(M) \subset \ell^\infty(M)$, and that when equipped with the ℓ^∞-norm, $C(M)$ is also a complete normed vector space.

Hint: Since by (i) $\ell^\infty(M)$ is complete, you can show completeness by proving that $C(M)$ is closed in $\ell^\infty(M)$.

Chapter 4

First order scalar quasilinear equations

We now consider first order quasilinear equations. These have the form

$$(4.1) \qquad a(x,y,u)\, u_x + b(x,y,u)\, u_y = c(x,y,u),$$

with a,b,c at least C^1, given real-valued functions. There is an immediate difference between semilinear and quasilinear equations at this point: since a and b depend on u, we cannot associate a vector field on $\mathbb{R}^2$ to the equation. We need to work on $\mathbb{R}^3$ at least to account for the (x,y,u) dependence.

To achieve this, we proceed as follows. We consider the graph $\mathcal{S}$ of u in $\mathbb{R}^2 \times \mathbb{R}$, given by $z = u(x,y)$. If we know the solution u, we know the graph, and conversely we can recover u if we find its graph, a surface in $\mathbb{R}^3$. So let

$$U(x,y,z) = u(x,y) - z;$$

we need to find the 0-set of U. Now, substituting U into the left-hand side of the PDE we get

$$(a(x,y,u(x,y))\partial_x + b(x,y,u(x,y))\partial_y)\, U(x,y,z)$$
$$= a(x,y,u(x,y))\, u_x(x,y) + b(x,y,u(x,y))\, u_y(x,y).$$

If the PDE holds, this equals $c(x,y,u(x,y))$, which is in turn equal to

$$-c(x,y,u(x,y))\partial_z U$$

as $\partial_z U \equiv -1$! Thus, our PDE implies that

$$(a(x,y,u(x,y))\partial_x + b(x,y,u(x,y))\partial_y + c(x,y,u(x,y))\partial_z)\, U(x,y,z) = 0$$

for all x, y, z. This equation states that the directional derivative of U along the vector field $(a, b, c) = (a(x, y, u(x, y)), b(x, y, u(x, y)), c(x, y, u(x, y)))$ vanishes. Note that $\partial_z U = -1$, i.e. never vanishes, so in fact at each point (x, y, z) the directional derivative can only vanish along a plane (and not in every direction). But along any level set $\mathcal{S}$ of U its directional derivative certainly vanishes, so applied to the 0-level set, where $u(x, y) = z$, the PDE states that the vector field

$$W(x, y, z) = (a(x, y, z), b(x, y, z), c(x, y, z))$$

is tangent to $\mathcal{S}$; see Figure 4.1. We can now rephrase our problem as follows: we want to find $\mathcal{S}$ as a union of integral curves of W. As before, integral curves of W, where we introduce a parameter r again, satisfy

$$\frac{\partial x}{\partial s}(r, s) = a(x(r, s), y(r, s), z(r, s)),$$

$$\frac{\partial y}{\partial s}(r, s) = b(x(r, s), y(r, s), z(r, s)),$$

$$\frac{\partial z}{\partial s}(r, s) = c(x(r, s), y(r, s), z(r, s)).$$

These are called the characteristic ODEs, and integral curves are called the *characteristics*, though perhaps it would be most appropriate to call them *lifted characteristics* in view of our terminology for semilinear equations. Note that although the integral curves are in $\mathbb{R}^3$, hence there is a two dimensional family of them, we only care about the ones which will give $\mathcal{S}$ as their union, i.e. a one dimensional family of integral curves. Our initial condition will again come from the initial condition of the PDE, which is of the form

$$u(\Gamma(r)) = \phi(r)$$

along a curve $\Gamma(r) = (\Gamma_1(r), \Gamma_2(r))$. The initial condition states that over the curve Γ, $\mathcal{S}$ is given by $z = \phi(r)$, i.e. that $\mathcal{S}$ goes through the curve

$$\tilde{\Gamma}(r) = (\Gamma_1(r), \Gamma_2(r), \phi(r)).$$

Our initial condition for the characteristic ODEs is then

$$x(r, 0) = \Gamma_1(r), \ \ y(r, 0) = \Gamma_2(r), \ \ z(r, 0) = \phi(r).$$

Once we solve the characteristic ODEs with these initial conditions, we again express (r, s) in terms of (x, y), namely $r = R(x, y)$, $s = S(x, y)$, by inverting the map $(r, s) \mapsto (x, y)$. Then the solution of our PDE is

$$u(x, y) = z(R(x, y), S(x, y)).$$

There are two possible geometric problems with this method, even near Γ (for large values of s of course the ODEs may not have a solution, as we already observed in the semilinear case). First, our vector field W may be tangent to $\tilde{\Gamma}$ in which case we do not have a smooth surface $\mathcal{S}$. Second, even

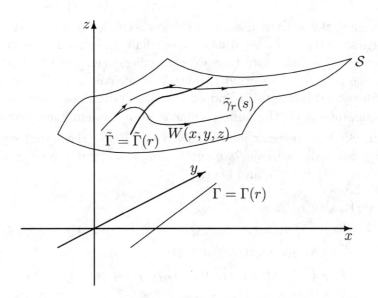

Figure 4.1. (Lifted) characteristic curves $\tilde{\gamma} = \tilde{\gamma}_r(s)$, shown by thin lines, for the vector field $W = W(x, y, z)$, shown by arrows, going through the lifted initial curve $\tilde{\Gamma} = \tilde{\Gamma}(r)$ inside the surface $\mathcal{S}$.

where the ODEs have a solution, it can happen that the union of the integral curves is not a graph above the (x, y)-plane (two points in $\mathcal{S}$ have the same (x, y)-coordinates), or it can happen in a differential sense, i.e. the vertical vectors $(0, 0, 1)$ may become tangent to $\mathcal{S}$. If neither of these happens, we get a solution of the PDE.

Now, the first condition is easy to check, as it only involves the initial conditions. Note that this property (that W is not tangent to $\tilde{\Gamma}$) now *depends on the initial condition, $\phi(r)$,* unlike those of semilinear equations.

The second problem cannot happen near $\tilde{\Gamma}$ (i.e. sufficiently close to Γ) provided that $(a(\Gamma(r), \phi(r)), b(\Gamma(r), \phi(r)))$ is *not* tangent to Γ, i.e. provided that Γ is *not characteristic* for this PDE *with these initial data*. Note this assumption means that (a, b) is *not* a multiple of $(\Gamma_1'(r), \Gamma_2'(r))$, so in particular (a, b, c) is not a multiple of $(\Gamma_1'(r), \Gamma_2'(r), \phi'(r))$, and so under our non-characteristic assumption $\mathcal{S}$ is a smooth surface near $\tilde{\Gamma}$ at least and the first problem is eliminated. To see our claim that the second problem cannot happen for non-characteristic Γ, we remark that in this case $(0, 0, 1)$ is *not* a tangent vector to $\mathcal{S}$ at $\tilde{\Gamma}$, since the tangent space of $\mathcal{S}$ is spanned by linear combinations of tangent vectors to $\tilde{\Gamma}$, one of which is $(\Gamma_1'(r), \Gamma_2'(r), \phi'(r))$, and of W, i.e. $(a(\Gamma(r), \phi(r)), b(\Gamma(r), \phi(r)), c(\Gamma(r), \phi(r)))$, non-trivial linear combinations of which cannot have vanishing first two components since $(\Gamma_1'(r), \Gamma_2'(r))$ and (a, b) are not parallel by the non-characteristic assumption.

Let's check the analytic meaning of the non-characteristic condition. We can always solve the characteristic ODE for small $|s|$, just as in the semilinear case. We need to make sure that we can express (r, s) in terms of (x, y), i.e. that the map $(r, s) \mapsto (x, y)$ is invertible, at least near Γ. But by the inverse function theorem this is indeed possible under the non-characteristic assumption, by exactly the same argument as in the semilinear setting.

To check that we have indeed solved the PDE (the initial condition is clearly satisfied), we calculate, with $u(x, y) = z(R(x, y), S(x, y))$, i.e. $z(r, s) = u(x(r, s), y(r, s))$, and hence

$$\partial_s u(x(r, s), y(r, s))$$
$$= (\partial_s x)(r, s)\, u_x(x(r, s), y(r, s)) + (\partial_s y)(r, s)\, u_y(x(r, s), y(r, s))$$
$$= a(x(r, s), y(r, s), z(r, s))\, u_x(x(r, s), y(r, s))$$
$$\quad + b(x(r, s), y(r, s), z(r, s))\, u_y(x(r, s), y(r, s)).$$

On the other hand,

$$\partial_s u(x(r, s), y(r, s)) = \partial_s z(r, s) = c(x(r, s), y(r, s), z(r, s)).$$

Thus,

$$a(x(r, s), y(r, s), z(r, s))\, u_x(x(r, s), y(r, s))$$
$$\quad + b(x(r, s), y(r, s), z(r, s))\, u_y(x(r, s), y(r, s))$$
$$= c(x(r, s), y(r, s), z(r, s)),$$

and substituting in $z(r, s) = u(x(r, s), y(r, s))$, writing $r = R(x, y)$, $s = S(x, y)$, we deduce that the PDE indeed holds.

Just as for semilinear equations, it is easy to generalize the method to more than two independent variables, i.e. replacing (x, y) by $x = (x_1, \ldots, x_n)$, so the PDE is

$$\sum_{j=1}^{n} a_j(x, u(x))\, \partial_{x_j} u(x) = c(x, u(x)).$$

In this case $r = (r_1, \ldots, r_{n-1})$ is an $n - 1$-dimensional variable and the characteristic ODEs are

$$\partial_s x_j(r, s) = a_j(x(r, s), z(r, s)), \quad x_j(r, 0) = \Gamma_j(r),$$
$$\partial_s z(r, s) = c(x(r, s), z(r, s)), \quad z(r, 0) = \phi(r).$$

To see how this method works, let's first work out what it gives for semilinear PDEs. Thus, assume that a and b are independent of u. Then the characteristic ODE's become

$$\frac{\partial x}{\partial s}(r,s) = a(x(r,s), y(r,s)),$$

$$\frac{\partial y}{\partial s}(r,s) = b(x(r,s), y(r,s)),$$

$$\frac{\partial z}{\partial s}(r,s) = c(x(r,s), y(r,s), z(r,s)),$$

with initial conditions

$$x(r,0) = \Gamma_1(r), \ y(r,0) = \Gamma_2(r), \ z(r,0) = \phi(r).$$

The first two ODEs (for x and y) then decouple from the third and are exactly the same equations as the ones we had for our (projected) characteristic curves for semilinear equations. This explains the term 'projected': the projected characteristics are just the projections of the lifted characteristics! Finally, the ODE for z is also the same as the transport ODE along the characteristic in the semilinear case, except we denoted the unknown function by v there.

Turning to a quasilinear equation now:

Example 4.1 (*Burgers' equation*). Solve $u_t + u u_x = 0$, $u(x,0) = \phi(x)$.

We write our independent variables as (x,t), i.e. t takes the role of y. Our initial curve is then $\Gamma(r) = (r,0)$, and as $x = r$ along Γ, the initial condition is $u(\Gamma(r)) = \phi(r)$. The equations for the (lifted) characteristic curves are

$$x_r'(s) = z, \ x_r(0) = r,$$

$$t_r'(s) = 1, \ t_r(0) = 0,$$

$$z_r'(s) = 0, \ z_r(0) = \phi(r).$$

We check that this is non-characteristic: tangent vectors of Γ are multiples of $(1,0)$ while $(a,b) = (z,1)$ has a non-vanishing second component, so (a,b) is not tangent to Γ.

We now solve the ODEs, starting with the z and t ODEs (since these do not involve x) and followed by the x ODE. The first two give

$$z(r,s) = \phi(r), \ t(r,s) = s.$$

The first of these again shows that z (i.e. u) is constant along the projected characteristic curves. Substituting into the x ODE we get

$$x(r,s) = \phi(r)s + r.$$

There is only one more step, that of expressing (r,s) in terms of (x,t). While $s = t$, the remaining equation is

(4.2) $$x = \phi(r)t + r,$$

which is hard to solve explicitly for r in general (except at $t = 0$!). Note that x depends on the initial data, ϕ. We can rewrite this slightly as

$$r = x - \phi(r)t = x - zt,$$

so we deduce that

$$u(x,t) = z(R(x,t), S(x,t)) = \phi(R(x,t)) = \phi(x - u(x,t)t),$$

or $u = \phi(x - ut)$ for short. While this is rather implicit, note that along the projected characteristics

$$\frac{dx}{dt} = x_r'(s)/t_r'(s) = z(r, s) = \phi(r),$$

and r is constant along these, so the slope of the projected characteristics is constant! Moreover, as already mentioned, z (hence u) is constant along these projected characteristics.

To see what sort of problems this causes, suppose that there is $r_1 < r_2$ such that $\phi(r_1) > \phi(r_2)$. Then the characteristic curves intersect somewhere with $t > 0$. But u is constant along the characteristic curves, so at the intersection point it should take both of the unequal values $\phi(r_1)$ and $\phi(r_2)$. Thus, singularities of solutions must form.

We solve the equation in a special case:

Example 4.2 (Burgers' equation with special initial data). Solve $u_t + u u_x = 0$, $u(x,0) = x$.

Solution. We have $\Gamma(r) = (r, 0)$ and $\phi(r) = r$. Thus, (4.2) becomes

$$x = rt + r = r(t + 1),$$

so $r = x/(t + 1)$. Thus,

$$u(x,t) = z(R(x,t), S(x,t)) = \phi(R(x,t)) = R(x,t) = \frac{x}{t + 1},$$

and one checks easily that this indeed solves the PDE. Notice that while for $t \geq 0$ (which is what one usually considers for Burgers' equation) the solution is smooth, for $t < 0$ it fails to be such: it blows up as $t \to -1$.

The same general method works for equations describing scalar conservation laws, namely

(4.3) $u_t + f(u)_x = 0, \quad u(x,0) = \phi(x),$

where we assume $f(0) = 0$ and f is C^1. This is called a conservation law because for C^1 functions u with sufficient decay as x goes to infinity, if they solve this PDE,

$$\frac{d}{dt}\int_{\mathbb{R}} u(x,t)\,dx = \int_{\mathbb{R}} u_t\,dx = -\int_{\mathbb{R}} f(u)_x\,dx = 0,$$

where the last equation follows from the fundamental theorem of calculus and the decay of $f(u)$ as $|x| \to \infty$. Thus, the total 'mass', $\int_{\mathbb{R}} u(t, x) \, dx$, is conserved, i.e. is independent of t. Burgers' equation is a special case with $f(z) = z^2/2$.

Since $f(u)_x = f'(u)u_x$, the characteristic equations for (4.3) are

$$x'_r(s) = f'(z), \ x_r(0) = r,$$
$$t'_r(s) = 1, \ t_r(0) = 0,$$
$$z'_r(s) = 0, \ z_r(0) = \phi(r).$$

We now solve the ODEs, starting with the z and t ODEs and followed by the x ODE. The first two give

$$z(r, s) = \phi(r), \ t(r, s) = s,$$

just as before. In particular, z (i.e. u) is constant along the projected characteristic curves. Substituting into the x ODE we get

$$x(r, s) = f'(\phi(r))s + r.$$

The projected characteristics are thus still straight lines, but now

$$\frac{dx}{dt} = x'_r(s)/t'_r(s) = f'(z(r, s)) = f'(\phi(r));$$

hence typically all the phenomena mentioned beforehand (colliding characteristics) occur.

Example 4.3. Solve

$$u_t + u^2 u_x = 0, \ u(x, 0) = x.$$

Solution. Notice that this is a conservation law with $f(z) = z^3/3$, i.e. $f'(z) = z^2$. Our initial condition, with the above parameterization, is $\phi(r) = r$. Thus,

$$x(r, s) = f'(\phi(r))s + r = \phi(r)^2 s + r = r^2 s + r.$$

Since $s = t$, this states that r solves

$$tr^2 + r - x = 0,$$

so

$$r = \frac{-1 \pm \sqrt{1 + 4xt}}{2t}.$$

For x bounded and t small, the square root behaves like $1 + 2xt$ by Taylor's theorem, so the formula gives that for t small r is approximately $(-1 \pm (1 + 2xt))/(2t)$. Since we want this to be finite, and equal to x for $t = 0$, we need to pick the $+$ sign, so

$$R(x, t) = r = \frac{-1 + \sqrt{1 + 4xt}}{2t}.$$

Hence, the solution is

$$u(x,t) = z(R(x,t), S(x,t)) = \phi(R(x,t)) = R(x,t) = \frac{-1 + \sqrt{1 + 4xt}}{2t},$$

defined where $xt > -1/4$. In particular, the time interval for which the equation has a solution shrinks as $x \to -\infty$.

Problems

Problem 4.1. Solve

$$u_t + uu_x = 0, \ \ u(x,0) = -x^2$$

for $|t|$ small.

Problem 4.2. Solve

$$e^t u_t - uu_x = 0, \ \ u(x,0) = x$$

for $|t|$ small.

Problem 4.3. Consider the PDE

$$(1 + u)u_x + uu_y = 0, \ \ (x,y) \in \mathbb{R}^2,$$
$$u(0,y) = y.$$

Solve the PDE in a neighborhood of the origin $(0,0)$. Can you solve the PDE in a neighborhood of the full y-axis? What happens near $x = 0, y = -1$? Justify your answer.

Problem 4.4. Consider the PDE

$$uu_x + uu_y = 1.$$

(i) Solve the PDE for small $|y|$ if $u(x,0) = 1$.

(ii) Solve the PDE for small $|y|$ near the positive x-axis $\{(x,0) : x > 0\}$ if $u(x,0) = x$. Compare the equations you get if you consider this as a quasilinear PDE vs. if you consider it as a semilinear PDE after division by u. (Note that for $x > 0$, $u \neq 0$ along the initial curve, so this is justified.)

Problem 4.5. Consider the PDE

$$u_t + uu_x = 0, \ \ u(x,0) = \phi(x).$$

Suppose that $\phi' \geq -C$, where $C > 0$. Show that the PDE has a C^1 solution on $\mathbb{R}_x \times [0, \frac{1}{C})_t$. Show also that for $t \in [0, \frac{1}{C})$, u_x satisfies the estimate

$$u_x(x,t) \geq \frac{1}{t - C^{-1}}.$$

(Note that the right-hand side is negative!) (*Hint:* Consider the difference quotients $\frac{u(\xi_2(t),t)-u(\xi_1(t),t)}{\xi_2(t)-\xi_1(t)}$, where $x = \xi_j(t)$ are the projected characteristic curves emanating from point x_j on the x-axis.)

Chapter 5

Distributions and weak derivatives

The basic motivation for distribution theory is that sometimes a PDE does not have a classical solution, i.e. there are mth order PDEs which do not have C^m solutions (with whatever conditions we want to impose). By going to a larger framework, namely that of distributions, we may find solutions to the PDE. As an example, we saw that solutions of quasilinear conservation laws may break down when the projected characteristics intersect. This makes the existence of continuous solutions impossible. However, discontinuous solutions may, and in fact do, exist, as some problems below show. These are called shock waves: airplanes breaking the speed of sound generate a sonic boom, which is an example of such a shock. Thus, our task is to make sense of what a PDE means for a discontinuous putative solution. We exceed this goal by going to an even larger framework, that of distributions, in which one can differentiate as many times as one wants to.

To see how this is built up, we start with a reasonable class of objects, such as continuous functions on $\mathbb{R}^n$, and embed them into a bigger space by a map ι. The bigger space is that of distributions, which we soon define. The idea is that ι is a one-to-one map, thus we may think of continuous functions as distributions by identifying $f \in C(\mathbb{R}^n)$ with $\iota(f)$. (Below we write $\iota(f) = \iota_f$ often, by analogy with the notation of a sequence, as a distribution itself will be a map, or function, on functions, so we need to write an expression like $\iota_f(\phi)$, which is nicer than $(\iota(f))(\phi)$.) An analogy is that letters of the English alphabet can be considered as numbers via their ASCII encoding; there are more ASCII codes than letters, but to every letter there corresponds a unique ASCII code. One can just think of letters then as numbers, e.g. one thinks of the letter 'A' as the decimal number 65 (i.e. the letters are thought of as a subset of the integers 0 through 255).

So on to distributions. Suppose V is a vector space over $\mathbb{F} = \mathbb{R}$ or $\mathbb{F} = \mathbb{C}$. The algebraic dual of V is the vector space $\mathcal{L}(V, \mathbb{F})$ consisting of linear functionals from V to $\mathbb{F}$. That is, elements of $f \in \mathcal{L}(V, \mathbb{F})$ are linear maps $f : V \to \mathbb{F}$ satisfying

$$f(v + w) = f(v) + f(w), \ f(cv) = cf(v), \ v, w \in V, \ c \in \mathbb{F}.$$

When V is infinite dimensional, we need additional information, namely continuity. So if V is a topological space (which means that one has a notion of *open* sets, which we do not emphasize, and thus of *convergence*, which we do), with the topology compatible with the vector space structure (namely the vector space operations are continuous, which we have already shown in the case of normed vector spaces), i.e. if V is a topological vector space, we define the dual space V^* as the space of *continuous* linear maps $f : V \to \mathbb{F}$. In the cases we are interested in, continuity is the same as *sequential continuity*. The latter means that we consider maps f with the property that if $\phi_j \to \phi$ in V, then $f(\phi_j) \to f(\phi)$ in $\mathbb{F}$.

For us, V is the class of 'very nice objects', and V^* will be the class of 'bad objects'. Of course, normally there is no way of comparing elements of V with those of V^*, so we will also need an injection (i.e. a one-to-one map)

$$\iota : V \to V^*$$

so that elements of V can be regarded as elements of V^* (by identifying $v \in V$ with $\iota(V)$). As we want to differentiate functions, as much as we desire, V will consist of infinitely differentiable functions. As we need to control behavior at infinity to integrate (even infinitely differentiable functions on $\mathbb{R}^n$ need not be integrable on all of $\mathbb{R}^n$ since they need not decay at infinity, e.g. polynomials), the elements of V will be *compactly supported*, i.e. they vanish outside $B_R(0) = \{x \in \mathbb{R}^n : |x| < R\}$ for sufficiently large R. So we define $V = \mathcal{C}_c^\infty(\mathbb{R}^n)$ to be the space of infinitely continuously differentiable functions of compact support, i.e. elements ϕ of $\mathcal{C}^\infty(\mathbb{R}^n)$ for which there is $R > 0$ such that ϕ vanishes outside $B_R(0)$.

In order to explain the terminology, we define the *support* $\operatorname{supp} \phi$ of a continuous function ϕ as the closure of the set where $\phi \neq 0$, so $\operatorname{supp} \phi$ is compact if and only if it is bounded (since it is automatically closed), i.e. if and only if it is the subset of $B_R(0)$ for some $R > 0$. Thus, if $\phi \in \mathcal{C}_c^\infty(\mathbb{R}^n)$ with the above definition, $\operatorname{supp} \phi$ is certainly compact, as $\{x \in \mathbb{R}^n : \phi(x) \neq 0\} \subset B_R(0)$, and thus its closure is in $\overline{B}_R(0) = \{x \in \mathbb{R}^n : |x| \leq R\} \subset B_{R+1}(0)$, and so is compact. Conversely, if ϕ is $\mathcal{C}^\infty$ and $\operatorname{supp} \phi$ is compact, then $\{x \in \mathbb{R}^n : \phi(x) \neq 0\} \subset \operatorname{supp} \phi \subset B_R(0)$ for some $R > 0$, so $\phi \in \mathcal{C}_c^\infty(\mathbb{R}^n)$. This in combination with the previous sentence justifies the terminology for elements of V as 'infinitely continuously differentiable functions of compact support'.

We also call elements of V *test functions* since V^* is defined as continuous linear maps from V to $\mathbb{F}$, i.e. we are applying elements of V^* to elements of V, so we are 'testing' elements of V^* on V.

As $\mathcal{C}_c^\infty(\mathbb{R}^n)$ is infinite dimensional (see Lemma 5.2 and its corollary), we also need to put a topology on it. Technically this means that we should define what open sets are. Rather than doing this (to avoid complexity) we define what convergence of a sequence ϕ_j of functions in $\mathcal{C}_c^\infty(\mathbb{R}^n)$ means.

Definition 5.1. Suppose $\{\phi_j\}_{j=1,2,\dots}$ is a sequence in $\mathcal{C}_c^\infty(\mathbb{R}^n)$, and $\phi \in \mathcal{C}_c^\infty(\mathbb{R}^n)$. We say that $\lim_{j\to\infty} \phi_j = \phi$ if

  (i) there is a compact set K such that $\phi_j \equiv 0$ outside K for all j,

  (ii) and all derivatives of ϕ_j converge uniformly to the corresponding derivatives of ϕ, i.e. for all multiindices α, $\sup_{\mathbb{R}^n} |\partial^\alpha(\phi_j - \phi)| \to 0$ as $j \to \infty$. Explicitly, for all $\alpha \in \mathbb{N}^n$ and $\epsilon > 0$ there exists N such that for $j \geq N$, $\sup_{\mathbb{R}^n} |\partial^\alpha(\phi_j - \phi)| < \epsilon$.

At this point we do not even know if $\mathcal{C}_c^\infty(\mathbb{R}^n)$ has any elements besides the identically zero function; our potential fears are however laid to rest by the following lemma, which gives a plentiful supply of elements of this space:

Lemma 5.2. *For all $x_0 \in \mathbb{R}^n$ and $\epsilon > 0$ there is a function $\phi \in \mathcal{C}_c^\infty(\mathbb{R}^n)$ such that $\phi(x_0) > 0$, $\phi \geq 0$ and $\operatorname{supp}\phi \subset \{x : |x - x_0| < \epsilon\}$.*

Such functions are often called *bump functions*.

Proof. First one checks that the function χ defined by

(5.1) $$\chi(t) = e^{-1/t},\ t > 0;\ \chi(t) = 0,\ t \leq 0,$$

is in $\mathcal{C}^\infty(\mathbb{R})$; see Problem 5.1. Then we let

$$\phi(x) = \chi\left(\frac{\epsilon^2}{2} - |x - x_0|^2\right);$$

see Figure 5.1. This has all the desired properties; in particular, being $\mathcal{C}^\infty$ follows from the chain rule. □

An immediate corollary is:

Corollary 5.3. *The space $\mathcal{C}_c^\infty(\mathbb{R}^n)$ is infinite dimensional.*

Proof. One just needs to vary x_0 and ϵ in the statement of the lemma. The resulting functions are certainly linearly independent for instance if a sequence $(x_0)_j$, $j = 1, 2, \dots$, is chosen so that the corresponding $\phi = \phi_{(x_0)_j}$ have disjoint supports. □

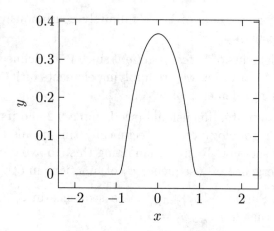

Figure 5.1. The plot of the function $\chi(\frac{\epsilon^2}{2} - |x - x_0|^2) = \chi(1 - x^2)$ with $x_0 = 0$, $\epsilon = \sqrt{2}$, with χ given by (5.1).

Another consequence of Lemma 5.2 is that a smooth function is *not* determined by its Taylor series at any one point; e.g. the zero function has the same Taylor series as ϕ in Lemma 5.2 at any $x \in \mathbb{R}^n$ with $|x - x_0| \geq \epsilon/\sqrt{2}$.

We can now define distributions:

Definition 5.4. A *distribution* $u \in \mathcal{D}'(\mathbb{R}^n)$ is a continuous linear functional on $\mathcal{C}_c^\infty(\mathbb{R}^n)$. That is, a distribution u is a map $u : \mathcal{C}_c^\infty(\mathbb{R}^n) \to \mathbb{F}$ such that

  (i) u is linear, $u(\phi + \psi) = u(\phi) + u(\psi)$, $u(c\phi) = cu(\phi)$ for $\phi, \psi \in \mathcal{C}_c^\infty(\mathbb{R}^n)$ and $c \in \mathbb{F}$,

  (ii) and u is continuous, so if ϕ_j is any sequence such that $\phi_j \to \phi$ in $\mathcal{C}_c^\infty(\mathbb{R}^n)$, then $u(\phi_j) \to u(\phi)$ in $\mathbb{F}$.

It is not hard to check that continuity of u is equivalent to the following: for all K compact there exists m and $C > 0$ such that for all $\phi \in \mathcal{C}_c^\infty(\mathbb{R}^n)$ with $\operatorname{supp} \phi \subset K$,

$$(5.2) \qquad |u(\phi)| \leq C \sum_{|\alpha| \leq m} \sup_{\mathbb{R}^n} |\partial^\alpha \phi|;$$

see Problem 9.8 in the setting of tempered distributions. Again, equivalently, one can specialize to balls for the compact sets, so the continuity of u is equivalent to the following: for all $R > 0$ there exist m and $C > 0$ such that for all $\phi \in \mathcal{C}_c^\infty(\mathbb{R}^n)$ with $\phi(x) = 0$ when $|x| > R$,

$$|u(\phi)| \leq C \sum_{|\alpha| \leq m} \sup_{\mathbb{R}^n} |\partial^\alpha \phi|.$$

The simplest distribution is the *delta distribution* at some point $a \in \mathbb{R}^n$; it is given by

$$\delta_a(\phi) = \phi(a).$$

One can also generate many similar examples, e.g. fixing $a, b \in \mathbb{R}^n$, $n \geq 2$; the map $u : \mathcal{C}_c^\infty(\mathbb{R}^n) \to \mathbb{C}$ given by

$$u(\phi) = (\partial_1 \phi)(a) - (\partial_2^2 \phi)(b)$$

is also a distribution. Indeed, it is linear:

$$u(\phi + \psi) = (\partial_1(\phi + \psi))(a) - (\partial_2^2(\phi + \psi))(b)$$
$$= (\partial_1 \phi)(a) + (\partial_1 \psi)(a) - (\partial_2^2 \phi)(b) - (\partial_2^2 \psi)(b) = u(\phi) + u(\psi),$$

with $u(c\phi) = cu(\phi)$ being similar, and it is continuous: if $\phi_j \to \phi$ in $\mathcal{C}_c^\infty(\mathbb{R}^n)$, then all derivatives of ϕ_j converge uniformly to the corresponding derivative of ϕ, and thus the derivatives evaluated at specific points certainly converge, so $(\partial_1 \phi_j)(a) \to (\partial_1 \phi)(a)$, $(\partial_2^2 \phi_j)(b) \to (\partial_2^2 \phi)(b)$, and thus $u(\phi_j) \to u(\phi)$. Note that in this example the linearity of u comes down to the linearity of differentiation, while the continuity of u again to that of differentiation. Note that from the perspective of the estimate (5.2), the continuity is also easily seen:

$$|u(\phi)| \leq |(\partial_1 \phi)(a)| + |(\partial_2^2 \phi)(b)| \leq \sum_{|\alpha| \leq 2} \sup_{\mathbb{R}^n} |\partial^\alpha \phi|,$$

since two of the summands on the right-hand side are $\sup_{\mathbb{R}^n} |\partial_1 \phi|$ and $\sup_{\mathbb{R}^n} |\partial_2^2 \phi|$, which are respectively upper bounds for $|(\partial_1 \phi)(a)|$ and $|(\partial_2^2 \phi)(b)|$.

A large class of distributions is obtained from (reasonable, such as piecewise continuous) functions, i.e. functions give rise to distributions, and indeed, the map from functions to distributions is one-to-one, i.e. to different functions correspond different distributions, so one can *identify* this class of functions with a *subset* of distributions, as mentioned at the very beginning of the chapter. Now, the natural class of functions to consider here is the set $L_{\text{loc}}^1(\mathbb{R}^n)$ of locally Lebesgue integrable functions. However, as this book does not assume any familiarity with measure theory, we do not consider this class officially, though for the sake of readers already familiar with it, we make remarks about it along the way. Thus, we work with continuous or piecewise continuous functions instead. In addition, we shall work with square integrable functions L^2 in Chapter 13, obtained as a completion. Completely similar techniques also construct L^1 without measure theory, and we do so in the additional material at the end of this chapter for the interested reader.

So suppose that f is (piecewise) continuous (or $f \in L^1_{\text{loc}}(\mathbb{R}^n)$ for those familiar with it). Then f defines a distribution $u = \iota_f$ as follows:

$$u(\phi) = \iota_f(\phi) = \int_{\mathbb{R}^n} f(x)\phi(x)\,dx, \qquad \phi \in \mathcal{C}^\infty_c(\mathbb{R}^n).$$

Note that the integral converges since ϕ has compact support. Certainly $\iota_f : \mathcal{C}^\infty_c(\mathbb{R}^n) \to \mathbb{C}$ is linear due to the linearity of the integral, e.g.

$$\iota_f(\phi + \psi) = \int_{\mathbb{R}^n} f(x)(\phi(x) + \psi(x))\,dx = \int_{\mathbb{R}^n} (f(x)\phi(x) + f(x)\psi(x))\,dx$$

$$= \int_{\mathbb{R}^n} f(x)\phi(x)\,dx + \int_{\mathbb{R}^n} f(x)\psi(x)\,dx = \iota_f(\phi) + \iota_f(\psi),$$

with a similar argument for $\iota_f(c\phi)$. The continuity of ι_f can be seen as follows. Suppose that $\phi_j \to \phi$ in $\mathcal{C}^\infty_c(\mathbb{R}^n)$. Then by the definition of convergence, there is $R > 0$ such that ϕ_j and ϕ vanish for $|x| > R$. Thus,

$$\left| \int_{\mathbb{R}^n} f(x)\phi_j(x)\,dx - \int_{\mathbb{R}^n} f(x)\phi(x)\,dx \right| = \left| \int_{\mathbb{R}^n} f(x)(\phi_j(x) - \phi(x))\,dx \right|$$

$$= \left| \int_{|x|\leq R} f(x)(\phi_j(x) - \phi(x))\,dx \right| \leq \int_{|x|\leq R} |f(x)|\,|\phi_j(x) - \phi(x)|\,dx$$

$$\leq (\sup_{|x|\leq R} |\phi_j(x) - \phi(x)|) \int_{B_R(0)} |f(x)|\,dx,$$

where $B_R(0) = \{x : |x| < R\}$ is the ball of radius R. Now, continuous (and piecewise continuous) f are bounded over compact sets, so

$$\int_{B_R(0)} |f(x)|\,dx \leq (\sup_{|x|\leq R} |f(x)|)\text{vol}(B_R(0)),$$

where vol denotes volume. Now, $\int_{B_R(0)} |f(x)|\,dx$ is a constant (independent of j), while $\sup_{|x|\leq R} |\phi_j(x) - \phi(x)| \to 0$ as $j \to \infty$ by the definition of convergence in $\mathcal{C}^\infty_c(\mathbb{R}^n)$. Thus, $\iota_f(\phi_j) \to \iota_f(\phi)$, showing the desired continuity.

The important fact is that we do not lose any information by thinking of continuous functions as distributions, i.e. the map ι is one-to-one.

Lemma 5.5. *The map $\iota : C(\mathbb{R}^n) \to \mathcal{D}'(\mathbb{R}^n)$ is injective.*

Remark 5.6. The argument below extends easily to piecewise continuous f; see Problem 5.4.

Proof. Since ι is linear, we just need to show that if $\iota_f = 0$, then $f = 0$. But, $\iota_f = 0$ means that $\int f\phi\,dx = 0$ for all $\phi \in \mathcal{C}^\infty_c(\mathbb{R}^n)$; we need to show that this implies $f = 0$. Equivalently, we need to show (and this is what we do) that if f is not the zero function (i.e. f does not vanish identically), then $\iota_f \neq 0$, i.e. $\iota_f(\phi) \neq 0$ for some $\phi \in \mathcal{C}^\infty_c(\mathbb{R}^n)$.

If $f \in C(\mathbb{R}^n)$ and f is not the zero function, so $f(x_0) \neq 0$ for some $x_0 \in \mathbb{R}^n$, then by the continuity of f, for $\epsilon > 0$ sufficiently small, $|f(x) - f(x_0)| < |f(x_0)|/2$ whenever $|x - x_0| < \epsilon$. Now let ϕ be as in Lemma 5.2, so $|f(x) - f(x_0)| < |f(x_0)|/2$ on supp ϕ. Thus,

$$\left| \int f\phi \, dx - f(x_0) \int \phi \, dx \right| \leq \int |f(x) - f(x_0)| \, \phi(x) \, dx \leq \frac{|f(x_0)|}{2} \int \phi \, dx,$$

so

$$\left| \int f\phi \, dx \right| \geq \left| f(x_0) \int \phi(x) \, dx \right| - \left| \int f\phi \, dx - f(x_0) \int \phi \, dx \right| \geq \frac{|f(x_0)|}{2} \int \phi \, dx,$$

so $\int f\phi \, dx \neq 0$ as $\int \phi \, dx > 0$. □

Because of this lemma, we can consider $C(\mathbb{R}^n)$ as a subset of $\mathcal{D}'(\mathbb{R}^n)$, via the identification ι.

For the sake of completeness, we now state without proof the L^1 version of the result, from which we can consider $L^1_{\text{loc}}(\mathbb{R}^n)$ as a subset of $\mathcal{D}'(\mathbb{R}^n)$, via the identification ι.

Lemma 5.7. *The map $\iota : L^1_{loc}(\mathbb{R}^n) \to \mathcal{D}'(\mathbb{R}^n)$ is injective.*

It is both useful and intuitively helpful to consider the convergence of sequences of distributions. One usually equips $\mathcal{D}'(\mathbb{R}^n)$ with the so-called *weak-* topology*, which means the following in terms of convergence:

Definition 5.8. One says that a sequence $u_j \in \mathcal{D}'(\mathbb{R}^n)$ *converges* to $u \in \mathcal{D}'(\mathbb{R}^n)$ if $u(\phi) = \lim_{j \to \infty} u_j(\phi)$ for all $\phi \in \mathcal{C}_c^\infty(\mathbb{R}^n)$.

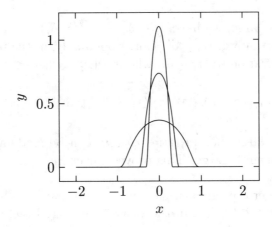

Figure 5.2. The functions $\delta_j^{-1}\phi(./\delta_j)$ with $\delta_j = 1, 1/2, 1/3$; the peaks become narrower but taller as $\delta_j \to 0$. The limit, however, is *not* a function, rather it is a multiple of the delta distribution.

Example 5.1. Fix ϵ (say $\epsilon = \sqrt{2}$, though the choice does not really make a difference) and let ϕ be as in Lemma 5.2 with $x_0 = 0$. Let δ_j be a sequence of positive constants with $\lim_{j \to \infty} \delta_j = 0$. Let u_j be the distribution given by $\delta_j^{-n} \phi(./\delta_j)$, i.e.

$$u_j(\psi) = \int \delta_j^{-n} \phi(x/\delta_j) \psi(x)\, dx.$$

Let $c = \int \phi\, dx$. Show that

$$\lim_{j \to \infty} u_j = c\delta_0.$$

Solution. By the continuity of ψ, given $\epsilon' > 0$, there is $\delta' > 0$ such that $|x| < \delta'$ implies $|\psi(x) - \psi(0)| < \epsilon'$. Then, noting that

$$c = \int \delta_j^{-n} \phi(x/\delta_j)\, dx$$

by a change of variables, we have

$$|u_j(\psi) - c\delta_0(\psi)| = \left| \int \delta_j^{-n} \phi(x/\delta_j) \psi(x)\, dx - \psi(0) \int \phi(x)\, dx \right|$$

$$= \left| \int \delta_j^{-n} \phi(x/\delta_j) \psi(x)\, dx - \int \delta_j^{-n} \phi(x/\delta_j) \psi(0)\, dx \right|$$

$$= \left| \int \delta_j^{-n} \phi(x/\delta_j)(\psi(x) - \psi(0))\, dx \right|$$

$$\leq \int \delta_j^{-n} \phi(x/\delta_j) |\psi(x) - \psi(x_0)|\, dx.$$

For j sufficiently large, we have $\delta_j < \delta'$, and $\phi(x/\delta_j) = 0$ if $|x|/\delta_j \geq 1$, i.e. if $|x| \geq \delta_j$, so certainly if $|x| \geq \delta'$. Correspondingly, in the integral, one can restrict the integration to $|x| \leq \delta_j$, where $|\psi(x) - \psi(0)| < \epsilon'$, to deduce that

$$|u_j(\psi) - c\delta_0(\psi)| \leq \epsilon' \int \delta_j^{-n} \phi(x/\delta_j)\, dx = c\epsilon'$$

for j sufficiently large; i.e. $|u_j(\psi) - c\delta_0(\psi)|$ can be made arbitrarily small by taking j sufficiently large. This proves our claim.

This is a rather typical example, and it is not hard to show that one can approximate any $u \in \mathcal{D}'(\mathbb{R}^n)$ in the weak-* topology by u_j which are given by $C_c^\infty(\mathbb{R}^n)$ functions, i.e. $u_j = \iota_{f_j}$, $f_j \in C_c^\infty(\mathbb{R}^n)$, and for all $\phi \in C_c^\infty(\mathbb{R}^n)$, $\lim_{j \to \infty} u_j(\phi) = u(\phi)$. Thus, $C_c^\infty(\mathbb{R}^n)$ is *dense* in $\mathcal{D}'(\mathbb{R}^n)$. See Problem 5.3 for a related argument showing the approximability of continuous functions of compact support by C^∞ functions of compact support.

We can now consider differentiation. The idea is that we already know what the derivative of a C^1 function is, so we should express it as a distribution in such a way that it easily extends to the class of all distributions. Now, for $f \in C^1(\mathbb{R}^n)$, the distribution associated to the function $\partial_j f$ satisfies

$$\iota_{\partial_j f}(\phi) = \int \partial_j f \, \phi \, dx = - \int f \, \partial_j \phi \, dx = -\iota_f(\partial_j \phi)$$

for all $\phi \in \mathcal{C}_c^\infty(\mathbb{R}^n)$. Motivated by this, we make the definition:

Definition 5.9. The *partial derivatives* of $u \in \mathcal{D}'(\mathbb{R}^n)$ are defined by

$$\partial_j u(\phi) = -u(\partial_j \phi).$$

Note that for $\phi \in \mathcal{C}_c^\infty(\mathbb{R}^n)$, $\partial_j \phi \in \mathcal{C}_c^\infty(\mathbb{R}^n)$, so this definition makes sense.

It is straightforward to check that $\partial_j u$ *is* a distribution, in particular is continuous as a map $\mathcal{C}_c^\infty(\mathbb{R}^n) \to \mathbb{C}$. Indeed, for $c \in \mathbb{F}$, $\phi \in \mathcal{C}_c^\infty(\mathbb{R}^n)$, we have

$$(\partial_j u)(c\phi) = -u(\partial_j(c\phi)) = -u(c\partial_j \phi) = -cu(\partial_j \phi) = c(\partial_j u)(\phi),$$

and similarly for $(\partial_j u)(\phi + \psi)$, giving linearity. On the other hand, if $\phi_k \to \phi$, as $k \to \infty$, in $\mathcal{C}_c^\infty(\mathbb{R}^n)$, then $\partial_j \phi_k \to \partial_j \phi$ in $\mathcal{C}_c^\infty(\mathbb{R}^n)$. Indeed, the supports of ϕ_k are all in a fixed compact set, and thus the same holds for the derivatives $\partial_j \phi_k$. Now, $\partial_j \phi_k \to \partial_j \phi$ in $\mathcal{C}_c^\infty(\mathbb{R}^n)$ means $\sup |\partial^\alpha(\partial_j \phi_k - \partial_j \phi)| \to 0$ as $k \to \infty$. However, $\sup |\partial^\alpha(\partial_j \phi_k - \partial_j \phi)| = \sup |\partial^\alpha \partial_j(\phi_k - \phi)| \to 0$ as $k \to \infty$ since $\partial^\alpha \partial_j$ is simply a derivative, (i.e. is ∂^β for an appropriate multiindex β), so the assumption $\phi_k \to \phi$, as $k \to \infty$, in $\mathcal{C}_c^\infty(\mathbb{R}^n)$ implies this, giving the desired continuity.

Note also that this is the only reasonable notion of a derivative as the map $u \mapsto \partial_j u$ is continuous (i.e. $u_k \to u$ implies $\partial_j u_k \to \partial_j u$) and $\mathcal{C}_c^\infty(\mathbb{R}^n)$, on which we already know ∂_j, is dense in $\mathcal{D}'(\mathbb{R}^n)$.

As an example, on $\mathbb{R}$,

$$\delta_a'(\phi) = -\delta_a(\phi') = -\phi'(a), \quad \phi \in \mathcal{C}_c^\infty(\mathbb{R}).$$

Also, if H is the Heaviside step function, so $H(x) = 1$ for $x \geq 0$, $H(x) = 0$ for $x < 0$, then

$$H'(\phi) = (\iota_H)'(\phi) = -\iota_H(\phi') = -\int_0^\infty \phi'(x) \, dx = \phi(0)$$

for all $\phi \in \mathcal{C}_c^\infty(\mathbb{R})$, where we used the fundamental theorem of calculus in the last step and where ϕ vanishes identically near infinity, so there is no other 'boundary term'. Therefore $H' = \delta_0$.

Note also that distributions u may be multiplied by C^∞ functions g. Indeed, we proceed again by analogy with ι_f where $f \in C(\mathbb{R}^n)$. In that case

$$\iota_{fg}(\phi) = \int (fg)\phi\,dx = \int f(g\phi)\,dx = \iota_f(g\phi),$$

so for arbitrary $u \in \mathcal{D}'(\mathbb{R}^n)$ we make the following definition.

Definition 5.10. For $g \in C^\infty(\mathbb{R}^n)$, $u \in \mathcal{D}'(\mathbb{R}^n)$, we define $gu \in \mathcal{D}'(\mathbb{R}^n)$ by

$$gu(\phi) = u(g\phi).$$

Note that for $\phi \in C_c^\infty(\mathbb{R}^n)$, $g\phi \in C_c^\infty(\mathbb{R}^n)$ since $g \in C^\infty(\mathbb{R}^n)$, so the definition makes sense.

As an example of calculation with distributions, consider the following:

Lemma 5.11. *Suppose that $u \in \mathcal{D}'(\mathbb{R})$ and $xu = 0$. Then there is a constant $c \in \mathbb{F}$ such that $u = c\delta_0$.*

Before the proof, let's recall Taylor's theorem with an integral remainder formula on $\mathbb{R}$ (see Problem 1.2): if f is a C^{k+1} function, then

$$f(x) = \sum_{j \le k} \frac{f^{(j)}(x_0)}{j!}(x-x_0)^j + (x-x_0)^{k+1} \int_0^1 \frac{(1-s)^k}{k!} f^{(k+1)}(x_0 + s(x-x_0))\,ds.$$

Notice that

$$\int_0^1 \frac{(1-s)^k}{k!} f^{(k+1)}(x_0 + s(x-x_0))\,ds$$

is a continuous function of x; if f is C^∞, it is in fact a C^∞ function of x. This formula can be seen in the $k = 0$ case (which is what we use below) by the fundamental theorem of calculus:

$$f(x) = f(x_0) + \int_{x_0}^x f'(t)\,dt = f(x_0) + (x - x_0)\int_0^1 f'(x_0 + s(x - x_0))\,ds,$$

where we wrote $t = x_0 + s(x-x_0)$, and changed variables, so $dt = (x-x_0)\,ds$. To continue one writes the integrand as $1 \cdot f'(x_0 + s(x - x_0))$ and integrates by parts, making the indefinite integral of 1 to be $s - 1$ (and after k steps, starting from the above expression, one gets $\frac{1}{k!}(s - 1)^k$ for this).

Proof. First note that if $u = c\delta_0$, then $xu = 0$. Indeed:

$$xu(\phi) = u(x\phi) = c\delta_0(x\phi) = c(x\phi)(0) = 0$$

for all $\phi \in C_c^\infty(\mathbb{R})$ since $x(0) = 0$, so $(x\phi)(0) = 0$.

Suppose now that $xu = 0$, i.e. $u(x\phi) = 0$ for all $\phi \in C_c^\infty(\mathbb{R})$. If ψ is a test function such that $\psi(0) = 0$, then Taylor's theorem allows one to write $\psi = x\phi$ with $\phi \in C_c^\infty(\mathbb{R})$ (namely $\phi = x^{-1}\psi$ for $x \ne 0$ extends to be C^∞ at 0; see Problem 1.2), so $u(\psi) = u(x\phi) = 0$.

So now suppose that $\psi \in \mathcal{C}_c^\infty(\mathbb{R})$. Let $\phi_0 \in \mathcal{C}_c^\infty(\mathbb{R})$ be such that $\phi_0(0) \neq 0$. We choose $\alpha \in \mathbb{F}$ such that $\psi - \alpha\phi_0$ vanishes at 0, i.e. let $\alpha = \frac{\psi(0)}{\phi_0(0)}$. Then by the argument of the previous paragraph, $u(\psi - \alpha\phi_0) = 0$. Thus,

$$u(\psi) = u((\psi - \alpha\phi_0) + \alpha\phi_0) = u(\psi - \alpha\phi_0) + \alpha u(\phi_0)$$

$$= \alpha u(\phi_0) = \frac{u(\phi_0)}{\phi_0(0)}\psi(0) = c\delta_0(\psi), \quad c = \frac{u(\phi_0)}{\phi_0(0)}.$$

This finishes the proof. $\qquad\qquad\qquad\qquad\qquad\qquad\qquad\qquad\square$

If L is a linear partial differential operator, so L is of the form

$$L = \sum_{|\alpha|\leq m} a_\alpha \partial^\alpha$$

and a_α are in $\mathcal{C}^\infty(\mathbb{R}^n)$, then for all $u \in \mathcal{D}'(\mathbb{R}^n)$, Lu makes sense as an element of $\mathcal{D}'(\mathbb{R}^n)$. In particular, we make the following definition:

Definition 5.12. Suppose f, u are piecewise continuous on $\mathbb{R}^n$ (or more generally $f \in L_{\text{loc}}^1(\mathbb{R}^n)$, $u \in L_{\text{loc}}^1(\mathbb{R}^n)$) and L is linear with $\mathcal{C}^\infty(\mathbb{R}^n)$ coefficients. We say that u is a *weak solution* of $Lu = f$ if $Lu = f$ in the sense of distributions.

Notice that explicitly, with L as above, this simply means that for all $\phi \in \mathcal{C}_c^\infty(\mathbb{R}^n)$,

$$\sum_{|\alpha|\leq m} \int_{\mathbb{R}^n} u(x)(-1)^{|\alpha|}\partial^\alpha(a_\alpha(x)\phi(x))\,dx = \int_{\mathbb{R}^n} f(x)\phi(x)\,dx,$$

i.e. that

$$(5.3) \qquad \int u(x)L^\dagger\phi(x)\,dx = \int_{\mathbb{R}^n} f(x)\phi(x)\,dx,$$

where $L^\dagger$ is the *transpose* of L:

$$(5.4) \qquad (L^\dagger\phi)(x) = \sum_{|\alpha|\leq m} (-1)^{|\alpha|}\partial^\alpha(a_\alpha(x)\phi(x)),$$

or simply

$$(5.5) \qquad L^\dagger = \sum_{|\alpha|\leq m} (-1)^{|\alpha|}\partial^\alpha a_\alpha,$$

where a_α are understood as multiplication operators. (So M_{a_α} would be a better notation, where $(M_{a_\alpha}\phi)(x) = a_\alpha(x)\phi(x)$, but one usually just abuses the notation and writes a_α for the multiplication operator.)

The reason for the term 'transpose' is that if $u \in C^m(\mathbb{R}^n)$, $\phi \in \mathcal{C}_c^\infty(\mathbb{R}^n)$, then

$$(5.6) \qquad \int_{\mathbb{R}^n} (Lu)(x)\phi(x)\,dx = \int_{\mathbb{R}^n} u(x)(L^\dagger\phi)(x)\,dx,$$

which is analogous to the role matrix transposes play: $Av \cdot w = v \cdot A^{\mathrm{T}} w$, where A is a linear map from $\mathbb{R}^k$ to $\mathbb{R}^p$ (so it is given by a p by k matrix), A^{T} its transpose (so given by a k by p matrix) and $\cdot$ is the usual dot (scalar) product, $v \in \mathbb{R}^k$, $w \in \mathbb{R}^p$.

We reiterate that (5.6) is the result of integrating by parts m times (where m is the order of L), which is justified when $u \in C^m(\mathbb{R}^n)$. Distribution theory makes sense of the left-hand side via the right-hand side even when u is piecewise continuous only, i.e. then the left-hand side is *defined by the right-hand side*, and *cannot be thought of as an integral* (i.e. the expression as a whole makes sense, as a distribution applied to ϕ, not its individual parts).

All these considerations go through unaffected if $\mathbb{R}^n$ is replaced by an *open* set $\Omega \subset \mathbb{R}^n$. In this case, we consider only test functions $\phi \in C_c^\infty(\mathbb{R}^n)$ with $\operatorname{supp} \phi \subset \Omega$.

Even if L is not linear, it *may* make sense to talk about weak solutions. For instance, for conservation laws,

$$L(u) = u_t + (f(u))_x,$$

and $u \in C(\mathbb{R}^2)$ or u has jump discontinuities and $f \in C(\mathbb{R})$, we say that u is a weak solution of $L(u) = 0$ if $Lu = 0$ in the sense of distributions. Note that in this case, $f(u)$ is (piecewise) continuous (so it is in $L_{\mathrm{loc}}^1(\mathbb{R}^2)$), so $f(u)$ can be considered as an element of $\mathcal{D}'(\mathbb{R}^2)$, and so the derivative makes sense.

One often wants solutions only for $t \geq 0$, with initial conditions, in which case it is helpful to reformulate the definition of a weak solution by rewriting the distributional derivatives explicitly. If $u \in C^1(\mathbb{R}_x \times [0, \infty)_t)$ and $\phi \in C_c^\infty(\mathbb{R}_x \times [0, \infty)_t)$, then

$$\int_{\mathbb{R}_x \times [0,\infty)_t} (u_t + f(u)_x)\phi \, dx \, dt$$

$$= -\int_{\mathbb{R}_x \times [0,\infty)_t} (u\,\phi_t + f(u)\phi_x) \, dx \, dt - \int_{\mathbb{R}_x} u(x,0)\,\phi(x,0) \, dx.$$

Definition 5.13. We say that a function u which is piecewise continuous on $\mathbb{R}_x \times [0, \infty)_t$ is a weak solution of

$$u_t + f(u)_x = 0, \; u(x,0) = v(x)$$

in $\mathbb{R}_x \times [0, \infty)_t$ if for all $\phi \in C_c^\infty(\mathbb{R}_x \times [0, \infty)_t)$,

$$\int_{\mathbb{R}_x \times [0,\infty)_t} (u\,\phi_t + f(u)\phi_x) \, dx \, dt + \int_{\mathbb{R}_x} v(x)\,\phi(x,0) \, dx = 0.$$

In particular, if ϕ is supported in $t > 0$, so $\phi(x,0) = 0$ for all x, then

$$\int_{\mathbb{R}_x \times [0,\infty)_t} (u\,\phi_t + f(u)\phi_x)\,dx\,dt = 0,$$

so $u_t + f(u)_x = 0$ in the sense of distributions on $\mathbb{R}_x \times (0,\infty)_t$, but we also have the initial condition encoded in this definition. In other words, a weak solution on $\mathbb{R}_x \times [0,\infty)_t$ is necessarily a weak solution on the smaller set $\mathbb{R}_x \times (0,\infty)_t$, but there is an additional criterion to fulfill since one tests by $\phi \in \mathcal{C}_c^\infty(\mathbb{R}_x \times [0,\infty)_t)$, not merely $\phi \in \mathcal{C}_c^\infty(\mathbb{R}_x \times (0,\infty)_t)$.

The problems at the end of the chapter discuss applications of distributions and weak solutions, such as discontinuous solutions (shocks) for Burgers' equation.

Finally we comment on the larger picture for distributions, some of which we can shed light on towards the end of the book. In general, linear PDE theory constructs solutions of PDE by duality arguments. To get a bit of feel for this, consider the following analogue. In finite dimensional vector spaces V, W, if one has a map $P : V \to W$ and $P^* : W^* \to V^*$ is the adjoint, defined by

$$(P^*\ell)(v) = \ell(Pv), \ \ell \in W^*, \ v \in V,$$

then P is onto if and only if P^* is one-to-one, and P is one-to-one if and only if P^* is onto. (This reduces to a statement about matrices by introducing bases; the adjoint is the transpose, or the conjugate transpose, depending on the definition used.) Thus, if we try to solve $P^*\ell = f$, where $f \in V^*$ is given, we just need to show that P is one-to-one. In infinite dimensions there are serious complications; for one thing instead of just the one-to-one nature, we want an estimate of the form

$$\|v\|_V \leq C\|Pv\|_W,$$

i.e. there is $C > 0$ such that for all $v \in V$ this estimate holds. Notice that this implies the statement that P is injective, but it is strictly stronger in general. The space $\mathcal{C}_c^\infty(\mathbb{R}^n)$ is not normed, which causes some complications; one typically works with a somewhat larger space instead. In any case, this gives distributional solutions to a PDE in the presence of some estimates for the adjoint operator; in the case of $P^* = L$ above, this amounts to estimates for P which are essentially (up to possibly some complex conjugates in the complex-valued setting) $L^\dagger$. We address this more explicitly at the end of the book; see Chapter 17. (Also, one would typically want to impose extra conditions, such as initial conditions or boundary conditions, so there are further subtleties.)

We also mention that weak solutions give rise to a numerical way of solving PDE. Namely, consider (5.3). Let's take a finite dimensional space X_N of *trial functions* which we take piecewise continuous, with say a basis

$\psi_1, \ldots, \psi_N$, and a finite dimensional space Y_N of *test functions* which we take C^m, with a basis $\phi_1, \ldots, \phi_N$. Let's assume that u can be approximated by a linear combination of the ψ_j, i.e. by $\sum_{j=1}^{N} c_j \psi_j$. If this combination actually solved the PDE, we would have

$$\int_{\mathbb{R}^n} \left(\sum_{j=1}^{N} c_j \psi_j(x) \right) (L^\dagger \phi)(x) \, dx = \int_{\mathbb{R}^n} f(x) \phi(x) \, dx,$$

for all $\phi \in C_c^\infty(\mathbb{R}^n)$. Since we have only finitely many constants c_j to vary, it is unlikely that we can solve the equation so that it is satisfied for all ϕ. However, let us demand that the equation is only satisfied for $\phi \in Y_N$, in which case it suffices to check it for the basis consisting of the ϕ_k, i.e. let us demand

$$\int_{\mathbb{R}^n} \left(\sum_{j=1}^{N} c_j \psi_j(x) \right) (L^\dagger \phi_k)(x) \, dx = \int_{\mathbb{R}^n} f(x) \phi_k(x) \, dx, \quad k = 1, \ldots, N.$$

This is N equations ($k = 1, \ldots, N$) for N unknowns (the c_j), so typically (when the corresponding matrix is invertible) we would expect that we can solve these equations, which are just linear equations. They are of the form

$$\sum_{j=1}^{N} A_{kj} c_j = g_k,$$

where $A_{kj} = \int_{\mathbb{R}^n} \psi_j(x)(L^\dagger \phi_k)(x) \, dx$ and $g_k = \int_{\mathbb{R}^n} f(x)\phi_k(x) \, dx$. Again, there are issues about additional conditions (which would influence the choice of X_N and Y_N), and one may want to write the equations in a different form. For instance, for a second order equation have one derivative each on ϕ_k and ψ_j, requiring them to be piecewise C^1. Such a rewriting, with an appropriate choice of ϕ_k and ψ_j, is the basis of the very important method of finite elements for numerically solving PDE.

Additional material: The space L^1

This section is separate from the main material of the book, but some readers might be interested in this extension of the text.

Although the Lebesgue theory of integration plays no role in this textbook, the set of Lebesgue integrable functions is a natural class to consider in the context of distribution theory, since functions which are integrable on compact sets are exactly the functions which give rise to distributions via the map ι:

(5.7) $\iota_f(\phi) = \int_{\mathbb{R}^n} f(x)\phi(x) \, dx, \quad \phi \in C_c^\infty(\mathbb{R}^n).$

While L^1 is usually defined through measure theory, it can also be defined via metric space techniques. The advantage of the former is that L^1 is *almost* a class of functions as seen from measure theory, namely one thinks of two functions as equivalent if their difference vanishes except on a set which has volume 0 in the measure theoretic sense, i.e. it has measure 0. However, it has significantly greater overhead, as one needs to develop measure theory, and with our abstract metric space-based L^1 we can actually do everything we might want in this book. We refer to the book of Stein and Shakarchi [5] for a measure theoretic treatment for the interested reader.

The key metric space notion is that of *completions*; first recall the notion of completeness of a metric space from the additional material in Chapter 3; see Definition 3.1. For a metric space (X, d) which is not complete, one can construct a new metric space (M, ρ) which is such, so that (X, d) 'sits inside' (M, ρ), namely there is a map $j : X \to M$ which satisfies $\rho(j(x), j(y)) = d(x, y)$ for $x, y \in X$ (so the distances of the images under j are the same as the distances of the original points), and such that $j(X)$ is dense in M, i.e. every $m \in M$ is the limit of images of points from X. Since $x \neq y$ means $d(x, y) \neq 0$, and thus $\rho(j(x), j(y)) \neq 0$, we have $j(x) \neq j(y)$, so j is necessarily one-to-one. This basically means that one can think of X as a subset of M (and indeed one often writes $X \subset M$), and j is just a relabelling, much like it happened with continuous functions sitting inside distributions via the map ι!

Theorem 5.14. *For any metric space (X, d) there is a complete metric space (M, ρ) (called the completion of (X, d)) and a map $j : X \to M$ such that*

$$(5.8) \qquad \rho(j(x), j(y)) = d(x, y), \ x, y \in X.$$

For the proof we refer to analysis texts, such as [3]. However, the idea is simple. As a set, M will be the set of equivalence classes of Cauchy sequences in X, under a certain equivalence relation. Namely, two Cauchy sequences $\{x_n\}_{n=1}^\infty, \{y_n\}_{n=1}^\infty$ in X are called equivalent, written as $\{x_n\}_{n=1}^\infty \sim \{y_n\}_{n=1}^\infty$, if $\lim_{n \to \infty} d(x_n, y_n) = 0$. Note that if say x_n converge to some $x \in X$, and the y_n are equivalent to the x_n, then y_n also converge to x, and conversely, two sequences converging to the same limit are equivalent. Thus, our notion of equivalence is just the right notion of generalizing the notion of 'having the same limit'. So then M will be the set of these Cauchy sequences, up to this equivalence relation, i.e. we do not distinguish between two equivalent sequences (we regard them as the same point in M). One defines j by $j(x)$ being the equivalence class of the constant sequence $x, x, x, x, \ldots$; this is certainly a Cauchy sequence. Then one defines for any two Cauchy sequences

$\{x_n\}_{n=1}^\infty, \{y_n\}_{n=1}^\infty$ in X the following distance function:

$$\rho(\{x_n\}_{n=1}^\infty, \{y_n\}_{n=1}^\infty) = \lim_{n\to\infty} d(x_n, y_n);$$

one of course needs to check first that this limit exists (which follows from the Cauchy property). Further, one checks that if $\{x_n\}_{n=1}^\infty \sim \{x'_n\}_{n=1}^\infty$ and $\{y_n\}_{n=1}^\infty \sim \{y'_n\}_{n=1}^\infty$, then

$$\rho(\{x_n\}_{n=1}^\infty, \{y_n\}_{n=1}^\infty) = \rho(\{x'_n\}_{n=1}^\infty, \{y'_n\}_{n=1}^\infty),$$

i.e. ρ does not change if one replaces a Cauchy sequence by an equivalent one, so it is well defined on M. Note that directly from the definition of j, (5.8) holds. Then one checks that ρ is indeed the metric (it inherits this property from d), and finally that ρ is complete, proving the theorem.

In fact, this is *exactly* the process by which the real numbers are constructed from the rationals, since the latter are not complete under the metric $d(x,y) = |x - y|$, $x, y \in \mathbb{Q}$. Thus, a real number is actually an equivalence class of Cauchy sequences of rationals!

While we did not state it, the completion is essentially unique, i.e. any other metric space $(\tilde{M}, \tilde{\rho})$ which has the properties of (M, ρ) simply relabels the points of M. This justifies the article 'the' in front of 'completion'.

Now, we can equip the set $C_c^0(\mathbb{R}^n)$ of continuous functions of compact support (we could work instead with continuous functions which decay sufficiently fast at infinity) with the so-called L^1-norm:

$$\|f\|_{L^1} = \int_{\mathbb{R}^n} |f(x)|\, dx.$$

This space is *not* complete with this norm. However, we can complete it! *The metric space $L^1(\mathbb{R}^n)$ is then the completion of $C_c^0(\mathbb{R}^n)$ with the L^1-norm.* Thus, elements of $L^1(\mathbb{R}^n)$ are equivalence classes of Cauchy sequences of continuous functions of compact support, such as $\{f_j\}_{j=1}^\infty$.

For instance (for $n = 1$), the sequence

$$f_j(x) = \begin{cases} 0, & |x| \geq 1, \\ j(1 - |x|), & 1 - 1/j \leq |x| \leq 1, \\ 1, & |x| \leq 1 - 1/j \end{cases}$$

is a Cauchy sequence of continuous functions of compact support; see Figure 5.3. For each $|x| < 1$, $f_j(x) \to 1$, while for each $|x| \geq 1$, $f_j(x) \to 0$. Thus, pointwise, f_j converges to a step function, which is certainly not a continuous function. On the other hand, the sequence is L^1-Cauchy: $f_j - f_k$ vanishes outside

$$[-1, -1 + \max(j^{-1}, k^{-1})] \cup [1 - \max(j^{-1}, k^{-1}), 1],$$

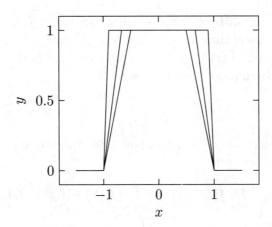

Figure 5.3. The functions f_j with $j = 2, 3, 10$ are continuous and Cauchy in $L^1(\mathbb{R})$, with the slope of the non-constant segments increasing with j. We think of the limit, however, *not* as a function, but rather as an element of $L^1(\mathbb{R})$, which is a completion.

and on these intervals it is bounded in absolute value by 2, so

$$\|f_j - f_k\|_{L^1} = \int_{\mathbb{R}} |f_j(x) - f_k(x)| \, dx$$

$$\leq \int_{-1}^{-1+\max(j^{-1}, k^{-1})} 2 \, dx + \int_{1-\max(j^{-1}, k^{-1})}^{1} 2 \, dx$$

$$= 4 \max(j^{-1}, k^{-1}) \to 0$$

as $j, k \to \infty$. In particular, $\{f_j\}_{j=1}^{\infty}$ represents an element of $L^1(\mathbb{R}^n)$ (i.e. it is one of the elements of an equivalence class). This way, the step function 'is' an element of L^1, though technically we cannot really think of the equivalence class of $\{f_j\}_{j=1}^{\infty}$ as a function; it is an abstract point in the metric space $L^1(\mathbb{R}^n)$. Thus, one can think of elements of $L^1(\mathbb{R}^n)$ as abstract objects which can, in the sense of the L^1-norm, be approximated by continuous functions of compact support.

The main reason this completion is useful is the following result, which essentially uses a special case of Lemma 17.10, though we prove it directly:

Theorem 5.15. *The Riemann integral $\int_{\mathbb{R}^n} : C_c^0(\mathbb{R}^n) \to \mathbb{R}$ has a unique continuous extension to $L^1(\mathbb{R}^n)$, i.e. there is a unique continuous map $I : L^1(\mathbb{R}^n) \to \mathbb{R}$ such that $I(f) = \int_{\mathbb{R}^n} f(x) \, dx$ when $f \in C_c^0(\mathbb{R}^n)$. Here $I(f)$ is shorthand for $I(j(f))$ when $f \in C_c^0(\mathbb{R}^n)$, i.e. j is surpressed (regarded as relabelling) as usual.*

One calls I the Lebesgue integral *and simply writes $\int$ or $\int_{\mathbb{R}^n}$.*

Proof. If $f \in L^1(\mathbb{R}^n)$, there exists a Cauchy sequence $f_j \in C_c^0(\mathbb{R}^n)$, Cauchy in the L^1-norm, such that $f_j \to f$ in L^1. (Namely, f is represented by the Cauchy sequence $\{f_j\}_{j=1}^\infty$.) We claim that $\int_{\mathbb{R}^n} f_j(x)\, dx$ converges (as a sequence of real numbers); then we want to let

$$I(f) = \lim_{j \to \infty} \int_{\mathbb{R}^n} f_j(x)\, dx.$$

Since $\mathbb{R}$ is complete, it suffices to show that the sequence $\int_{\mathbb{R}^n} f_j(x)\, dx$ is Cauchy. But this is easy:

$$\left| \int_{\mathbb{R}^n} f_j(x)\, dx - \int_{\mathbb{R}^n} f_k(x)\, dx \right| = \left| \int_{\mathbb{R}^n} (f_j(x) - f_k(x))\, dx \right|$$

$$\leq \int_{\mathbb{R}^n} |f_j(x) - f_k(x)|\, dx = \|f_j - f_k\|_{L^1},$$

so $\{f_j\}_{j=1}^\infty$ being L^1-Cauchy shows that $\int_{\mathbb{R}^n} f_j(x)\, dx$ is $\mathbb{R}$-Cauchy.

This is not enough. We still need to show that if we take a *different* Cauchy sequence representing f, say $\{g_j\}_{j=1}^\infty$, then we get the same result, i.e. that

$$\lim_{j \to \infty} \int_{\mathbb{R}^n} f_j(x)\, dx = \lim_{j \to \infty} \int_{\mathbb{R}^n} g_j(x)\, dx.$$

This can be seen in different ways. One is that the alternating sequence

$$f_1, g_1, f_2, g_2, f_3, \ldots$$

is still Cauchy, as one easily checks (using that f_j, g_j are equivalent); thus the integrals of elements of this alternating sequence are still convergent by the previous paragraph, and hence all subsequences, such as the integrals of the f_j and those of the g_j, converge to the very same limit.

The Riemann integral is a linear map, so I inherits this property as one easily checks using

$$\int_{\mathbb{R}^n} (af_j(x) + bg_j(x))\, dx = a \int_{\mathbb{R}^n} f_j(x)\, dx + b \int_{\mathbb{R}^n} g_j(x)\, dx.$$

Next, for $f, g \in L^1(\mathbb{R}^n)$ represented by sequences $\{f_j\}_{j=1}^\infty$, resp. $\{g_j\}_{j=1}^\infty$,

$$\left| \int_{\mathbb{R}^n} f_j(x)\, dx - \int_{\mathbb{R}^n} g_j(x)\, dx \right| = \left| \int_{\mathbb{R}^n} (f_j(x) - g_j(x))\, dx \right|$$

$$\leq \int_{\mathbb{R}^n} |f_j(x) - g_j(x)|\, dx = \|f_j - g_j\|_{L^1},$$

so

$$|I(f) - I(g)| = \lim_{j \to \infty} \left| \int_{\mathbb{R}^n} f_j(x)\, dx - \int_{\mathbb{R}^n} g_j(x)\, dx \right| \leq \lim_{j \to \infty} \|f_j - g_j\|_{L^1} = \|f - g\|,$$

which shows the continuity of I.

Finally, if $\tilde{I}$ has the same properties of I, then for any $f \in L^1(\mathbb{R}^n)$ represented by a Cauchy sequence $\{f_j\}_{j=1}^\infty$, we must have

$$\tilde{I}(f) = \lim_{j \to \infty} \int_{\mathbb{R}^n} f_j(x)\,dx,$$

so we have $\tilde{I}(f) = I(f)$, completing the proof. $\qquad\qquad\square$

For $f \in L^1(\mathbb{R})$ and $\phi \in C^0(\mathbb{R}^n)$ bounded (in particular for $\phi \in C_c^0(\mathbb{R}^n)$), we can also make sense of $f\phi$ as an element of $L^1(\mathbb{R}^n)$; see Problem 5.13. As a result, ι_f is well defined in (5.7).

Note that elements of $L^1(\mathbb{R}^n)$ do not actually have values at points $x \in \mathbb{R}^n$ (thus they are not functions); in the above approximation of the step function one might think that 0 is a reasonable value at $x = \pm 1$, but one can just as well write an equivalent Cauchy sequence which would give 1 as the pointwise limit. Of course, the step function is still continuous away from these two points, so it makes sense to talk about its values away from these two points!

However, it does make sense to talk about restricting elements of $L^1(\mathbb{R}^n)$, e.g. to balls, or more generally open sets Ω. For this, one can just define $L^1(\Omega)$ as the completion of continuous functions of compact support included in Ω. Then it is not hard to see that a Cauchy sequence in $L^1(\mathbb{R}^n)$ of elements of $C_c^0(\mathbb{R}^n)$ (whose equivalence class is an element f of $L^1(\mathbb{R}^n)$ by definition) gives rise to a Cauchy sequence of compactly supported functions with support in Ω, e.g. by a cutoff procedure as for the step function considered above, with the resulting Cauchy sequence (or more precisely its equivalence class) giving an element $\tilde{f}$ of $L^1(\Omega)$ which we define to be the restriction of f to Ω.

The space $L^1_{\mathrm{loc}}(\mathbb{R}^n)$ then is the space of locally L^1 functions. This is a bit awkward to define in our present setting, though the idea is that this consists of (almost) functions which are L^1 on compact sets. The reason it is awkward to say is that one needs a larger framework (which in the measure theoretic setting would consist of measurable functions) within which to define it. One way to say this without measure theory is that $L^1_{\mathrm{loc}}(\mathbb{R}^n)$ consists of a sequence f_k of elements of $L^1(B(k))$ (so $f_k \in L^1(B(k))$), with $B(k)$ the ball of radius k around the origin, such that the restriction of f_k to $B(j)$ (which makes sense by the open set discussion above), when $j < k$, is given by f_j. For example, the function $e^{|x|^2}$ is in $L^1_{\mathrm{loc}}(\mathbb{R}^n)$, since for each k, the function $f_k(x) = e^{|x|^2}$ for $|x| < k$ is in $L^1(B(k))$ (by considering an approximating Cauchy sequence of continuous functions of compact support, as above for the step function), and has the desired property. Thus, the f_k are the 'same' where this statement makes sense.

Problems

Problem 5.1. Show that the function defined by

$$\chi(t) = e^{-1/t}, \ t > 0; \ \chi(t) = 0, \ t \le 0,$$

is in $\mathcal{C}^\infty(\mathbb{R})$.

Note that being $\mathcal{C}^\infty$ in $\{t < 0\}$ is automatic (the function is constant); in $\{t > 0\}$, follows from the chain rule and the differentiability of the exponential, so the issue is the behavior near $t = 0$.

Hint: It is standard from basic calculus that for all N, $\lim_{x \to +\infty} e^{-x}x^N = 0$; i.e. 'exponential growth beats polynomial growth' in the sense that e^x is much larger than x^N for x sufficiently large. So the continuity of χ, its differentiability at 0 with derivative 0, as well as the continuity of χ', with χ' given by $t^{-2}e^{-1/t}$ for $t > 0$, 0 for $t \le 0$, follow. Proceed by induction, showing in particular that the derivatives of χ in $t > 0$ are of the form $P(t)t^{-k}e^{-1/t}$ for a suitable polynomial P and $k \ge 1$ integer.

Problem 5.2. One can also use the function χ from Problem 5.1 to construct partitions of unity. For instance, suppose we want to localize smoothly to the intervals $I_+ = (-1, +\infty)$ and $I_- = (-\infty, 1)$ in $\mathbb{R}$; i.e. we want to write a $\mathcal{C}^\infty$ function u as $u = u_+ + u_-$, where u_+ is supported in I_+ and u_- is supported in I_-. We can achieve this if we find $\mathcal{C}^\infty$ functions $\phi_\pm$ such that $\phi_+ + \phi_- = 1$ (identically 1) and $\operatorname{supp}\phi_\pm \subset I_\pm$ (the subscripts of both sides are $+$ or both sides are $-$, so $\operatorname{supp}\phi_+ \subset I_+$, and similarly for the $-$ sign), for then $u_+ = \phi_+u$, $u_- = \phi_-u$ (i.e. $u_\pm = \phi_\pm u$) does the job. To get such $\phi_\pm$, simply let

$$\tilde{\phi}_+ = \chi(t + 1/2), \ \tilde{\phi}_- = \chi(1/2 - t), \ \phi_+ = \frac{\tilde{\phi}_+}{\tilde{\phi}_+ + \tilde{\phi}_-}, \ \phi_- = \frac{\tilde{\phi}_-}{\tilde{\phi}_+ + \tilde{\phi}_-};$$

it is easy to see that this satisfies all criteria.

Show that if instead we let $I_+ = (1, +\infty)$, $I_0 = (-\infty, -1)$, $I_0 = (-2, 2)$, then there still exist $\phi_+, \phi_-, \phi_0 \in \mathcal{C}^\infty(\mathbb{R})$ such that $\phi_+ + \phi_0 + \phi_- = 1$ (identically 1). One calls ϕ_+, ϕ_0, ϕ_- a *partition of unity* subordinate to the cover I_+, I_0, I_- of $\mathbb{R}$.

Problem 5.3. Let ϕ, c be as in Example 5.1 (so $c > 0$), and let $\delta_j \to 0$. For $y \in \mathbb{R}^n$, let $f_{y,j}(x) = c^{-1}\delta_j^{-n}\phi((x - y)/\delta_j)$.

    (i) Show that the distribution $u_{y,j} = \iota_{f_{y,j}}$ given by $f_{y,j}$ converges to δ_y as $j \to \infty$, i.e.

$$\lim_{j \to \infty} \int_{\mathbb{R}^n} f_{y,j}(x)\psi(x)\,dx = \psi(y).$$

(ii) Show that even if ψ is merely a continuous function of compact support, the function

$$\psi_j(y) = \int_{\mathbb{R}^n} f_{y,j}(x)\psi(x)\,dx = \int_{\mathbb{R}^n} c^{-1}\delta_j^{-n}\phi((x-y)/\delta_j)\psi(x)\,dx$$

is actually a $\mathcal{C}^\infty$ function of compact support.

(iii) Show that if ψ is a continuous function of compact support, then $\psi_j \to \psi$ uniformly, i.e.

$$\sup_{y\in\mathbb{R}^n} |\psi_j(y) - \psi(y)| \to 0$$

as $j \to \infty$.

Hint: ψ continuous of compact support implies that ψ is uniformly continuous, i.e. for all $\epsilon > 0$ there exists $\delta > 0$ such that $|\psi(x) - \psi(y)| < \epsilon$ if $|x - y| < \delta$.

Note: This shows that continuous functions of compact support can be approximated, in the natural norm on continuous functions, by $\mathcal{C}^\infty$ functions of compact support.

Problem 5.4. A real-valued piecewise continuous function f in $\mathbb{R}^2_{x,y}$, say with discontinuity given by the curve $y = F(x)$ for a continuous function F, is a function $f : \mathbb{R}^2 \to \mathbb{R}$ for which in each of the two regions

$$\Omega_+ = \{(x,y) \in \mathbb{R}^2 : y > F(x)\}, \quad \Omega_- = \{(x,y) \in \mathbb{R}^2 : y < F(x)\}$$

there exist functions $f_+ \in C(\overline{\Omega_+}), f_- \in C(\overline{\Omega_-})$ such that $f|_{\Omega_\pm} = f_\pm|_{\Omega_\pm}$ (with all signs being $+$ or all signs being $-$). Thus, $f_\pm$ are continuous to the curve $y = F(x)$, but they typically take different values on this curve. Note that the values of f along this curve are ignored since we only ask for equality to $f_\pm$ away from the curve; only f in $\Omega_\pm$ matters. Correspondingly, one says $f = g$ for two such functions f,g if $f|_{\Omega_+\cup\Omega_-} = g|_{\Omega_+\cup\Omega_-}$, i.e. if $f_+ = g_+$ and $f_- = g_-$.

Show that every such piecewise continuous function f gives rise to a distribution ι_f, and $f \neq g$ implies $\iota_f \neq \iota_g$.

Problem 5.5. Show that the only solution $u \in \mathcal{D}'(\mathbb{R})$ of $u' = 0$ is $u = c$, with c a constant function.

Hint: $u' = 0$ means that $u(\phi') = 0$ for all $\phi \in \mathcal{C}_c^\infty(\mathbb{R})$. You need to show that there is a constant c such that $u(\psi) = \int c\psi\,dx$ for all $\psi \in \mathcal{C}_c^\infty(\mathbb{R})$. To do so, consider when $\psi \in \mathcal{C}_c^\infty(\mathbb{R})$ is of the form $\psi = \phi'$, $\phi \in \mathcal{C}_c^\infty(\mathbb{R})$, paying particular attention to the issue of compact supports. Then write an arbitrary ψ as a linear combination of a fixed $\phi_0 \in \mathcal{C}_c^\infty(\mathbb{R})$ and the derivative ϕ' of some $\phi \in \mathcal{C}_c^\infty(\mathbb{R})$.

Problem 5.6. Consider the PDE

$$au_x + u_y = 0.$$

We already know that the C^1 solutions are of the form $u(x, y) = f(x - ay)$, $f \in C^1(\mathbb{R})$.

(i) Show that if f is merely piecewise continuous (or if you wish locally integrable), then the so-defined u still solves the PDE in the sense of distributions.

(ii) Suppose now that f is a distribution, e.g. $f = \delta_0$. Can you make sense of the formula $u(x, y) = f(x - ay)$? That is, find a procedure giving a distribution u when you start with $f \in \mathcal{D}'(\mathbb{R})$ which depends continuously on f (i.e. if $f_j \to f$, then the corresponding distributions satisfy $u_j \to u$) and make sure that if f is a piecewice continuous function, then you end up with the function $u(x, y) = f(x - ay)$.

(iii) Check that the distribution you defined still solves the PDE.

Problem 5.7. In $t \geq 0$, consider the conservation law

$$u_t + f(u)_x = 0, \quad u(x, 0) = \phi(x).$$

Suppose u is C^1, except that it has a jump discontinuity along the C^1-curve given by $x = \xi(t)$, and $u|_{t=0} = \phi$. Let

$$\Omega_+ = \{(x, t) : \ x > \xi(t)\}, \quad \Omega_- = \{(x, t) : \ x < \xi(t)\},$$

and let $u_\pm = u|_{\Omega_\pm}$, so $u_\pm$ are C^1 on $\overline{\Omega_\pm}$. Show that u is a weak solution of the PDE if and only if u_+ and u_- solve the PDE in the classical sense, i.e. the PDE holds pointwise in $\Omega_+ \cup \Omega_-$ and the initial condition holds there at $t = 0$, and

$$(u_-(\xi(t), t) - u_+(\xi(t), t))\xi'(t) = f(u_-(\xi(t), t)) - f(u_+(\xi(t), t)).$$

This is called the *Rankine-Hugoniot jump condition*.

Using this, in $t \geq 0$, find a weak solution of Burgers' equation

$$u_t + uu_x = 0, \quad u(x, 0) = \phi(x)$$

with

$$\phi(x) = \begin{cases} 1, & x < 0, \\ 0, & x > 0, \end{cases}$$

which is constant to the left, resp. the right, of a curve $x = \xi(t)$. This is called a *shock wave*.

Problem 5.8. In $t \geq 0$, consider Burgers' equation

$$u_t + uu_x = 0, \quad u(x, 0) = \phi(x),$$

with initial condition

$$\phi(x) = \begin{cases} 0, & x < 0, \\ \frac{x}{\epsilon}, & 0 < x < \epsilon, \\ 1, & x > \epsilon, \end{cases}$$

$\epsilon > 0$. Solve the equation, and show that as $\epsilon \to 0$, it converges to a weak solution of Burgers' equation with initial condition given by the Heaviside step function $\phi(x) = H(x)$ (so $\phi = 1$ on $(0, \infty)$, $\phi = 0$ on $(-\infty, 0)$). This solution is called a *rarefaction wave*.

Find another solution of Burgers' equation with initial condition given by the Heaviside step function $\phi(x) = H(x)$, namely find a solution which is piecewise constant (cf. the previous problem).

Show that this second solution does not satisfy the *entropy condition* along the shock:

$$f'(u_+(\xi(t), t)) \le \frac{f(u_-(\xi(t), t)) - f(u_+(\xi(t), t))}{u_-(\xi(t), t) - u_+(\xi(t), t)} \le f'(u_-(\xi(t), t)),$$

while the solution of Problem 5.7 satisfies it. Note that the entropy condition states that the speed of waves (which, recall, is $\frac{dx}{dt} = f'(u)$) to the right of the shock is less than the speed to the left, and that the speed of the shock (which is given by the Rankine-Hugoniot condition) is in between these values (i.e. the shock is 'forced' by colliding waves), which is a physical justification of preferring the rarefaction wave to a shock in this problem.

Problem 5.9. Let $\psi \in C(\mathbb{R})$ be given by

$$\psi(x) = \begin{cases} 0, & x < -1, \\ 1 + x, & -1 < x < 0, \\ 1 - x, & 0 < x < 1, \\ 0, & x > 1, \end{cases}$$

so $\psi \ge 0$, $\psi(x) = 0$ if $|x| \ge 1$ and $\int_{\mathbb{R}} \psi(x)\, dx = 1$. Let $\psi_j(x) = j\psi(jx)$, so $\psi_j(x) = 0$ if $|x| \ge 1/j$, and $\int \psi_j(x)\, dx = 1$ for all j.

(i) Show that $\psi_j \to \delta_0$ in $\mathcal{D}'(\mathbb{R})$ (i.e., to be pedantic, $\iota_{\psi_j} \to \delta_0$). (*Hint:* This is essentially written up in the notes on distributions, and was done in the chapter!)

(ii) Let $\phi \in C_c^\infty(\mathbb{R})$ be such that $\phi(x) = 1$ for $|x| < 1$. Show that $\{\iota_{\psi_j^2}(\phi)\}_{j=1}^\infty$ is *not* a convergent sequence in $\mathbb{R}$. Use this to conclude that $\{\iota_{\psi_j^2}\}_{j=1}^\infty$ does not converge to any distribution.

(iii) Show that there is no continuous extension of the map $Q : f \mapsto f^2$ on $C(\mathbb{R})$ (so $Q : C(\mathbb{R}) \to C(\mathbb{R})$) to $\mathcal{D}'(\mathbb{R})$; i.e. there is no map $\tilde{Q} : \mathcal{D}'(\mathbb{R}) \to \mathcal{D}'(\mathbb{R})$ such that
  • $\tilde{Q}(\iota_f) = \iota_{Qf}$ for every $f \in C(\mathbb{R})$ and
  • $u_j \to u$ in $\mathcal{D}'(\mathbb{R})$ implies $\tilde{Q}u_j \to \tilde{Q}u$ in $\mathcal{D}'(\mathbb{R})$.
  This means that one cannot square (and more generally multiply) distributions in general.

Problem 5.10. Consider the conservation law

$$u_t + (f(u))_x = 0, \ u(x,0) = \phi(x),$$

with $f \in C^2(\mathbb{R})$. Let $v = f'(u)$. Show that if $f'' \neq 0$ and u is continuous and is C^1 apart from jump discontinuities *in its first derivatives*, then v has the same properties and satisfies Burgers' equation. (Note that the Rankine-Hugoniot condition is vacuous: there are no shocks.) If f is strictly convex, i.e. $f'' > 0$, conclude that one can reduce general scalar conservation laws to Burgers' equation, i.e. that one can find u by solving for v first.

Is the same true if u has jump discontinuities, i.e. is it true that if u satisfies the Rankine-Hugoniot conditions, then v does as well?

Problem 5.11. Consider Burgers' equation

$$u_t + u u_x = 0, \ u(x,0) = \phi(x),$$

with initial condition

$$\phi(x) = \begin{cases} 0, & x < -1, \\ -1 - x, & -1 < x < 0, \\ -1 + x, & 0 < x < 1, \\ 0, & x > 1. \end{cases}$$

(i) Find the weak solution u that satisfies the entropy condition (see Problem 5.8).

(ii) Show that $\int u(x,t)\,dx$ is constant (independent of t).

(iii) Show that $\int u(x,t)^3\,dx$ is constant until the time when a shock develops. What happens for later times? Explain this in terms of the results of Problem 5.10. (*Hint:* For the last part: Consider a conservation law $v_t + g(v)_x = 0$, so $\int v(x,t)\,dx$ is conserved; find g so you can use this.)

Problem 5.12. This and the following problems concern the additional material on L^1.

For $d < n$ make sense of

$$f(x) = \begin{cases} |x|^{-d}, & |x| \leq 1, \\ 0, & |x| > 1 \end{cases}$$

as an element of $L^1(\mathbb{R}^n)$ by showing that the sequence

$$f_j(x) = \begin{cases} j^d, & |x| \leq 1/j, \\ |x|^{-d}, & 1/j \leq |x| \leq 1, \\ 1 - j(|x| - 1), & 1 \leq |x| \leq 1 + 1/j, \\ 0, & |x| \geq 1 + 1/j \end{cases}$$

is an L^1-Cauchy sequence of continuous functions. Sketch some elements of the sequence and the function f.

Problem 5.13. Show that if $\phi \in C^0(\mathbb{R}^n)$ is bounded, then for $f \in L^1(\mathbb{R}^n)$, $f\phi$ makes sense as an element of $L^1(\mathbb{R}^n)$, extending the notion of pointwise multiplication of continuous functions.

Hint: You just need to show that if $\{f_j\}_{j=1}^{\infty}$ is a Cauchy sequence representing f, then $f_j\phi$ (which makes sense as a product of continuous functions) is also L^1-Cauchy, and any other Cauchy sequence $\{g_j\}_{j=1}^{\infty}$ representing f gives an equivalent Cauchy sequence $g_j\phi$.

Chapter 6

Second order constant coefficient PDE: Types and d'Alembert's solution of the wave equation

1. Classification of second order PDE

We consider second order constant coefficient scalar linear PDEs on $\mathbb{R}^n$. These have the form

$$\mathcal{L}u = f,$$

$$\mathcal{L} = \sum_{i,j=1}^{n} a_{ij}\partial_{x_i}\partial_{x_j} + \sum_{i=1}^{n} b_i\partial_{x_i} + c$$

where a_{ij}, b_i and c are (complex) constants, and f is given. As a general principle, the leading, i.e. second order, terms are the most important in analyzing the PDE. The second order terms then take the form

$$L = \sum_{i,j=1}^{n} a_{ij}\partial_{x_i}\partial_{x_j},$$

here we are interested in the case when the a_{ij} are real constants. Since $\partial_{x_i}\partial_{x_j} = \partial_{x_j}\partial_{x_i}$, we might as well assume that $a_{ij} = a_{ji}$ (otherwise replace both by $\frac{1}{2}(a_{ij} + a_{ji})$, which does not change L), i.e. that A is symmetric.

We rewrite L in a different form. With

$$\nabla_x = \begin{bmatrix} \partial_{x_1} \\ \partial_{x_2} \\ \cdots \\ \partial_{x_n} \end{bmatrix}$$

and

$$A = \begin{bmatrix} a_{11} & \cdots & a_{1n} \\ & \cdots & \\ a_{n1} & \cdots & a_{nn} \end{bmatrix},$$

we can write

$$L = \nabla_x^{\mathrm{T}} A \nabla_x,$$

where ∇_x^{T} is the transpose of ∇_x:

$$\nabla_x^{\mathrm{T}} = \begin{bmatrix} \partial_{x_1} & \cdots & \partial_{x_n} \end{bmatrix}.$$

Technically, there is no need for the x in the subscripts, and one should write e.g. $\nabla^{\mathrm{T}} = \begin{bmatrix} \partial_1 & \cdots & \partial_n \end{bmatrix}$. However, the subscripts are very useful below, as they enable us to compress (and abuse) the notation in the change of variables formula. Note that this T is the transpose only in the *matrix sense*, but *not* in the differential operator sense discussed in Chapter 5 in (5.4)-(5.5); hence we are using T rather than †. (In the full, joint matrix and differential operator sense, transpose would be the negative of this in accordance with (5.5), as the operator is first order: $\partial_{x_j}^{\dagger} = -\partial_{x_j}$.)

We would like to bring it to a simpler form; the simpler form we wish for may depend on the intended use. As a first step we would like to have L brought into a diagonal form, i.e. A made diagonal. We can achieve this by changing coordinates. Namely, if we let $\xi = \Xi(x)$, i.e. $\xi_j = \Xi_j(x_1, \ldots, x_n)$, $j = 1, \ldots, n$, with inverse function X, i.e. $x = X(\xi)$, meaning that $X \circ \Xi$ and $\Xi \circ X$ are both the identity maps (both X and Ξ are assumed to be C^1), then by the chain rule for any C^1 function u we have

$$\partial_{x_k}(u \circ \Xi) = \sum_{\ell=1}^{n} \frac{\partial \Xi_\ell}{\partial x_k}(\partial_{\xi_\ell} u) \circ \Xi.$$

We can rewrite this in a matrix form:

$$(6.1) \qquad \begin{bmatrix} \partial_{x_1} \\ \cdots \\ \partial_{x_n} \end{bmatrix} = \begin{bmatrix} \frac{\partial \Xi_1}{\partial x_1} & \cdots & \frac{\partial \Xi_n}{\partial x_1} \\ & \cdots & \\ \frac{\partial \Xi_1}{\partial x_n} & \cdots & \frac{\partial \Xi_n}{\partial x_n} \end{bmatrix} \begin{bmatrix} \partial_{\xi_1} \\ \cdots \\ \partial_{\xi_n} \end{bmatrix},$$

and so, in particular, if ξ is a linear function of x, i.e. $\xi = Bx$, $B = \begin{bmatrix} b_{ij} \end{bmatrix}_{i,j=1}^{n}$, so $\xi_i = \sum_{j=1}^{n} B_{ij} x_j$, then

$$(6.2) \qquad\qquad\qquad \nabla_x = B^{\mathrm{T}} \nabla_\xi.$$

Thus,
$$L = \nabla_x^{\mathrm{T}} A \nabla_x = \nabla_\xi^{\mathrm{T}} B A B^{\mathrm{T}} \nabla_\xi.$$

Hence, in order to make L be given by a diagonal matrix in the ξ coordinates, we need that BAB^{T} is diagonal. Recalling that A is symmetric, we proceed as follows.

Any real symmetric matrix, such as A, can be diagonalized by conjugating it by an orthogonal matrix. That is, there is an orthogonal matrix O (recall that orthogonal means that $OO^{\mathrm{T}} = O^{\mathrm{T}}O = \mathrm{Id}$, i.e. $O^{\mathrm{T}} = O^{-1}$) and a diagonal matrix Λ such that
$$A = O^{\mathrm{T}} \Lambda O,$$

where the diagonal entries λ_j of Λ are the eigenvalues of A and the rows of O, i.e. the columns of O^{T}, are corresponding (unit length) eigenvectors e_j. (Thus, the orthogonality of O is a consequence of the orthogonality of the e_j.) Choosing $B = O$, we deduce that
$$BAB^{\mathrm{T}} = OO^{\mathrm{T}} \Lambda OO^{\mathrm{T}} = \Lambda$$

is indeed diagonal.

In fact, we can go a bit further. We write $\mathrm{sgn}\, t$ for the sign of a real number t, so $\mathrm{sgn}\, t = 1$ if $t > 0$, $\mathrm{sgn}\, t = -1$ if $t < 0$ and $\mathrm{sgn}\, t = 0$ if $t = 0$. One can factor a diagonal matrix with real entries, such as Λ, by writing its diagonal entries as $\lambda_j = |\lambda_j|^{1/2} (\mathrm{sgn}\, \lambda_j) |\lambda_j|^{1/2}$. It turns out that if $\lambda_j = 0$ for some j, it is convenient to simply write $\lambda_j = 1 \cdot \mathrm{sgn}\, \lambda_j \cdot 1$. Thus, if D is the diagonal matrix with diagonal entries $d_j = |\lambda_j|^{1/2}$ when $\lambda_j \neq 0$, $d_j = 1$ if $\lambda_j = 0$, and P is the diagonal matrix with diagonal entries $p_j = \mathrm{sgn}\, \lambda_j$, then $\Lambda = DPD$, so
$$A = O^{\mathrm{T}} DPDO.$$

Now, D is invertible since it is diagonal and none of its diagonal entries vanish. We let $B = D^{-1}O$, and hence
$$BAB^{\mathrm{T}} = (D^{-1}O)(O^{\mathrm{T}} DPDO)(O^{\mathrm{T}} D^{-1}) = P.$$

We deduce that for any L there is a change of coordinates, $\xi = Bx$, such that in the ξ coordinates
$$L = \nabla_\xi^{\mathrm{T}} P \nabla_\xi = \sum_{j=1}^{n} p_j \partial_{\xi_j}^2.$$

Now, the overall sign of L is not important, so e.g. if all the $p_j < 0$, the behavior of solutions of $Lu = f$ is the same as those of $-Lu = g$ (simply let $g = -f$), and the corresponding coefficients of $-L$ are $-p_j > 0$. Also, the case when some p_j vanish is *degenerate* and first order terms influence the behavior (and existence) of solutions of $\mathcal{L}u = f$ tremendously. As an example, the heat operator, $\partial_x^2 - \partial_t$, behaves very differently from the 'backward heat operator', $\partial_x^2 + \partial_t$.

We thus make the following classification of the non-degenerate setting, i.e. when $\det A \neq 0$, so each p_j is 1 or -1:

(i) If all p_j (i.e. λ_j) have the same sign, L (and $\mathcal{L}$) is called *elliptic*.

(ii) If all but one p_j have the same sign, L is called *hyperbolic*.

(iii) If there are at least two p_j of each sign, then L is called *ultrahyperbolic*.

In certain ways ultrahyperbolic equations are like hyperbolic equations (though not as far as initial conditions on hypersurfaces are concerned!); the biggest difference is between elliptic equations and the other two classes.

We conclude that by a change of coordinates and possible multiplication by -1, every non-degenerate second order linear constant coefficient PDE with real coefficients in its leading part can be brought to the following form:

(i) If $\mathcal{L}$ is elliptic, then

$$\mathcal{L} = \sum_{j=1}^{n} \partial_{\xi_j}^2 + \sum_{j=1}^{n} b_j \partial_{\xi_j} + c.$$

(ii) If $\mathcal{L}$ is hyperbolic, then

$$\mathcal{L} = -\sum_{j=1}^{n-1} \partial_{\xi_j}^2 + \partial_{\xi_n}^2 + \sum_{j=1}^{n} b_j \partial_{\xi_j} + c.$$

(iii) If $\mathcal{L}$ is ultrahyperbolic, then for some $2 \leq k \leq n-2$,

$$\mathcal{L} = -\sum_{j=1}^{n-k} \partial_{\xi_j}^2 + \sum_{j=n-k+1}^{n} \partial_{\xi_j}^2 + \sum_{j=1}^{n} b_j \partial_{\xi_j} + c.$$

The model examples, when b_j and c vanish, are:

(i) If $\mathcal{L}$ is elliptic, then

$$\Delta = \sum_{j=1}^{n} \partial_{\xi_j}^2,$$

the Laplacian.

(ii) If $\mathcal{L}$ is hyperbolic, then

$$\Box = -\sum_{j=1}^{n-1} \partial_{\xi_j}^2 + \partial_{\xi_n}^2,$$

the wave operator or d'Alembertian.

We can of course change our notation and work with $\Delta = \sum_{j=1} \partial_{x_j}^2$ and $\Box = -\sum_{j=1}^{n-1} \partial_{x_j}^2 + \partial_{x_n}^2$. For $\Box$, one often wants to think of x_n as 'time' and write it as t, and write x for $(x_1, \ldots, x_{n-1})$; then $\Box_{\mathbb{R}^n} = -\Delta_{\mathbb{R}^{n-1}} + \partial_t^2$.

We remark that this classification of second order equations remains valid even if the coefficients are not constant, by simply fixing the coefficients at a point. Namely taking some $x_0 \in \mathbb{R}^n$ and replacing

$$L = \sum_{i,j=1}^{n} a_{ij}(x) \partial_{x_i} \partial_{x_j} + R,$$

where R is a first order differential operator by

$$\mathcal{L} = \mathcal{L}_{x_0} = \sum_{i,j=1}^{n} a_{ij}(x_0) \partial_{x_i} \partial_{x_j},$$

which has constant coefficients, and assigning to L a type at x_0 based on $\mathcal{L}_{x_0}$. Thus, the type can vary from point to point; for instance for the operator

$$L = u_{xx} - x u_{yy}$$

on $\mathbb{R}^2_{x,y}$, the operator is elliptic at points with $x < 0$, hyperbolic at points with $x > 0$, and degenerate at points with $x = 0$. This is called *Tricomi's operator*. Another type-changing operator (with quite different behavior) is

$$L = x u_{xx} + u_{yy}.$$

2. Solving second order hyperbolic PDE on $\mathbb{R}^2$

We start our study of second order equations by studying hyperbolic equations on $\mathbb{R}^2$. The advantage of these is that they can be factored (disregarding the first order terms) into the product of two first order operators, which we already know how to analyze. Thus, with the slightly more general d'Alembertian

$$\Box = \partial_t^2 - c^2 \partial_x^2,$$

we factor it as

$$\Box = (\partial_t - c \partial_x)(\partial_t + c \partial_x).$$

With such a factorization it is easy to solve $\Box u = 0$. Namely, we first let $v = (\partial_t + c \partial_x)u$, so the PDE states

$$(\partial_t - c \partial_x)v = 0.$$

This is a first order linear PDE, which we solve easily (indeed we solved this in Chapter 3, with slightly different notation) to obtain

$$v(x, t) = h(x + ct)$$

for some h. (Explicitly, we could impose initial conditions say at $t = 0$, namely $v(x, 0) = h(x)$, and then we get $v(x, t) = h(x + ct)$.) Next, we need to solve

$$(\partial_t + c\partial_x)u = v, \text{ with } v(x, t) = h(x + ct).$$

We use the method of characteristics again, so we have characteristic equations with initial data, e.g. at the x axis,

$$\frac{dx}{ds} = c, \ x(r, 0) = r,$$

(6.3)
$$\frac{dt}{ds} = 1, \ t(r, 0) = 0,$$

$$\frac{dz}{ds} = h(x + ct), \ u(r, 0) = \phi(r).$$

We obtain that $t = s$, $x = cs + r$, hence $x + ct = 2cs + r$ and $r = x - ct$, so the ODE for z is

$$\frac{dz}{ds} = h(2cs + r), \ z(r, 0) = \phi(r).$$

Integrating from $s = 0$, we deduce that

$$z = \int_0^s h(2cs + r) \, ds + \phi(r).$$

Letting $H(s) = \int_r^s h(s') \, ds'$ (so $H' = h$ and $H(r) = 0$), we deduce that

(6.4)
$$z = \frac{1}{2c} H(2cs + r) + \phi(r)$$

and

(6.5)
$$u(x, t) = \frac{1}{2c} H(x + ct) + \phi(x - ct),$$

so u is of the form

(6.6)
$$u(x, t) = f(x + ct) + g(x - ct).$$

Conversely, it is easy to check that every u of this form with f, g in C^2 solves the PDE, so this is our general solution of the wave equation. Note that $x + ct = C$ and $x - ct = C$, with C a constant, are the *characteristics* of the two factors $\partial_t - c\partial_x$ and $\partial_t + c\partial_x$, respectively (recall that r is constant along the characteristics in our method to solve first order PDEs). Thus, the general solution is the sum of two 'waves', each of which is constant along the characteristics corresponding to one of the factors.

We now find the solution with initial conditions:

$$u(x, 0) = \phi(x), \ u_t(x, 0) = \psi(x).$$

Directly from (6.6),

$$f(x) + g(x) = \phi(x),$$
$$cf'(x) - cg'(x) = \psi(x).$$

Differentiating the first line gives $f'(x) + g'(x) = \phi'(x)$. Combining with the second line and solving for f' and g' yields

$$f'(x) = \frac{1}{2}(\phi'(x) + c^{-1}\psi(x)), \quad g'(x) = \frac{1}{2}(\phi'(x) - c^{-1}\psi(x)).$$

Integrating yields

$$f(x) = \frac{1}{2}\phi(x) + \frac{1}{2c}\int_0^x \psi(\sigma)\,d\sigma + A,$$

$$g(x) = \frac{1}{2}\phi(x) - \frac{1}{2c}\int_0^x \psi(\sigma)\,d\sigma + B,$$

for some constants A and B — substitution into $f + g = \phi$ gives $B = -A$. In summary,

$$u(x,t) = f(x+ct) + g(x-ct) = \frac{1}{2}(\phi(x+ct) + \phi(x-ct)) + \frac{1}{2c}\int_{x-ct}^{x+ct}\psi(\sigma)\,d\sigma.$$

This is *d'Alembert's solution* of the wave equation. Notice that if ϕ is C^2 and ψ is C^1, then this indeed satisfies the wave equation in the classical sense, and it also satisfies the initial conditions.

This method was somewhat ad hoc in that we ignored the initial conditions at first. We could have gone through the whole calculation enforcing the initial conditions. The key issue is to find initial conditions for v. But this is easy: $v(x,0) = u_t(x,0) + cu_x(x,0) = \psi(x) + c\phi'(x)$, i.e. the initial condition is $v(x,0) = h(x)$, $h(r) = \psi(r) + c\phi'(r)$. Then our solution is

$$v(x,t) = h(x+ct) = \psi(x+ct) + c\phi'(x+ct).$$

Then in (6.4),

$$H(s) = \int_r^s h(s')\,ds' = \int_r^s (\psi(s') + c\phi'(s'))\,ds' = \int_r^s \psi(s')\,ds' + c(\phi(s) - \phi(r)),$$

and so

$$z(r,s) = \frac{1}{2c}\int_r^{2cs+r}\psi(s')\,ds' + \frac{1}{2}(\phi(2cs+r) - \phi(r)) + \phi(r);$$

hence

$$u(x,t) = \frac{1}{2c}\int_{x-ct}^{x+ct}\psi(s')\,ds' + \frac{1}{2}(\phi(x+ct) + \phi(x-ct)),$$

in agreement with the previous calculation.

If we are given a second order hyperbolic linear PDE,

$$a_{11}u_{\xi\xi} + 2a_{12}u_{\xi\eta} + a_{22}u_{\eta\eta} = 0,$$

by a change of coordinates we could always bring it to the form $\partial_t^2 - \partial_x^2$, which we just analyzed; hence we can solve this. Instead, we can also proceed by directly factoring this equation. Note that hyperbolicity means in this case exactly that $\det A < 0$, i.e. that $a_{11}a_{22} - a_{12}^2 < 0$. If $a_{11} = a_{22} = 0$, then it is very easy to factor the equation, it is just $2a_{12}\partial_\xi\partial_\eta u = 0$. Otherwise either a_{11} or a_{22} is non-zero; assume it is the former. Then we factor the quadratic polynomial

$$a_{11}\mu^2 + 2a_{12}\mu + a_{22} = a_{11}(\mu - \mu_+)(\mu - \mu_-),$$

where $\mu_\pm$ are the roots of this polynomial, so

$$\mu_\pm = \frac{-2a_{12} \pm \sqrt{4a_{12}^2 - 4a_{11}a_{22}}}{2a_{11}}.$$

Note that hyperbolicity is exactly the condition where the quantity whose square root we are taking is positive. Then

$$a_{11}\alpha^2 + 2a_{12}\alpha\beta + a_{22}\beta^2 = a_{11}(\alpha - \mu_+\beta)(\alpha - \mu_-\beta)$$

for all α and β (note that for $\beta = 0$ this is automatic, and for $\beta \neq 0$ divide through by β^2), so

$$a_{11}\partial_\xi^2 + 2a_{12}\partial_\xi\partial_\eta + a_{22}\partial_\eta^2 = a_{11}(\partial_\xi - \mu_+\partial_\eta)(\partial_\xi - \mu_-\partial_\eta).$$

For instance, if $a_{12} = 0$, $a_{11} = 1$, $a_{22} = -c^2$, this gives $\mu_\pm = \pm c$ and the factoring

$$\partial_\xi^2 - c^2\partial_\eta^2 = (\partial_\xi + c\partial_\eta)(\partial_\xi - c\partial_\eta);$$

exactly the factoring we had for the wave operator beforehand.

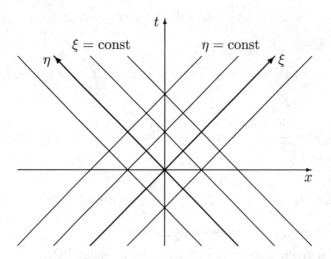

Figure 6.1. Characteristic coordinates ξ, η for the wave equation. Note that the ξ-axis is the line $\eta = 0$, and the η-axis is the line $\xi = 0$.

A slightly different method is to change the hyperbolic equation to a different model form, i.e. another form rather than $\partial_t^2 - c^2\partial_x^2$, namely $\partial_\xi\partial_\eta$.

As already mentioned, $L = \partial_\xi \partial_\eta$ is already factored, and the general solution of $Lu = 0$ can be read immediately:

$$u(\xi, \eta) = f(\xi) + g(\eta).$$

Note that this change of coordinates can always be done for a hyperbolic second order operator on $\mathbb{R}^2$, for they can all (including $\partial_\xi \partial_\eta$) be changed to the standard wave operator. Hence by composing one of these changes with the inverse of another one, they can be changed into each other. The coordinates (ξ, η), shown on Figure 6.1, are called *characteristic coordinates*. To see why, note that for the wave operator we want $\partial_\xi = \partial_t + c\partial_x$, $\partial_\eta = \partial_t - c\partial_x$, or vice versa, i.e.

$$\begin{bmatrix} \partial_\xi \\ \partial_\eta \end{bmatrix} = \begin{bmatrix} 1 & c \\ 1 & -c \end{bmatrix} \begin{bmatrix} \partial_t \\ \partial_x \end{bmatrix},$$

which in view of (6.1)-(6.2) means that (with the role of ξ and x switched) we need

$$\begin{bmatrix} t \\ x \end{bmatrix} = \begin{bmatrix} 1 & 1 \\ c & -c \end{bmatrix} \begin{bmatrix} \xi \\ \eta \end{bmatrix},$$

i.e. $t = \xi + \eta$, $x = c\xi - c\eta$ or

$$\xi = \frac{1}{2c}(x + ct), \quad \eta = -\frac{1}{2c}(x - ct),$$

so the coordinate lines $\xi = C$ and $\eta = C$ are exactly the characteristics. (If we did not mind a constant in front of $\partial_\xi \partial_\eta$, we could have chosen $\xi = x + ct$, $\eta = x - ct$.)

While so far have we worked with classical, C^2, solutions of the wave equation, one can immediately generalize this using Problem 5.6 in Chapter 5. Namely, for any *distributions* $f, g \in \mathcal{D}'(\mathbb{R})$, the distribution given by $u(x, t) = f(x + ct) + g(x - ct)$ (in the same sense as the problem just mentioned) solves

$$(\partial_t^2 - c^2 \partial_x^2)u = 0$$

in the sense of distributions. This can be read since

$$(\partial_t^2 - c^2 \partial_x^2)u = (\partial_t + c\partial_x)(\partial_t - c\partial_x)f(x + ct) + (\partial_t - c\partial_x)(\partial_t + c\partial_x)g(x - ct)$$
$$= (\partial_t + c\partial_x)0 + (\partial_t - c\partial_x)0 = 0,$$

where we used the result of Problem 5.6 in Chapter 5. So, for instance, the function $u(x, t) = H(x - ct)$, where H is now the step function, solves the wave equation, as does $u(x, t) = \delta_0(x + ct)$.

The converse of our claim is also true, i.e. every distribution u satisfying

$$(\partial_t^2 - c^2 \partial_x^2)u = 0$$

is of the form $u(x, t) = f(x + ct) + g(x - ct)$. This is particularly easy to read in characteristic coordinates, where one has to solve equations such as $\partial_\xi v = 0$, $\partial_\eta u = v$, using a slightly modified version of Problem 5.5 in Chapter 5, namely with the extension of this problem to $\mathbb{R}^2$.

Problems

Problem 6.1. Find the type (elliptic, hyperbolic, degenerate) of the following PDEs:

    (i) $u_{xx} - u_{xy} - 2u_{yy} = 0$.

    (ii) $u_{xx} - 2u_{xy} + u_{yy} = 0$.

    (iii) $u_{xx} + 2u_{xy} + 2u_{yy} = 0$.

Problem 6.2. Find the type of the following variable coefficient PDE:

    (i) $\nabla \cdot (c(x)^2 \nabla u) - q(x)u = f(x)$, where c, q, f are given functions on $\mathbb{R}_x^n$, and $c > 0$.

    (ii) $\nabla \cdot (c(x,t)^2 \nabla u) - q(x,t)u = u_{tt}$, where c, q are given functions on $\mathbb{R}_x^n \times \mathbb{R}_t$, and $c > 0$.

Problem 6.3.

    (i) Find the general C^2 solution of the PDE

$$u_{xx} - u_{xt} - 6u_{tt} = 0$$

    by reducing it to a system of first order PDEs.

    (ii) Show that if $f, g \in \mathcal{D}'(\mathbb{R})$, and we define new distributions $v, w \in \mathcal{D}'(\mathbb{R}^2)$ as in Problem 5.6, i.e. formally $v(x,t) = f(3x+t)$, $w(x,t) = g(-2x+t)$, then $u = v + w$ solves the PDE in (1). (*Hint:* Use the result of Problem 5.6, and factor our second order operator. This should only take a few lines.)

Problem 6.4. Solve $u_{xx} - 2u_{xy} - 3u_{yy} = 0$, $u(1,y) = 0$, $u_x(1,y) = y^2$.

Problem 6.5. Solve $u_{xx} + 3u_{xy} - 4u_{yy} = xy$, $u(x,x) = \sin x$, $u_x(x,x) = 0$, $-\infty < x, y < \infty$.

Problem 6.6. On $\mathbb{R}_x \times \mathbb{R}_y$, solve

$$u_{xx} + u_{xy} - 6u_{yy} = 0,$$

with initial conditions $u(x,0) = 0$, $u_y(x,0) = (\sin x) - x$.

Problem 6.7. On $\mathbb{R}^2$, consider the following two PDE:

$$u_{xx} - 2u_{xy} - 3u_{yy} = 0, \qquad x, y \in \mathbb{R},$$
$$u_{xx} + 2u_{xy} + 3u_{yy} = 0, \qquad x, y \in \mathbb{R}.$$

State the type of each of these two PDE, and solve the hyperbolic one of them with the additional conditions

$$u(x, 0) = 0, \qquad\qquad x \in \mathbb{R},$$
$$u_y(x, 0) = x^2, \qquad\qquad x \in \mathbb{R}.$$

Properties of solutions of second order PDE: Propagation, energy estimates and the maximum principle

1. Properties of solutions of the wave equation: Propagation phenomena

We have solved the initial value problem for the wave equation

$$(\partial_t^2 - c^2 \partial_x^2)u = 0, \ u(x,0) = \phi(x), \ u_t(x,0) = \psi(x),$$

namely we showed that the solution is

$$u(x,t) = \frac{1}{2}(\phi(x+ct) + \phi(x-ct)) + \frac{1}{2c} \int_{x-ct}^{x+ct} \psi(\sigma) \, d\sigma.$$

There are a few facts that can be read immediately from this expression. We consider $t > 0$ here; $t < 0$ is similar. First, for $t_0 > 0$, $u(x_0, t_0)$ depends on the initial data ϕ just at the two points $x_0 \pm ct_0$, while it depends on the values of ψ in the whole interval $[x_0 - ct_0, x_0 + ct_0]$. Thus, we call the interval $[x_0 - ct_0, x_0 + ct_0]$ the *domain of dependence* of (x_0, t_0): if the initial conditions vanish there, the solution vanishes at (x_0, t_0). Note that the straight lines $x - ct = x_0 - ct_0$ and $x + ct = x_0 + ct_0$ which go through (x_0, t_0) and $(x_0 \pm ct_0, 0)$ are characteristics.

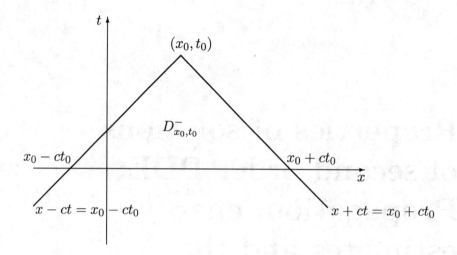

Figure 7.1. The domain of dependence of the point (x_0, t_0).

In fact, it is convenient (for reasons that will be more clear when we solve the *inhomogeneous* wave equation, $\Box u = f$) to consider the domain of dependence of (x, t) to be the whole region

$$D^-_{x_0, t_0} = \{(x, t) : \ t \leq t_0, \ |x - x_0| \leq c(t_0 - t)\};$$

see Figure 7.1. This is the *backward characteristic triangle* from (x_0, t_0): its sides are the characteristics $x - ct = x_0 - ct_0$ and $x + ct = x_0 + ct_0$. With this definition, if the initial data are imposed at $t = T$ instead, where $T < t_0$, then the solution u at (x_0, t_0) depends on the initial data in the interval $[x_0 - c(t_0 - T), x_0 + c(t_0 - T)]$, which is just the segment in which the line $t = T$ intersects the backward characteristic triangle. Indeed, a change of variables (replacing t by $\tau = t - T$) shows that the solution of

$$(\partial_t^2 - c^2 \partial_x^2)u = 0, \ u(x, T) = \phi_T(x), \ u_t(x, T) = \psi_T(x)$$

is

$$u(x, t) = \frac{1}{2}(\phi_T(x + c(t - T)) + \phi_T(x - c(t - T))) + \frac{1}{2c} \int_{x - c(t - T)}^{x + c(t - T)} \psi_T(\sigma) \, d\sigma.$$

On the flipside, we may ask at which points (x_0, t_0) do, or do not, the initial data at $(x, 0)$ influence the solution. More precisely, we can ask at which points (x_0, t_0) is the solution guaranteed to be unaffected if we change the initial data at $(x, 0)$. This happens exactly if $(x, 0)$ is *not* in the backward characteristic triangle from (x_0, t_0), $D^-_{x_0, t_0}$, i.e. (keeping in mind $t_0 > 0$) if $|x - x_0| > ct_0$. Conversely, if $|x - x_0| \leq ct_0$, then the solution may change (and in general does change) if ϕ and ψ are changed at $(x, 0)$. Thus, we call *the forward characteristic triangle*

$$D_{x,0}^+ = \{(x_0, t_0) : t_0 \geq 0, \ |x - x_0| \leq ct_0\}$$

the *domain of influence* of $(x, 0)$; see Figure 7.2.

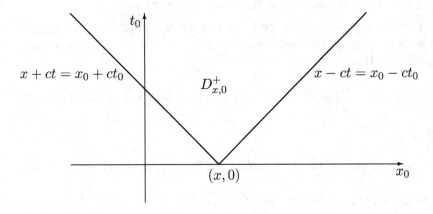

Figure 7.2. The domain of influence of the point $(x, 0)$.

More generally, for initial data at $t = T$, the (forward) domain of influence of (x, T) is

$$D_{x,T}^+ = \{(x_0, t_0) : t_0 \geq T, \ |x - x_0| \leq c(t_0 - T)\}.$$

The size of the intersection of the domain of influence with the line $t = t_0$ increasing at speed c is called *Huygens' principle*, and it is also valid for wave equations with variable coefficients: waves propagate (at most) as fast as c, so c is reasonably called the speed of waves.

Although the solution at (x_0, t_0) depends on the initial data everywhere inside its domain of dependence, its dependence on the data may not be very significant. One way to think about this is the following. 'Information' is carried by singularities of solutions to the PDE, e.g. one flips a switch, and gets a jump in the amplitude of the electromagnetic field. It is this jump that carries the information that the switch was flipped. There are different ways of measuring singularities. The simplest is if we say that a function u is non-singular at a point $y_0 \in \mathbb{R}^n$ if u is C^∞ near y_0, or equivalently, if there is a function $\phi \in C_c^\infty(\mathbb{R}^n)$ such that $\phi(y_0) \neq 0$ and $\phi u \in C^\infty(\mathbb{R}^n)$. (Here, one thinks mostly of functions ϕ which are 0 outside a small neighborhood of y_0.) We make this into a definition, which even works for distributions:

Definition 7.1. Suppose $u \in \mathcal{D}'(\mathbb{R}^n)$. We say that $y_0 \notin \operatorname{singsupp} u$, i.e. y_0 is not in the *singular support* of u, if there exists $\phi \in C_c^\infty(\mathbb{R}^n)$ such that $\phi(y_0) \neq 0$ and $\phi u \in C^\infty(\mathbb{R}^n)$. (Recall that, to be pedantic, $\phi u \in C^\infty(\mathbb{R}^n)$ means that there is $f \in C^\infty(\mathbb{R}^n)$ such that $\phi u = \iota_f$.)

We claim that the singular support of u is at most affected by the singular support of ϕ and ψ spreading along characteristics. In particular, the singular support cannot 'linger', i.e. it always 'goes away': for fixed $x \in \mathbb{R}$ in the singular support of ϕ or ψ, the singular support of u is only affected along the characteristics $\{(x_0, t_0) : x_0 \pm ct_0 = x\}$.

This is easy to see. Indeed, for ϕ even the solution u at (x_0, t_0) (and not merely its singularities!) only depends on ϕ at $x_0 \pm ct_0$, so it suffices to consider ψ. But

$$u_t(x, t) = \frac{c}{2}(\phi'(x + ct) - \phi'(x - ct)) + \frac{1}{2}(\psi(x + ct) + \psi(x - ct))$$

and

$$u_x(x, t) = \frac{c}{2}(\phi'(x + ct) + \phi'(x - ct)) + \frac{1}{2}(\psi(x + ct) - \psi(x - ct)),$$

so it follows that u is C^∞ near (x_0, t_0) if ϕ and ψ are such near $x_0 \pm ct_0$. (This is because if the derivative of a function (or even a distribution) is C^∞, then so is the function, directly from the definition of being C^∞!)

This is an example of the *propagation of singularities*, which occurs even for variable coefficient wave equations, and is one of the most fundamental facts about wave propagation.

As a further excercise, suppose that the initial data ϕ, ψ vanish for $|x| > R$. Then consider the domain of influence of the interval $[-R, R]$:

$$D^+_{[-R,R],0} = \bigcup_{x \in [-R,R]} D^+_{x,0} = \{(x_0, t_0) : t_0 \geq 0, \ -R - ct_0 \leq x_0 \leq R + ct_0\};$$

see Figure 7.3. We know that the solution u can only be singular in the strips

$$-R - ct_0 \leq x_0 \leq R - ct_0 \text{ and } -R + ct_0 \leq x_0 \leq R + ct_0.$$

We can ask though just what it is in between these strips (where it need not vanish; outside the strips it vanishes in view of the domain of influence). This is easy to see: if $R - ct_0 < x_0 < -R + ct_0$, then $x_0 - ct_0 < -R < R < x_0 + ct_0$, so the interval of integration for ψ includes the whole interval $[-R, R]$. Since ψ vanishes outside this interval, inside this region u is constant, namely

$$u(x, t) = \frac{1}{2c} \int_{\mathbb{R}} \psi(\sigma) \, d\sigma.$$

Thus, we see that for such initial data, for any point x_0, if we wait long enough (i.e., eventually) the solution will be constant: independent of (x, t) as long as x is close enough to x_0.

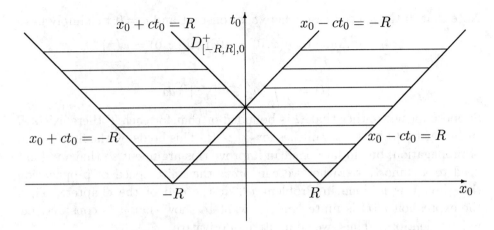

Figure 7.3. The domain of influence of the interval $[-R, R]$. The shaded strips are the only region in $t_0 \geq 0$ in which the solution might be singular. In the triangular region above the two strips the solution u is constant.

2. Energy conservation for the wave equation

While the finite speed of propagation (i.e. Huygens' principle) and the propagation of singularities are very general phenomena, i.e. they hold for variable coefficient equations as well, so far we could only justify them for our special constant coefficient equation. (Though note that our change of variables argument shows that this works for *all* hyperbolic PDE in $\mathbb{R}^2$!) We next consider the conservation of energy, which is more amenable to analysis even in general, without having to develop further tools.

Consider the variable coefficient (real-valued, i.e. u is real) wave equation

$$(7.1) \qquad u_{tt} - \nabla \cdot (c^2 \nabla u) + qu = 0, \;\; u(x,0) = \phi(x), \;\; u_t(x,0) = \psi(x),$$

where $c, q \geq 0$ depend on x only. Since it does not cause serious issues, we can also work on a domain Ω in $\mathbb{R}^n$, in which case we impose Dirichlet or Neumann boundary conditions, i.e. assume either that

$$(7.2) \qquad\qquad\qquad u|_{\partial\Omega} = 0$$

or

$$(7.3) \qquad\qquad\qquad \frac{\partial u}{\partial n}\bigg|_{\partial\Omega} = 0.$$

(If $\Omega = \mathbb{R}^n$, these conditions are vacuous. The notation for normal derivatives is that $\frac{\partial u}{\partial n} = \hat{n} \cdot \nabla u$, where $\hat{n}$ is the outward pointing unit normal.) We define the total energy at time t of a C^2 solution u as

$$E(t) = \frac{1}{2} \int_\Omega (u_t^2 + c(x)^2 |\nabla u|^2 + q(x)u^2) \, dx.$$

Note that in the special case where c is constant and $q = 0$ we simply have

$$u_{tt} - c^2 \Delta u = 0, \ u(x,0) = \phi(x), \ u_t(x,0) = \psi(x)$$

and

$$E(t) = \frac{1}{2} \int_\Omega (u_t^2 + c^2 |\nabla u|^2) \, dx.$$

Suppose for now either that Ω is bounded or that for each T there exists R such that for $|t| \leq |T|$, u vanishes for $|x| > R$. This holds by the finite speed of propagation, but in fact one can improve this argument so that we don't need to assume this, rather we can *prove* the finite speed of propagation this way; this is done in Problem 7.3 at the end of the chapter. Then the expression $E(t)$ is finite, and we want to show that it is constant, i.e. independent of t. Thus, we compute its derivative,

$$\frac{dE}{dt}(t) = \int_\Omega (u_t u_{tt} + c(x)^2 \nabla u \cdot \nabla u_t + q(x) u u_t) \, dx.$$

We want to rewrite this using the PDE. The terms $u_t u_{tt}$ and $q u u_t$ are promising: they are u_t times corresponding terms in the PDE; we need to bring the middle term into this form. To do so, we 'integrate by parts', which is also known as 'using the divergence theorem'. Thus, for each t we use the divergence theorem in the region Ω applied to the vector field $c(x)^2 u_t \nabla u$ (this is a vector field in $\mathbb{R}^n$!) to deduce that

$$\int_\Omega \mathrm{div}(c(x)^2 u_t \nabla u) \, dx = \int_{\partial \Omega} c^2 u_t \hat{n} \cdot \nabla u \, dS(x),$$

where $\hat{n}$ is the outward pointing unit normal of Ω and $dS(x)$ is the area form of $\partial \Omega$. Now

- if u satisfies Dirichlet boundary conditions, then $u_t|_{\partial\Omega} = 0$,
- if u satisfies Neumann boundary conditions, then $\hat{n} \cdot \nabla u|_{\partial\Omega} = \frac{\partial u}{\partial n}\big|_{\partial\Omega} = 0$;

of course, if $\Omega = \mathbb{R}^n$, then there is no boundary to talk about—so the boundary integral vanishes. Expanding the left-hand side we deduce that

$$0 = \int_\Omega \mathrm{div}(c(x)^2 u_t \nabla u) \, dx = \int_\Omega \nabla u_t \cdot c(x)^2 \nabla u \, dx + \int_\Omega u_t \nabla \cdot (c(x)^2 \nabla u) \, dx.$$

Thus, returning to $\frac{dE}{dt}$,

$$\frac{dE}{dt}(t) = \int_\Omega (u_t u_{tt} - u_t \nabla \cdot (c(x)^2 \nabla u) + q(x) u u_t) \, dx$$

$$= \int_\Omega (u_{tt} - \nabla \cdot (c(x)^2 \nabla u) + q(x) u) \, u_t \, dx = 0,$$

where in the last step we used the PDE, proving our claim. Thus, energy of solutions is *conserved*.

An immediate consequence is that there is at most one solution of the
PDE with the given initial conditions (and boundary conditions!), at least
if either Ω is bounded or for each T there exists R such that for $|t| \leq |T|$, u
vanishes for $|x| > R$. Indeed, if u_1, u_2 solve the wave equation (7.1) with the
same initial conditions and with either (7.2) or (7.3), then $u = u_1 - u_2$ solves
the wave equation with vanishing initial conditions, i.e. $\phi, \psi = 0$. Thus, the
energy of u satisfies $E(0) = 0$, and hence by the conservation of energy
$E(t) = 0$ for all t, so

$$\int_\Omega (u_t^2 + c(x)^2 |\nabla u|^2 + q(x)u^2) \, dx = 0$$

for all t. As the quantities inside are all non-negative, we conclude that
$u_t \equiv 0$ and $\nabla u \equiv 0$; if $q > 0$ we even conclude $u \equiv 0$. In general, if $q \geq 0$
merely, we use that $u(x,t) = u(x,0) + \int_0^t u_t(x,\tau) \, d\tau = 0$ in view of the initial
conditions and the vanishing of u_t. We thus deduce that $u \equiv 0$, and hence
$u_1 \equiv u_2$, giving uniqueness.

In fact, note that the uniqueness argument even works for *inhomogeneous*
boundary conditions, i.e.

(7.4) $u|_{\partial\Omega} = h$

or

(7.5) $\left.\frac{\partial u}{\partial n}\right|_{\partial\Omega} = h,$

for $u_1 - u_2$ then still satisfies the homogeneous boundary condition (i.e. (7.2)
or (7.3)).

We also have stability, i.e. if the initial data are close, then the solutions
are close (assuming they exist), since if $u = u^{(j)}$ solves

$$u_{tt} - \nabla \cdot (c^2 \nabla u) + qu = 0, \quad u(x,0) = \phi_j(x), \quad u_t(x,0) = \psi_j(x), \quad j = 1, 2,$$

then

$$\int_\Omega ((u_t^{(1)} - u_t^{(2)})^2 + c(x)^2 |\nabla u^{(1)} - \nabla u^{(2)}|^2 + q(x)(u^{(1)} - u^{(2)})^2) \, dx$$

$$= \int_\Omega ((\psi_1 - \psi_2)^2 + c(x)^2 |\nabla \phi_1 - \nabla \phi_2|^2 + q(x)(\phi_1 - \phi_2)^2) \, dx.$$

It is important to realize that the role of boundary and initial conditions
is not the same: this is a second order PDE, so just as for second order
ODE we may expect that we can specify two initial conditions (for u and
its derivative in some direction not tangential to the initial surface) and still
have both existence and uniqueness for the solution of the PDE. Indeed,
have these (ϕ and ψ), but we only need to know either $u|_{\partial\Omega}$ or $\frac{\partial u}{\partial n}|_{\partial\Omega}$ for
uniqueness. (We have not proved it, but in fact, the solution actually exists
under these conditions.)

3. The maximum principle for Laplace's equation and the heat equation

We now investigate analogous issues for Laplace's equation and the heat equation. In this case the simplest method is the *maximum principle*. We start with Laplace's equation.

Proposition 7.2. *Suppose $\Omega \subset \mathbb{R}^n$ is a bounded region and $u \in C^2(\Omega) \cap C^0(\bar{\Omega})$ solves $\Delta u = 0$. Then u attains its maximum on $\partial\Omega$, i.e.*

$$\sup_{x \in \Omega} u(x) = \sup_{x \in \partial\Omega} u(x).$$

Remark 7.3. What is assumed here is that u is C^2 in Ω (so all partial derivatives of order ≤ 2 are continuous in Ω), but not necessarily up to the boundary of Ω. Indeed, only u itself, not its partial derivatives, are assumed to be continuous up to $\partial\Omega$.

Proof. Since $\bar{\Omega}$ is closed and bounded, the continuous function u attains its maximum on $\bar{\Omega}$; we need to prove that this happens on $\partial\Omega$. To see why this should be the case note first that if y is a local maximum for u in Ω, then $(\partial_{x_j} u)(y) = 0$ for all j and $(\partial^2_{x_j} u)(y) \leq 0$ for all j — this follows from one variable calculus (thinking of u as a function of x_j only, fixing the other variables). Thus,

$$\Delta u(y) = \sum_{j=1}^n (\partial^2_{x_j} u)(y) \leq 0.$$

If the inequality were strict, this would contradict the PDE, $\Delta u = 0$. It is not, so we do not have a contradiction, though we can at least conclude that any local maximum in Ω is degenerate in each x_j separately. (This does not help, but it is nice to know.)

To get around this problem, for $\epsilon > 0$ we consider the function

$$u_\epsilon(x) = u(x) + \epsilon |x|^2.$$

Then

$$\Delta u_\epsilon = \Delta u + 2\epsilon n = 2\epsilon n > 0,$$

so by our previous calculation u_ϵ cannot have a local maximum in Ω (as $\Delta u_\epsilon \leq 0$ at such a local maximum). In particular, the global maximum cannot be attained in Ω, so it must be attained at $\partial\Omega$, i.e. we have proved that

$$\sup_{x \in \partial\Omega} u_\epsilon = \sup_{x \in \Omega} u_\epsilon.$$

But this gives for all $y \in \Omega$ that

$$\sup_{x \in \partial\Omega} u_\epsilon = \sup_{x \in \Omega} u_\epsilon \geq u_\epsilon(y) = u(y) + \epsilon |y|^2 \geq u(y),$$

so we deduce that for all $y \in \Omega$ and all $\epsilon > 0$,

$$u(y) \leq \sup_{x \in \partial\Omega} u_\epsilon \leq \sup_{x \in \partial\Omega} u + \epsilon R^2,$$

where $R = \sup_{x \in \partial\Omega} |x| < \infty$ since Ω is bounded. As this is true for every $\epsilon > 0$, letting $\epsilon \to 0$ we deduce that

$$u(y) \leq \sup_{x \in \partial\Omega} u,$$

and thus as $y \in \Omega$ is arbitrary,

$$\sup_{y \in \Omega} u(y) \leq \sup_{x \in \partial\Omega} u,$$

proving the maximum principle. $\qquad\qquad\qquad\qquad\qquad\qquad\qquad\square$

We can also apply the maximum principle to $-u$ (or run a similar argument directly for $\inf u$) to deduce that

$$\inf_{x \in \Omega} u(x) = \inf_{x \in \partial\Omega} u(x).$$

In particular, if $u|_{\partial\Omega} = 0$, then we deduce that for $y \in \Omega$, $0 \leq u(y) \leq 0$, i.e. $u \equiv 0$. By linearity, considering $u = u_1 - u_2$, this gives uniqueness for solutions of the Dirichlet problem for Laplace's equation:

$$\Delta u = f, \ u|_{\partial\Omega} = \phi.$$

We also get the following stability result: suppose that

$$\Delta u_j = f, \ u_j|_{\partial\Omega} = \phi_j, \ j = 1, 2.$$

Then

$$\sup_\Omega |u_1 - u_2| \leq \sup_{\partial\Omega} |\phi_1 - \phi_2|.$$

Indeed, to see this, just note that $u = u_1 - u_2$ satisfies $\Delta u = 0$ and $u|_{\partial\Omega} = \phi_1 - \phi_2$, so

$$\inf_{\partial\Omega}(\phi_1 - \phi_2) \leq u_1(y) - u_2(y) \leq \sup_{\partial\Omega}(\phi_1 - \phi_2)$$

for all $y \in \Omega$.

There is a complete analogue of this for the heat equation, $u_t = k\Delta u$, $k > 0$.

Proposition 7.4. *Suppose $\Omega \subset \mathbb{R}^n$ is a bounded region, $T > 0$, and*

$$u \in C^2(\Omega \times (0, T]) \cap C^0(\bar{\Omega} \times [0, T])$$

solves $u_t = k\Delta u$ on $\Omega \times (0, T)$. Then u attains its maximum on

$$(\partial\Omega \times [0, T]) \cup (\Omega \times \{0\}),$$

i.e. either for $x \in \partial\Omega$ or at $t = 0$, so

$$\sup_{(x,t)\in\Omega\times[0,T]} u(x,t) = \max\left(\sup_{(x,t)\in\partial\Omega\times[0,T]} u(x), \sup_{x\in\Omega} u(x,0)\right);$$

see Figure 7.4.

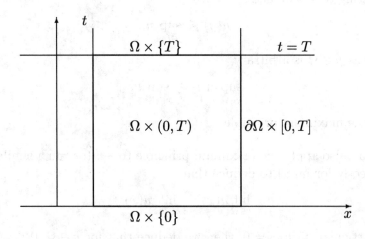

Figure 7.4. The maximum principle for the heat equation. The maximum of u over $\overline{\Omega} \times [0,T]$ is attained either on the sides, $\partial\Omega \times [0,T]$, or at the bottom, $\Omega \times \{0\}$.

Proof. We observe as above that if u had a local maximum at y in $\Omega\times(0,T)$, then $k\Delta u(y) - u_t(y) \le 0$ there. If u has a local maximum at y in $\Omega \times \{T\}$, then we still have $\partial_{x_j} u(y) = 0$ and $\partial_{x_j}^2 u(y) \le 0$, and hence $\Delta u(y) \le 0$ as above, by exactly the same argument as before, but now we can only conclude that $\partial_t u(y) \ge 0$ there (since we are at the upper boundary of the interval $[0,T]$, so the derivative need not vanish even at a local maximum). Thus, we conclude that

$$k\Delta u(y) - u_t(y) \le 0.$$

This does not contradict the heat equation, but we can modify the argument as before by replacing u by

$$u_\epsilon(x,t) = u(x,t) + \epsilon|x|^2,$$

which gives that the maximum of u_ϵ on $\overline{\Omega} \times [0,T]$ is attained either at $\partial\Omega$ or at $t = 0$. A limiting argument as before now gives the full maximum principle. $\square$

We again have an immediate uniqueness result as a consequence. The solution of the Dirichlet problem for the heat equation,

(7.6) $k\Delta u = u_t, \ u|_{\partial\Omega} = h, \ u(x,0) = \phi(x),$

is unique. We also have a stability estimate for solutions u_j of

(7.7) $\qquad k\Delta u_j = \partial_t u_j, \ u_j|_{\partial\Omega} = h_j, \ u_j(x,0) = \phi_j, \ j = 1,2,$

namely

(7.8) $\qquad \sup_{\Omega\times[0,T]} |u_1 - u_2| \leq \max(\sup_{\partial\Omega\times[0,T]} |h_1 - h_2|, \sup_{\Omega} |\phi_1 - \phi_2|).$

Note that there is a major qualitative difference with the 'backward heat equation' $k\Delta u = -u_t$: in that case, the roles of $t = T$ and $t = 0$ are reversed in the argument. This is an example of how first order terms can make a major difference for degenerate second order PDE.

4. Energy for Laplace's equation and the heat equation

In physical applications of the heat equation, u is temperature or concentration, and the corresponding conserved quantity is the total heat or mass,

$$M(t) = \int_\Omega u(x,t)\,dx,$$

at least if there is no heat flux/flow through $\partial\Omega$, i.e. if $\frac{\partial u}{\partial n} = 0$ (the Neumann boundary condition). This is easy to check:

$$\frac{dM}{dt}(t) = \int_\Omega u_t(x,t)\,dx = \int_\Omega k\Delta u\,dx = \int_\Omega k\nabla\cdot(\nabla u)\,dx$$

$$= \int_{\partial\Omega} k\hat{n}\cdot\nabla u\,dS(x) = 0,$$

where the last step used the boundary condition. If $\Omega = \mathbb{R}^n$ and there is sufficient decay of u at infinity, then there is no need for the boundary condition (one can think of the decay as the boundary condition), and one again concludes that M is conserved. The main issue here is that just because this integral vanishes, we cannot conclude that u vanishes, unlike in the case of the energy for the wave equation.

There is, however, an unphysical 'energy' akin to the energy for the wave equation, namely

$$E(t) = \frac{1}{2}\int_\Omega u(x,t)^2\,dx.$$

Then

$$\frac{dE}{dt} = \int_\Omega u u_t\,dx = \int_\Omega u\,k\Delta u\,dx = k\int_\Omega u\,\nabla\cdot\nabla u\,dx$$

$$= k\int_{\partial\Omega} u\,\hat{n}\cdot\nabla u\,dS(x) - k\int_\Omega \nabla u\cdot\nabla u\,dx$$

$$= k\int_{\partial\Omega} u\frac{\partial u}{\partial n}\,dS(x) - k\int_\Omega |\nabla u|^2\,dx,$$

where in the second equality we used the PDE and in the third we used the divergence theorem applied to the vector field $u\nabla u$, namely that

$$\int_\Omega \nabla \cdot (u\nabla u)\,dx = \int_{\partial\Omega} \hat{n} \cdot (u\nabla u)\,dS(x).$$

Since $\int_\Omega |\nabla u|^2\,dx \geq 0$, we deduce that for either homogeneous Dirichlet or Neumann boundary conditions,

$$\frac{dE}{dt} \leq 0.$$

Thus, this 'energy' is not conserved. However it is decreasing (in the sense of being non-increasing), so in particular if $E(0) = 0$, then $E(t) = 0$ for $t > 0$. We deduce another proof of uniqueness for the heat equation, i.e. that there is at most one solution of (7.6), and this also works for the Neumann boundary condition:

$$(7.9) \qquad k\Delta u = u_t, \quad \left.\frac{\partial u}{\partial n}\right|_{\partial\Omega} = h, \; u(x,0) = \phi(x).$$

Further, we have a new stability result: any solutions u_1 and u_2 of

$$(7.10) \qquad k\Delta u_j = \partial_t u_j, \; u_j|_{\partial\Omega} = h, \; u_j(x,0) = \phi_j, \; j = 1,2,$$

satisfy

$$\int_\Omega (u_1(x,t) - u_2(x,t))^2\,dx \leq \int_\Omega (\phi_1 - \phi_2)^2\,dx.$$

Here we had matching boundary conditions so that $u_1 - u_2$ has vanishing boundary conditions, thus satisfying the criteria in our energy calculation. The same estimate also applies to the Neumann problem:

$$(7.11) \qquad k\Delta u_j = \partial_t u_j, \quad \left.\frac{\partial u_j}{\partial n}\right|_{\partial\Omega} = h, \; u_j(x,0) = \phi_j, \; j = 1,2.$$

We finally work out an *energy estimate* for a generalized version of Laplace's equation, $\nabla \cdot (c(x)^2 \nabla u) - qu = f$ with a homogeneous Dirichlet or Neumann boundary condition. Then let

$$E = \int_\Omega (c(x)^2 |\nabla u|^2 + qu^2)\,dx.$$

Note that E is now a real number for each u. We may also write $E = E(u)$; it is sometimes called the Dirichlet form. Now note that

$$\begin{aligned}
(7.12) \quad E &= \int_\Omega (c(x)^2 |\nabla u|^2 + qu^2)\,dx \\
&= \int_{\partial\Omega} \hat{n} \cdot c(x)^2 \nabla u\, u\, dS(x) - \int_\Omega (\nabla \cdot (c(x)^2 \nabla u))\, u\, dx + \int_\Omega qu^2\,dx \\
&= -\int_\Omega (\nabla \cdot (c(x)^2 \nabla u) - qu)\, u\, dx = -\int_\Omega f u\,dx,
\end{aligned}$$

where in the last step we used the PDE.

Let us assume for now that $q \geq q_0 > 0$ and $c \geq c_0 > 0$. Then we can use the Cauchy-Schwarz inequality, which we will discuss in general when talking about inner product spaces in Chapter 13. For now we only state: for all $f, g \in C^0(\bar{\Omega})$ real-valued,

$$\left| \int_\Omega f(x) g(x) \, dx \right| \leq \left(\int_\Omega f^2 \, dx \right)^{1/2} \left(\int_\Omega g^2 \, dx \right)^{1/2},$$

while in the complex-valued case one has

$$\left| \int_\Omega f(x) g(x) \, dx \right| \leq \left(\int_\Omega |f|^2 \, dx \right)^{1/2} \left(\int_\Omega |g|^2 \, dx \right)^{1/2}.$$

(Note that in the complex-valued case one can replace g by $\bar{g}$ on the left hand side; as $|g| = |\bar{g}|$, the right-hand side is unchanged. This is the more standard version of Cauchy-Schwarz in that case.)

Thus, in our setting, by the Cauchy-Schwarz inequality,:

(7.13)
$$\left| \int_\Omega f u \, dx \right| \leq \left(\int_\Omega f^2 \, dx \right)^{1/2} \left(\int_\Omega u^2 \, dx \right)^{1/2}$$
$$\leq (q_0/4) \left(\int_\Omega u^2 \, dx \right) + (1/q_0) \left(\int_\Omega f^2 \, dx \right),$$

where the last inequality was an application of
(7.14)
$$ab \leq (a^2 + b^2)/2 \text{ with } a = \sqrt{q_0/2} \left(\int_\Omega u^2 \, dx \right)^{1/2}, \; b = \sqrt{2/q_0} \left(\int_\Omega f^2 \, dx \right)^{1/2}.$$

Thus, we deduce that

$$\int_\Omega (c_0^2 |\nabla u|^2 + (q_0/2) u^2) \, dx \leq (1/q_0) \left(\int_\Omega f^2 \, dx \right).$$

In particular, if we have two solutions u_j, $j = 1, 2$, of

$$\nabla \cdot (c(x)^2 \nabla u_j) - q u_j = f,$$

with either the *same inhomogeneous* Dirichlet or Neumann boundary conditions, then $u = u_1 - u_2$ solves

$$\nabla \cdot (c(x)^2 \nabla u) - q u = 0$$

with *homogeneous* Dirichlet or Neumann boundary conditions, so

$$\int_\Omega (c_0^2 |\nabla u|^2 + (q_0/2) u^2) \, dx \leq 0$$

by our energy estimate. But the left-hand side is non-negative, so this states that it vanishes; hence ∇u and u vanish identically, so the solution of the PDE is unique. Indeed, we get stability again, this time for the PDE

$$\nabla \cdot (c(x)^2 \nabla u_j) - q u_j = f_j,$$

with either the *same inhomogeneous* Dirichlet or Neumann boundary conditions, since we have

$$\int_\Omega (c_0^2 |\nabla u_1 - \nabla u_2|^2 + (q_0/2)(u_1 - u_2)^2)\, dx \le (1/q_0)\Big(\int_\Omega (f_1 - f_2)^2\, dx\Big).$$

The only somewhat serious assumption here was that $q \ge q_0 > 0$ since we might want to solve Laplace's equation itself. It turns out that we can merely assume $q \ge 0$, at least if we impose the Dirichlet boundary condition, $u|_{\partial\Omega} = 0$. This is due to the *Poincaré inequality*:

Lemma 7.5 (Poincaré inequality). *For a bounded domain Ω, there is a constant $C > 0$ such that for functions u with $u|_{\partial\Omega} = 0$,*

(7.15)
$$\int_\Omega u^2\, dx \le C \int_\Omega |\nabla u|^2\, dx.$$

Assuming this estimate, and using (7.14) but with c_0^2/C in place of $q_0/2$, we get in place of (7.13) that

(7.16)
$$\left|\int_\Omega f\, u\, dx\right| \le \Big(\int_\Omega f^2\, dx\Big)^{1/2} \Big(\int_\Omega u^2\, dx\Big)^{1/2}$$
$$\le (c_0^2/2C)\Big(\int_\Omega u^2\, dx\Big) + (C/2c_0^2)\Big(\int_\Omega f^2\, dx\Big).$$

Thus, (7.12) yields

$$\int_\Omega c_0^2 |\nabla u|^2\, dx \le E \le (c_0^2/2C)\Big(\int_\Omega u^2\, dx\Big) + (C/2c_0^2)\Big(\int_\Omega f^2\, dx\Big)$$
$$\le (c_0^2/2)\Big(\int_\Omega |\nabla u|^2\, dx\Big) + (C/2c_0^2)\Big(\int_\Omega f^2\, dx\Big),$$

so subtracting the first term of the extreme right-hand side,

$$\int_\Omega \frac{c_0^2}{2} |\nabla u|^2\, dx \le (C/2c_0^2)\Big(\int_\Omega f^2\, dx\Big).$$

The left-hand side is also an upper bound for

$$\frac{c_0^2}{2C} \int_\Omega u^2\, dx$$

by the Poincaré inequality, so we deduce that

$$\frac{c_0^2}{2} \int_\Omega (C^{-1}u^2 + |\nabla u|^2)\, dx \le \frac{C}{c_0^2} \int_\Omega f^2\, dx.$$

This gives uniqueness and stability, just under the assumption $q \ge 0$, for the Dirichlet problem for $\nabla \cdot (c^2 \nabla) - q$, just as before.

We now sketch a proof of the Poincaré inequality for bounded regions Ω, Lemma 7.5; we assume that $\Omega \subset [-R, R]^n$. We obtain this by an integration by parts argument for the function $u\, x_j \partial_{x_j} u$. Thus,

$$\int_\Omega u^2\, dx = \int_\Omega u(\partial_{x_j}(x_j u) - x_j \partial_{x_j} u)\, dx = -\int_\Omega ((x_j \partial_{x_j} u)u + u(x_j \partial_{x_j} u))\, dx$$

$$\leq 2 \int_\Omega |x_j|\, |\partial_{x_j} u|\, |u|\, dx \leq 2R \int_\Omega |\partial_{x_j} u|\, |u|\, dx$$

$$\leq 2R \left(\int_\Omega |\partial_{x_j} u|^2\, dx \right)^{1/2} \left(\int_\Omega |u|^2\, dx \right)^{1/2}$$

$$\leq \frac{1}{2} \left(\int_\Omega |u|^2\, dx \right) + 2R^2 \left(\int_\Omega |\partial_{x_j} u|^2\, dx \right),$$

where the last inequality was an analogue of (7.14). Here the second equality used the fact that $u = 0$ at $\partial\Omega$; hence the boundary term one would get in integration by parts vanishes. Rearranging and multiplying through by 2 yields

$$\int_\Omega u^2\, dx \leq 4R^2 \left(\int_\Omega |\partial_{x_j} u|^2\, dx \right) \leq 4R^2 \left(\int_\Omega |\nabla u|^2\, dx \right),$$

proving the inequality.

There are some simple generalizations of the energy estimate. For instance, if $c(x)$ is replaced by a symmetric positive definite matrix $A(x) = (A_{ij}(x))_{i,j=1}^n$, so our generalized Laplace equation is

$$\nabla \cdot (A(x)\nabla u) - qu = f;$$

see Problem 7.8.

Another generalization is for systems of PDE. For instance, suppose one has a system of PDE for functions u_j, $j = 1, \ldots, N$, given by

$$\nabla \cdot (c(x)^2 \nabla u_j) - \sum_{k=1}^N q_{jk}(x)u_k(x) = f_j(x),$$

and $q(x) = (q_{jk})_{j,k=1}^N$ is a symmetric matrix, with q positive in the sense that for all $x \in \overline{\Omega}$, $q(x)v \cdot v \geq 0$ for all $x \in \overline{\Omega}$ and $c(x) \geq c_0 > 0$ for all $x \in \overline{\Omega}$, and let us assume homogeneous Dirichlet boundary conditions. Then

$$E = \sum_{j=1}^N \int_\Omega c(x)^2 |\nabla u_j|^2\, dx + \sum_{j,k=1}^N \int_\Omega q_{jk}(x)u_j(x)u_k(x)\, dx$$

$$\geq \sum_{j=1}^N \int_\Omega c(x)^2 |\nabla u_j|^2\, dx \geq c_0^2 \sum_{j=1}^N \int_\Omega |\nabla u_j|^2\, dx \geq 0$$

by the positivity of q_{jk}. Further,

(7.17)

$$E = \int_\Omega \left(\sum_{j=1}^N c(x)^2 |\nabla u_j|^2 + \sum_{j,k=1}^N q_{jk} u_j u_k \right) dx$$

$$= \sum_{j=1}^N \int_{\partial\Omega} \hat{n} \cdot c(x)^2 \nabla u_j \, u_j \, dS(x) - \sum_{j=1}^N \int_\Omega (\nabla \cdot (c(x)^2 \nabla u_j)) \, u_j \, dx$$

$$+ \int_\Omega \sum_{j,k=1}^N q_{jk} u_j u_k \, dx$$

$$= - \int_\Omega \left(\sum_{j=1}^N \nabla \cdot (c(x)^2 \nabla u_j) - \sum_{j,k=1}^N q_{jk} u_k \right) u_j \, dx = - \sum_{j=1}^N \int_\Omega f_j \, u_j \, dx,$$

where in the last step we used the PDE. Then

$$c_0^2 \sum_{j=1}^N \int_\Omega |\nabla u_j(x)|^2 \, dx \leq E \leq \sum_{j=1}^N \frac{c_0^2}{2C} \int_\Omega u_j^2 \, dx + \sum_{j=1}^N \frac{C}{2c_0^2} \int_\Omega f_j^2 \, dx,$$

with C as in the Poincaré inequality, so

$$c_0^2 \sum_{j=1}^N \int_\Omega |\nabla u_j(x)|^2 \, dx \leq \sum_{j=1}^N \frac{c_0^2}{2} \int_\Omega |\nabla u_j|^2 \, dx + \sum_{j=1}^N \frac{C}{2c_0^2} \int_\Omega f_j^2 \, dx,$$

and thus

$$\sum_{j=1}^N \frac{c_0^2}{2} \int_\Omega |\nabla u_j(x)|^2 \, dx \leq \frac{C}{2c_0^2} \sum_{j=1}^N \int_\Omega f_j^2 \, dx.$$

As above, this gives

$$\sum_{j=1}^N \frac{c_0^2}{2} \int_\Omega (C^{-1} u^2 + |\nabla u_j(x)|^2) \, dx \leq \frac{C}{c_0^2} \sum_{j=1}^N \int_\Omega f_j^2 \, dx,$$

giving uniqueness and stability for the Dirichlet problem for the system as before in the case of a single equation.

Problems

Problem 7.1. Solve the wave equation on the line:

$$u_{tt} - c^2 u_{xx} = 0, \quad u(x,0) = \phi(x), \quad u_t(x,0) = \psi(x),$$

with

$$\phi(x) = \begin{cases} 0, & x < -1, \\ 1 + x, & -1 < x < 0, \\ 1 - x, & 0 < x < 1, \\ 0, & x > 1, \end{cases}$$

and

$$\psi(x) = \begin{cases} 0, & x < -1, \\ 2, & -1 < x < 1, \\ 0, & x > 1. \end{cases}$$

Also describe in $t > 0$ where the solution vanishes and where it is C^∞, and compare it with the general results discussed in the chapter (Huygens' principle and propagation of singularities).

Problem 7.2. Consider the PDE

$$u_{tt} - \nabla \cdot (c^2 \nabla u) + qu = 0, \ u(x,0) = \phi(x), \ u_t(x,0) = \psi(x),$$

on $\Omega_x \times \mathbb{R}_t$, Ω bounded, where $c, q \geq 0$ depend on x only, and with either a homogeneous Dirichlet ($u|_{\partial\Omega} = 0$) or Neumann ($\frac{\partial u}{\partial n}|_{\partial\Omega} = 0$) boundary condition. Suppose that u is $C^2(\bar\Omega \times \mathbb{R})$ and *complex-valued*. Show that

$$E(t) = \frac{1}{2} \int_\Omega (|u_t|^2 + c(x)^2 |\nabla u|^2 + q(x)|u|^2) \, dx \geq 0$$

is still conserved; note that here $|\nabla u|^2 = \sum_j |\partial_j u|^2$. Use this to show that the solution of the complex-valued wave equation is unique. (*Hint:* $|u_t|^2 = u_t \overline{u_t}$, etc.)

Problem 7.3. In this problem we prove the *finite speed of propagation* for solutions of variable coefficient wave equations.

Consider the PDE

(7.18) $u_{tt} - \nabla \cdot (c^2 \nabla u) + qu = 0, \ u(x,0) = \phi(x), \ u_t(x,0) = \psi(x),$

where $c, q \geq 0$ depend on x only and c is bounded between positive constants, i.e. for some $c_1, c_2 > 0$, $c_1 \leq c(x) \leq c_2$ for all $x \in \mathbb{R}^n$. Assume that u is C^2 throughout this problem, and u is real-valued. (All calculations would go through if one wrote $|u_t|^2$, etc., in the complex-valued case, see Problem 7.2.)

(i) Fix $x_0 \in \mathbb{R}^n$ and $R_0 > 0$, and for $t < \frac{R_0}{c_2}$, let

$$E(t) = \int_{|x-x_0| < R_0 - c_2 t} (u_t^2 + c(x)^2 |\nabla u|^2 + q(x)u^2) \, dx.$$

Show that E is decreasing with t (i.e. non-increasing). (*Hint:* To make sure you don't forget anything in the calculation, do it first on the line, when $n = 1$.)

(ii) Suppose that $\operatorname{supp}\phi, \operatorname{supp}\psi \subset \{|x| \leq R\}$, i.e. are 0 outside this ball. Show that $u(x,t) = 0$ if $t \geq 0$, $|x| > R + c_2 t$, i.e. the wave indeed propagates at speed $\leq c_2$.

(iii) Show that there is at most one real-valued C^2 solution of (7.18).

Problem 7.4. Consider the (real-valued) *damped wave equation* on $[0, \ell]_x \times [0, \infty)_t$ with *Robin boundary conditions*:

$$u_{tt} + a(x)u_t = (c(x)^2 u_x)_x, \ u_x(0, t) = \alpha u(0, t), \ u_x(\ell, t) = -\beta u(\ell, t),$$

where $\alpha, \beta \geq 0$ are constants, $a \geq 0$ and $c > 0$ depend on x only, and there are constants $c_1, c_2 > 0$ such that $c_1 \leq c(x) \leq c_2$ for all x. (Note that if $\alpha = 0$ and $\beta = 0$, then this is just the Neumann boundary condition! In general, this BC would hold for example for a string if its ends were attached to springs. Also, $a(x)u_t$ is the damping term; if $a \equiv 0$, there is no damping.) Assume throughout that u is C^2. Let

$$E(t) = \frac{1}{2} \int_0^\ell \left(u_t(x, t)^2 + c(x)^2 u_x(x, t)^2 \right) dx + \frac{1}{2} \left(c(0)^2 \alpha u(0, t)^2 + c(\ell)^2 \beta u(\ell, t)^2 \right).$$

(i) Show that if $a \equiv 0$, then E is constant.

(ii) Show that if $a \geq 0$, then E is a decreasing (i.e. non-increasing) function of t, and that the solution of the damped wave equation (under the conditions mentioned above) with given initial condition is unique.

Problem 7.5. Consider the wave equation on $\mathbb{R}^n$:

$$u_{tt} - c^2 \Delta u = f, \ u(x, 0) = \phi(x), \ u_t(x, 0) = \psi(x),$$

and write $x = (x', x_n)$, where $x' = (x_1, \ldots, x_{n-1})$.

(i) Show that if

$$f(x', x_n, t) = f(x', -x_n, t), \ \phi(x', x_n) = \phi(x', -x_n), \ \psi(x', x_n) = \psi(x', -x_n)$$

for all x and t, i.e. if f, ϕ, ψ are all even functions of x_n, then u is an even function of x_n as well. (*Hint:* Consider $u(x', x_n, t) - u(x', -x_n, t)$ and show that it solves the homogeneous wave equation with 0 initial conditions.)

(ii) Show that if

$$f(x', x_n, t) = -f(x', -x_n, t), \ \phi(x', -x_n) = -\phi(x', -x_n), \ \psi(x', x_n)$$
$$= -\psi(x', -x_n)$$

for all x and t, i.e. if f, ϕ, ψ are all odd functions of x_n, then u is an odd function of x_n as well.

(iii) If u is continuous and is an odd function of x_n, show that $u(x', 0, t) = 0$ for all x' and t.

(iv) If u is a C^1 and is an even function of x_n, show that $\partial_{x_n} u(x', 0, t) = 0$ for all x' and t.

These facts will enable us to solve the wave equation in the half space $x_n > 0$ with Dirichlet or Neumann boundary conditions later in Chapter 10.

Problem 7.6. Use the maximum principle for Laplace's equation on $\mathbb{R}^n$ to show the following statement: Suppose that $u \in C^2(\mathbb{R}^n)$ and $\Delta u = 0$. Suppose moreover that $u(x) \to 0$ at infinity uniformly in the following sense:

$$\sup_{|x|>R} |u(x)| \to 0$$

as $R \to \infty$. Then $u(x) = 0$ for all $x \in \mathbb{R}^n$. (*Hint:* Apply the maximum principle shown in the chapter for the ball $\Omega = \{x : |x| < R\}$ and for both u and $-u$.)

Use this to show that the solution of Laplace's equation on $\mathbb{R}^n$,

$$\Delta u = f,$$

with f given, is unique in the class of functions u such that $u \in C^2(\mathbb{R}^n)$ and $u(x) \to 0$ at infinity uniformly.

Problem 7.7. Show the following maximum principle for the heat equation on $\mathbb{R}^n$. Suppose that $u \in C^2(\mathbb{R}^n \times (0,T)) \cap C^0(\mathbb{R}^n \times [0,T])$ and $u_t = k\Delta u$, $k > 0$. Suppose moreover that $u(x,t) \to 0$ at infinity uniformly in the following sense:

$$\sup_{|x|>R,\ 0 \le t \le T} |u(x,t)| \to 0$$

as $R \to \infty$. Then

$$\sup_{(x,t) \in \mathbb{R}^n \times [0,T]} u(x,t) = \max(0, \sup_{x \in \mathbb{R}^n} u(x,0)).$$

(Note that you can think of 0 as the 'boundary value of u at infinity', in analogy with the case $\Omega \times [0,T]$ discussed in this chapter, where max would be the maximum of the initial and boundary values.)

Use this to show that the solution of the initial value problem for the heat equation on $\mathbb{R}^n \times [0,T]$:

$$u_t = k\Delta u, \ u(x,0) = \phi(x),$$

with ϕ given, is unique in the class of functions u such that

$$u \in C^2(\mathbb{R}^n \times (0,T]) \cap C^0(\mathbb{R}^n \times [0,T])$$

and $u(x,t) \to 0$ at infinity uniformly.

Problem 7.8. Suppose Ω is a bounded domain and $A(x) = (A_{ij}(x))_{i,j=1}^n$ is symmetric and positive definite in the sense that there is $c_0 > 0$ such that $A(x)v \cdot v \ge c_0|v|^2$ for all $v \in \mathbb{R}^n$, $x \in \overline{\Omega}$, and $q \ge 0$, with A_{ij} being C^1, q, continuous on $\overline{\Omega}$. Show that there is $C > 0$ such that solutions $u \in C^2(\overline{\Omega})$ of the PDE

$$\nabla \cdot (A(x)\nabla u) - qu = f$$

with Dirichlet boundary conditions $u|_{\partial\Omega} = 0$ satisfy

$$\int_\Omega (|\nabla u|^2 + u^2)\,dx \leq C \int_\Omega f^2\,dx.$$

Problem 7.9. We consider the setting of Problem 7.8 but for systems. Thus, suppose Ω is a bounded domain and $A(x) = (A_{ij}(x))_{i,j=1}^n$ is symmetric and positive definite in the sense that there is $c_0 > 0$ such that $A(x)v \cdot v \geq c_0|v|^2$ for all $v \in \mathbb{R}^n$, $x \in \overline{\Omega}$, and $q = (q_{k\ell}(x))$ is positive in the sense that $q(x)v \cdot v \geq 0$ for all $v \in \mathbb{R}^n$ and $x \in \overline{\Omega}$, with A_{ij} being C^1, $q_{k\ell}$ is continuous on $\overline{\Omega}$. Show that there is $C > 0$ such that solutions $u_1, \ldots, u_N \in C^2(\overline{\Omega})$ of the PDE system

$$\nabla \cdot (A(x)\nabla u_k) - \sum_{k=1}^N q_{k\ell}(x)u_\ell = f_k, \quad k = 1, \ldots, N,$$

with Dirichlet boundary conditions $u_k|_{\partial\Omega} = 0$, $k = 1, \ldots, N$, satisfy

$$\sum_{k=1}^N \int_\Omega (|\nabla u_k|^2 + u_k^2)\,dx \leq C \sum_{k=1}^N \int_\Omega f_k^2\,dx.$$

Can you extend the result to Neumann boundary conditions under stronger assumptions on q?

Chapter 8

The Fourier transform: Basic properties, the inversion formula and the heat equation

1. The definition and the basics

The Fourier transform is the basic and most powerful tool in studying constant coefficient PDE on $\mathbb{R}^n$. It is based on the following simple observation: For $\xi \in \mathbb{R}^n$, the functions

$$v_\xi(x) = e^{ix \cdot \xi} = e^{ix_1 \xi_1} \cdots e^{ix_n \xi_n}$$

are joint eigenfunctions of the operators ∂_{x_j}, namely for each j,

(8.1) $$\partial_{x_j} v_\xi = i\xi_j v_\xi.$$

It would thus be desirable to decompose an 'arbitrary' function u as an (infinite) linear combination of the v_ξ; namely, write it as

(8.2) $$u(x) = (2\pi)^{-n} \int_{\mathbb{R}^n} \hat{u}(\xi) e^{ix \cdot \xi} \, d\xi,$$

where $\hat{u}(\xi)$ is the 'amplitude' of the harmonic $e^{ix \cdot \xi}$ in u. (The factor $(2\pi)^{-n}$ here is due to a convention; it could also be moved to other places.) It turns out that this identity, (8.2), holds provided that we define

$$\hat{u}(\xi) = \int_{\mathbb{R}^n} e^{-ix \cdot \xi} u(x) \, dx;$$

(8.2) is then the *Fourier inversion formula*.

Rather than showing this at once, we start with a step-by-step approach. We first *define* the *Fourier transform* as

$$(\mathcal{F}u)(\xi) = \int_{\mathbb{R}^n} e^{-ix\cdot\xi} u(x)\, dx$$

for $u \in C(\mathbb{R}^n)$ with $|x|^N |u|$ bounded for some $N > n$ (i.e. $|u(x)| \le M|x|^{-N}$ for some M in say $|x| > 1$, the point being that in this case u is absolutely integrable as $\int_{|x|>1} M|x|^{-N}\, dx$ converges). One could instead assume simply that $u \in L^1(\mathbb{R}^n)$ if one is familiar with the Lebesgue integral (but actually just working with the special case mentioned above gives everything we want). Note that for such functions

$$|(\mathcal{F}u)(\xi)| \le \int_{\mathbb{R}^n} |e^{-ix\cdot\xi}|\,|u(x)|\, dx = \int_{\mathbb{R}^n} |u(x)|\, dx,$$

so $\mathcal{F}u$ is actually a *bounded continuous* function.

We can similarly define the *inverse Fourier transform*

$$(\mathcal{F}^{-1}\psi)(x) = (2\pi)^{-n} \int_{\mathbb{R}^n} e^{ix\cdot\xi} \psi(\xi)\, d\xi;$$

then $\mathcal{F}^{-1}$ maps $u \in C(\mathbb{R}^n)$ with $|x|^N |u|$ bounded for some $N > n$ (or indeed $u \in L^1(\mathbb{R}^n)$) to bounded continuous functions.

With this definition it is of course not clear whether $\mathcal{F}^{-1}$ is indeed the inverse of $\mathcal{F}$ and, worse, it not even clear whether $\mathcal{F}^{-1}\mathcal{F}\phi$ makes sense for $\phi \in L^1(\mathbb{R}^n)$, since $\mathcal{F}\phi$ is then a bounded continuous function, which is not sufficient to ensure that the integral defining $\mathcal{F}^{-1}$ actually converges!

We thus proceed to study the properties of $\mathcal{F}$ and $\mathcal{F}^{-1}$. First, we note a property of $\mathcal{F}$ which is the main reason for its usefulness in studying PDE and which is an immediate consequence of (8.1). Namely, suppose that $\phi \in C^1(\mathbb{R}^n)$ and both ϕ and all first derivatives $\partial_j\phi$, $j = 1,\ldots,n$, decay at infinity in the same sense as above (so $|x|^N \partial_j \phi$ is bounded for some $N > n$). Then integration by parts gives

$$(\mathcal{F}(\partial_{x_j}\phi))(\xi) = \int_{\mathbb{R}^n} e^{-ix\cdot\xi} \partial_{x_j}\phi(x)\, dx = -\int_{\mathbb{R}^n} \partial_{x_j}(e^{-ix\cdot\xi})\phi(x)\, dx$$

$$= i\xi_j \int_{\mathbb{R}^n} e^{-ix\cdot\xi}\phi(x)\, dx = i\xi_j(\mathcal{F}\phi)(\xi).$$

In other words, the operators $\mathcal{F}$, ∂_{x_j} and multiplication by ξ_j (usually just written as ξ_j) satisfy

$$\mathcal{F}\partial_{x_j} = i\xi_j\mathcal{F}.$$

In order to remove the factor of i, we let

$$D_{x_j} = \frac{1}{i}\partial_{x_j},$$

so

$$\mathcal{F}D_{x_j} = \xi_j \mathcal{F}.$$

Note, in particular, that this gives that for ϕ as above,

$$\xi_j \mathcal{F}\phi(\xi) = \mathcal{F}D_{x_j}\phi$$

is bounded for all j, so as $\mathcal{F}\phi$ is also bounded so

$$\left(1 + \sum_{j=1}^{n} \xi_j^2\right)|\mathcal{F}\phi(\xi)|^2$$

is bounded, we deduce that

(8.3) $$|\mathcal{F}\phi(\xi)| \leq C/(1+|\xi|^2)^{1/2},$$

i.e. the Fourier transform of ϕ actually decays and is not merely bounded. This gives us some hope that perhaps under some additional assumptions $\mathcal{F}^{-1}(\mathcal{F}\phi)$ actually makes sense.

Before proceeding, we note that $\frac{(1+|\xi|^2)^{1/2}}{1+|\xi|}$ is bounded from below and above by positive constants—indeed, it is certainly a positive continuous function, and as $|\xi| \to \infty$, it converges to 1 (since the summand 1 is negligible in the limit in both the numerator and the denominator). Thus, (8.3) is equivalent to, for some $C' > 0$,

$$|\mathcal{F}\phi(\xi)| \leq C'/(1+|\xi|).$$

There is an analogous formula for $\mathcal{F}(x_j\phi)$ if we instead assume that $\phi \in C(\mathbb{R}^n)$ and $|x|^N|\phi|$ is bounded for $N > n+1$; namely,

$$\mathcal{F}(x_j\phi) = \int_{\mathbb{R}^n} e^{-ix\cdot\xi} x_j \phi(x)\, dx = \int_{\mathbb{R}^n} (x_j e^{-ix\cdot\xi})\phi(x)\, dx$$

$$= \int_{\mathbb{R}^n} (i\partial_{\xi_j} e^{-ix\cdot\xi})\phi(x)\, dx$$

$$= i\partial_{\xi_j}\left(\int_{\mathbb{R}^n} e^{-ix\cdot\xi}\phi(x)\, dx\right) = -D_{\xi_j}(\mathcal{F}\phi)(\xi),$$

where $D_{\xi_j} = \frac{1}{i}\partial_{\xi_j}$. In operator notation,

$$\mathcal{F}x_j = -D_{\xi_j}\mathcal{F}.$$

In particular, this tells us that if $\phi \in C(\mathbb{R}^n)$ and $|x|^N|\phi|$ is bounded for $N > n+1$, then $\mathcal{F}\phi$ is continuously differentiable and its derivatives $D_{\xi_j}\mathcal{F}\phi$ are bounded.

In summary, the Fourier transform interchanges differentiation and multiplication by the coordinate functions (up to a $-$ sign), and correspondingly it interchanges differentiability and decay at infinity. If we only care about differentiation, the natural class of 'very nice' functions is C^∞, since we can differentiate its elements arbitrarily many times. In view of the properties

of the Fourier transform, the relevant class of 'very nice' functions consists
of functions which are C^∞ and decay arbitrarily rapidly, as measured by
powers of $|x|$, at infinity.

Definition 8.1. The set $\mathcal{S} = \mathcal{S}(\mathbb{R}^n)$, called the set of *Schwartz functions*,
consists of functions $\phi \in C^\infty(\mathbb{R}^n)$ such that for all $N \geq 0$ and all multiindices
$\alpha \in \mathbb{N}^n$, $|x|^N D^\alpha \phi$ is bounded on $\mathbb{R}^n$.

Here recall the multiindex notation:

$$D^\alpha = D_{x_1}^{\alpha_1} \dots D_{x_n}^{\alpha_n}.$$

The functions $\phi \in \mathcal{S}(\mathbb{R}^n)$ *decay rapidly at infinity* with all derivatives.

We can put this into a more symmetric form by noting that it suffices
to consider N even, and indeed merely ask if $(1 + |x|^2)^N D^\alpha \phi$ is bounded for
all N and α. Expanding the first term, using $|x|^2 = x_1^2 + \dots + x_n^2$, one easily
sees that this in turn is equivalent to the statement that for all multiindices
$\alpha, \beta \in \mathbb{N}^n$, $x^\alpha D^\beta \phi$ is bounded. Here we wrote

$$x^\alpha = x_1^{\alpha_1} x_2^{\alpha_2} \dots x_n^{\alpha_n},$$

in analogy with the notation for D^β. Note that by Leibniz' rule (i.e. the
product rule for differentiation), one can write $D^\beta x^\alpha \phi$ as a finite sum of
powers $\leq \alpha$ of x times derivatives of order $\leq \beta$ of ϕ, and conversely, so in
fact $x^\alpha D^\beta \phi$ being bounded for all multiindices α, β is equivalent to $D^\beta x^\alpha \phi$
being bounded for all multiindices α, β. For instance, the case $|\alpha| = |\beta| = 1$
takes the following form:

$$D_k(x_j \phi) = x_j(D_k \phi) + (D_k x_j)\phi,$$

with $D_k x_j$ being a constant, so the right-hand side is a linear combination
(with constant coefficients) of terms of the form $x^\mu D^\nu \phi$ with $|\mu| \leq 1 = |\alpha|$,
$|\nu| \leq 1 = |\beta|$, while

$$x_j(D_k \phi) = D_k(x_j \phi) - (D_k x_j)\phi,$$

and now the right-hand side is a sum of terms of the form $D^\nu(x^\mu \phi)$ with
$|\mu| \leq 1 = |\alpha|$, $|\nu| \leq 1 = |\beta|$. The general case proceeds iteratively and
bringing the powers of x to the left (front) in the first case, to the right
(back) in the second, and commuting them through one derivative at a
time.

With this definition, using the properties above, we conclude that if
$\phi \in \mathcal{S}(\mathbb{R}^n)$ then $\mathcal{F}\phi \in \mathcal{S}(\mathbb{R}^n)$ as well. Indeed,

$$\xi^\alpha D_\xi^\beta \mathcal{F}\phi = (-1)^{|\beta|} \mathcal{F} D_x^\alpha x^\beta \phi,$$

and $D_x^\alpha x^\beta \phi \in \mathcal{S}(\mathbb{R}^n) \subset L^1(\mathbb{R}^n)$ if $\phi \in \mathcal{S}(\mathbb{R}^n)$, so the right-hand side is indeed
bounded since this holds for the Fourier transform of an L^1 function.

Similar calculations show that the inverse Fourier transform satisfies

(8.4) $$\mathcal{F}^{-1} D_{\xi_j} \psi = -x_j \mathcal{F} \psi, \ D_{x_j} \mathcal{F}^{-1} \psi = \mathcal{F}^{-1}(\xi_j \psi),$$

so
$$\mathcal{F} : \mathcal{S} \to \mathcal{S}, \ \mathcal{F}^{-1} : \mathcal{S} \to \mathcal{S}.$$

In particular, $\mathcal{F}\mathcal{F}^{-1} : \mathcal{S} \to \mathcal{S}$ and $\mathcal{F}^{-1}\mathcal{F} : \mathcal{S} \to \mathcal{S}$; the Fourier inversion formula states that these are both the identity map on $\mathcal{S}(\mathbb{R}^n)$.

Before proceeding, let's see how we can use the Fourier transform to solve a constant coefficient PDE. Suppose that $a_\alpha \in \mathbb{C}$ and

$$P = \sum_{|\alpha| \le m} a_\alpha D_x^\alpha$$

is an mth order constant coefficient differential operator, and consider the PDE

$$Pu = f, \ f \in \mathcal{S}(\mathbb{R}^n).$$

Then for $u \in \mathcal{S}$ (for now),

$$\mathcal{F} Pu = \mathcal{F}\Big(\sum_{|\alpha| \le m} a_\alpha D_x^\alpha u \Big) = \sum_{|\alpha| \le m} a_\alpha \xi^\alpha \mathcal{F} u(\xi) = p(\xi) \mathcal{F} u,$$

where we let

$$p(\xi) = \sum_{|\alpha| \le m} a_\alpha \xi^\alpha$$

be the *full symbol* of P. Thus, if p never vanishes, then

$$\mathcal{F} u = \frac{\mathcal{F} f}{p(\xi)},$$

which is in $\mathcal{S}(\mathbb{R}^n)$ provided p has a lower bound like $|p(\xi)| \ge C(1+|\xi|)^{-N}$ for some N and $C > 0$; see Problem 8.7. Hence we get (assuming the Fourier inversion formula, which we soon prove)

$$u = \mathcal{F}^{-1}\left(\frac{\mathcal{F} f}{p(\xi)} \right),$$

solving the PDE. There are some issues we would like to understand better, e.g. the non-vanishing of p and also whether we really need $u, f \in \mathcal{S}$, but before getting further into this we need to investigate the Fourier inversion formula. To give an indication of what we'll see though, note the following examples:

- *Laplace's equation:* $P = \sum_{j=1}^{n} \partial_{x_j}^2$. Then $p(\xi) = -|\xi|^2$, so p vanishes at just one point, $\xi = 0$. Note that near infinity (well, for say $|\xi| > 1$), though, $|p(\xi)| > C(1 + |\xi|^2)$, for some $C > 0$.

- *Helmholtz equation:* $P = \sum_{j=1}^{n} \partial_{x_j}^2 + \lambda$. Then $p(\xi) = -|\xi|^2 + \lambda$, so if $\lambda < 0$, then p never vanishes, and indeed $|p(\xi)| \geq C(1 + |\xi|^2)$, for some $C > 0$.

- *Wave equation:* $P = -\sum_{j=1}^{n-1} \partial_{x_j}^2 + \partial_{x_n}^2$. Then $p(\xi) = |\xi'|^2 - \xi_n^2$, where $\xi' = (\xi_1, \ldots, \xi_{n-1})$, so p vanishes on the (light) cone $|\xi'| = |\xi_n|$.

- *Heat equation:* $P = -\sum_{j=1}^{n-1} \partial_{x_j}^2 + \partial_{x_n}$. Then $p(\xi) = |\xi'|^2 + i\xi_n$, so p only vanishes at the origin. Moreover, for $|\xi| \geq 1$, $|p(\xi)| > C(1 + |\xi|^2)^{1/2}$, for some $C > 0$. This is a *weaker* estimate than the one for Laplace's equation.

For a local result, i.e. whether one can solve a PDE locally, without regard to the behavior of the solution at infinity, what matters is whether $p(\xi)$ vanishes for large ξ; this is a reflection of the fact that the Fourier transform interchanges differentiability and decay. Thus, elliptic PDE, i.e. PDE of order m such that for sufficiently large $|\xi|$, $|p(\xi)| > C(1 + |\xi|^2)^{m/2}$ for some $C > 0$, are the best behaved PDE. Parabolic PDE like the heat equation where a weaker estimate holds are in certain aspects almost as well behaved, while hyperbolic PDE are the most interesting!

2. The inversion formula

As a first step towards the inversion formula, we calculate the Fourier transform of the Gaussian $\phi(x) = e^{-a|x|^2}$, $a > 0$, on $\mathbb{R}^n$ (note that $\phi \in \mathcal{S}$!) by writing it as

$$\hat{\phi}(\xi) = \left(\int_{\mathbb{R}} e^{-ax_1^2} e^{-ix_1\xi_1} \, dx_1 \right) \cdots \left(\int_{\mathbb{R}} e^{-ax_n^2} e^{-ix_n\xi_n} \, dx_n \right),$$

and hence reducing it to one-dimensional integrals which can be calculated by a change of variable and shift of contours if one is familiar with complex analysis. Since we do not assume the latter, we can also proceed as follows without complex analysis—actually we really prove a special case of the complex analysis result, Cauchy's theorem, below. Write x for the one-dimensional variable, ξ for its Fourier transform variable for simplicity, and $\psi(x) = e^{-ax^2}$,

$$\hat{\psi}(\xi) = \int_{\mathbb{R}} e^{-ix\xi} e^{-ax^2} \, dx = e^{-\xi^2/4a} \int_{\mathbb{R}} e^{-a(x+i\xi/(2a))^2} \, dx,$$

where we simply completed the square. We wish to show that

$$f(\xi) = \int_{\mathbb{R}} e^{-a(x+i\xi/(2a))^2} \, dx$$

is a constant, i.e. is independent of ξ, and in fact it is equal to $\sqrt{\pi/a}$. But that is easy: differentiating f, we obtain

$$f'(\xi) = -i \int_{\mathbb{R}} (x + i\xi/(2a)) e^{-a(x+i\xi/(2a))^2} \, dx.$$

The integrand is the derivative of $(-1/(2a)) e^{-a(x+i\xi/(2a))^2}$ with respect to x, so by the fundamental theorem of calculus,

$$f'(\xi) = (i/(2a)) e^{-a(x+i\xi/(2a))^2} |_{x=-\infty}^{+\infty} = 0,$$

due to the rapid decay of the Gaussian at infinity. This says that f is a constant, so for all ξ,

$$f(\xi) = f(0) = \int_{\mathbb{R}} e^{-ax^2} \, dx,$$

which can be evaluated by the usual polar coordinate trick, giving $\sqrt{\pi/a}$. Returning to $\mathbb{R}^n$, the final result is thus that

(8.5) $$\hat{\phi}(\xi) = (\pi/a)^{n/2} e^{-|\xi|^2/4a},$$

which is hence another Gaussian. A similar calculation shows that for such Gaussians $\mathcal{F}^{-1}\hat{\phi} = \phi$; i.e. for such Gaussians $T = \mathcal{F}^{-1}\mathcal{F}$ is the identity map. Indeed with $\psi(\xi) = e^{-b|\xi|^2}$, $b > 0$,

(8.6) $$\mathcal{F}^{-1}\psi(x) = (2\pi)^{-n} (\pi/b)^{n/2} e^{-|x|^2/4b} = (4\pi b)^{-n/2} e^{-|x|^2/4b},$$

so

$$\mathcal{F}^{-1}(\hat{\phi})(x) = (\pi/a)^{n/2} (4\pi/(4a))^{-n/2} e^{-4a|x|^2/4} = e^{-a|x|^2} = \phi(x).$$

Now we can show that T is the identity map on all Schwartz functions using the following lemma, which is due to Hörmander. (A different proof is given at the end of the chapter using the heat kernel.)

Lemma 8.2. *Suppose $T : \mathcal{S} \to \mathcal{S}$ is linear and commutes with x_j and D_{x_j}. Then T is a scalar multiple of the identity map, i.e. there exists $c \in \mathbb{C}$ such that $Tf = cf$ for all $f \in \mathcal{S}$.*

Proof. Let $y \in \mathbb{R}^n$. We show first that if $\phi(y) = 0$ and $\phi \in \mathcal{S}$ then $(T\phi)(y) = 0$.

Indeed, we can write, essentially by Taylor's theorem (see Problem 1.3),

(8.7) $$\phi(x) = \sum_{j=1}^{n} (x_j - y_j)\phi_j(x), \text{ with } \phi_j \in \mathcal{S} \text{ for all } j.$$

In one dimension this is just a statement that if ϕ is Schwartz and $\phi(y) = 0$, then $\phi_1(x) = \phi(x)/(x-y) = (\phi(x) - \phi(y))/(x-y)$ is Schwartz. Smoothness near y follows from Taylor's theorem (see Problem 1.2), while the rapid decay with all derivatives follows from $\phi_1(x) = \phi(x)/(x-y)$.

For the multi-dimensional version, Taylor's theorem as in Problem 1.3 guarantees that $\phi(x) = \sum_{j=1}^{n}(x_j - y_j)\phi_j^\sharp(x)$, with $\phi_j^\sharp \in \mathcal{C}^\infty$ for all j. However, it does not give the decay of $\phi_j^\sharp$ at infinity directly. On the other hand, one can take $\phi_j^\flat(x) = (x_j - y_j)\phi(x)/|x-y|^2$ for $|x-y| \geq 1$. One can paste the two together using a $\mathcal{C}^\infty$ function χ (called a cutoff, or bump, function) that is identically 1 for $|x-y| < 4/3$ and identically 0 for $|x-y| > 5/3$, say (see Problem 8.6), so

$$\phi_j = \chi\phi_j^\sharp + (1-\chi)\phi_j^\flat.$$

Indeed, then $\chi(x)\phi(x) = \sum_j(x_j - y_j)\chi(x)\phi_j^\sharp(x)$, with a similar formula for $(1-\chi)\phi$, and

$$\begin{aligned}
\phi(x) &= \chi(x)\phi(x) + (1-\chi(x))\phi(x) \\
&= \sum_j(x_j - y_j)\chi(x)\phi_j^\sharp(x) + \sum_j(x_j - y_j)(1-\chi(x))\phi_j^\flat(x) \\
&= \sum_j(x_j - y_j)\phi_j(x).
\end{aligned}$$

Further, $\chi\phi_j^\sharp$ has compact support and is $\mathcal{C}^\infty$, so it is Schwartz, while $(1-\chi)\phi_j^\flat$ is Schwartz directly from its definition (division by 0 not being an issue due to the vanishing of $1-\chi(x)$ when x is near y), so ϕ_j is Schwartz, proving our claim, (8.7).

Thus,

$$T\phi = \sum_{j=1}^{n}(x_j - y_j)(T\phi_j),$$

where we used the fact that T is linear and commutes with multiplication by x_j for all j. Substituting in $x = y$ yields $(T\phi)(y) = 0$ indeed.

Thus, fix $y \in \mathbb{R}^n$ and some $g \in \mathcal{S}$ such that $g(y) = 1$. Let $c(y) = (Tg)(y)$. We claim that for $f \in \mathcal{S}$, $(Tf)(y) = c(y)f(y)$. Indeed, let $\phi(x) = f(x) - f(y)g(x)$, so $\phi(y) = f(y) - f(y)g(y) = 0$. Thus, $0 = (T\phi)(y) = (Tf)(y) - f(y)(Tg)(y) = (Tf)(y) - c(y)f(y)$, proving our claim.

We have thus shown that there exists $c : \mathbb{R}^n \to \mathbb{C}$ such that for all $f \in \mathcal{S}$, $y \in \mathbb{R}^n$, $(Tf)(y) = c(y)f(y)$, i.e. $Tf = cf$. Taking $f \in \mathcal{S}$ such that f never vanishes (e.g. a Gaussian as above) shows that $c = Tf/f$ is $\mathcal{C}^\infty$, since Tf and f are such.

We have not yet used the fact that T commutes with D_{x_j}. But

$$\begin{aligned}
c(y)(D_{x_j}f)(y) = T(D_{x_j}f)(y) &= D_{x_j}(Tf)|_{x=y} = D_{x_j}(c(x)f(x))|_{x=y} \\
&= (D_{x_j}c)(y)f(y) + c(y)(D_{x_j}f)(y),
\end{aligned}$$

where the product rule is used in the last equality. Comparing the two sides and taking f such that f never vanishes yields

$$(D_{x_j} c)(y) = 0$$

for all y and for all j. Since all partial derivatives of c vanish, c is a constant, proving the lemma. $\qquad\qquad\qquad\qquad\qquad\qquad\qquad\qquad\qquad\qquad\qquad\qquad$ $\square$

The actual value of c can be calculated by applying T to a single Schwartz function, e.g. a Gaussian, and then the explicit calculation from above shows that $c = 1$, so $\mathcal{F}^{-1}\mathcal{F} = \mathrm{Id}$ indeed, i.e. the *Fourier inversion formula* holds.

3. The heat equation and convolutions

We already saw a use of the inversion formula in solving $\Delta u - u = f$. For PDEs with initial or boundary conditions, it is often best to use the partial Fourier transform. This is defined as follows. Let $\mathbb{R}^n = \mathbb{R}^m \times \mathbb{R}^k$ and write $\mathbb{R}^n \ni x = (y, z) \in \mathbb{R}^m \times \mathbb{R}^k$. (The variable y is a parameter and could instead be in an open subset of $\mathbb{R}^m$ or in $[0, \infty)$, say.) Suppose that $f \in C^1(\mathbb{R}^m \times \mathbb{R}^k)$ and $|z|^K f$, $|z|^K \partial_{x_j} f$ are bounded for all $j = 1, \ldots, n$, and $K > k$. Define the *partial Fourier transform* of f by

$$(\mathcal{F}_z f)(y, \zeta) = \int_{\mathbb{R}^k} e^{-iz \cdot \zeta} f(y, z) \, dz, \ y \in \mathbb{R}^m, \ \zeta \in \mathbb{R}^k,$$

i.e. for fixed y we perform a Fourier transform in z. By arguments as for the (full) Fourier transform, one can show easily (see the problems at the end of the chapter) that

(i) $(\mathcal{F}_z D_{z_j} f)(y, \zeta) = \zeta_j (\mathcal{F}_z f)(y, \zeta)$.

(ii) $(\mathcal{F}_z D_{y_j} f)(y, \zeta) = (D_{y_j} (\mathcal{F}_z f))(y, \zeta)$.

Similarly, as for the full Fourier transform, we have that if $f \in C^0(\mathbb{R}^m \times \mathbb{R}^k)$ and $|z|^K f$ is bounded for some $K > k + 1$, then

$$\mathcal{F}_z(z_j f) = -D_{\zeta_j} \mathcal{F}_z f, \ \mathcal{F}_z(y_j f) = y_j \mathcal{F}_z f.$$

Analogous formulae also hold for

$$(\mathcal{F}_\zeta^{-1} \psi)(y, z) = (2\pi)^{-k} \int_{\mathbb{R}^k} e^{iz \cdot \zeta} \psi(y, \zeta) \, dz.$$

An iterated application of these results also shows that

$$\mathcal{F}_z, \mathcal{F}_\zeta^{-1} : C^\infty(\mathbb{R}^m; \mathcal{S}(\mathbb{R}^k)) \to C^\infty(\mathbb{R}^m; \mathcal{S}(\mathbb{R}^k)),$$

where

(8.8) $$C^\infty(\mathbb{R}^m; \mathcal{S}(\mathbb{R}^k))$$

stands for C^∞ functions on $\mathbb{R}^m$ with values in $\mathcal{S}(\mathbb{R}^k)$, which means that its elements are C^∞ functions f on $\mathbb{R}^n = \mathbb{R}^m \times \mathbb{R}^k$ such that locally in y (i.e.

for $|y| < R$, $R > 0$ is arbitrary) $|z|^N D_x^\alpha f$ is bounded for all $N \geq 0$ and all $\alpha \in \mathbb{N}^n$.

As an application of these results, let's solve the heat equation on $(0, \infty)_t \times \mathbb{R}_x^n$:

$$u_t = k\Delta u, \ u(0, x) = \phi(x),$$

with $\phi \in \mathcal{S}(\mathbb{R}^n)$ given. Taking the partial Fourier transform in x and writing $\mathcal{F}_x u(t, \xi) = \hat{u}(t, \xi)$ gives

$$\frac{\partial \hat{u}}{\partial t}(t, \xi) = -k|\xi|^2 \hat{u}(t, \xi), \ \hat{u}(0, \xi) = (\mathcal{F}\phi)(\xi).$$

Solving the ODE for each fixed ξ yields

$$\hat{u}(t, \xi) = e^{-k|\xi|^2 t}(\mathcal{F}\phi)(\xi),$$

and hence

(8.9) $$u(t, x) = \mathcal{F}_\xi^{-1}\left(e^{-k|\xi|^2 t}(\mathcal{F}\phi)(\xi)\right).$$

We would like to rewrite this to have a more explicit expression for u in terms of ϕ. This can be done via convolutions.

Suppose first that f, g are continuous on $\mathbb{R}^n$ and $|x|^N f$, $|x|^N g$ are bounded for some $N > n$. (Again, $f \in L^1(\mathbb{R}^n)$ would do.) Then $\mathcal{F}f, \mathcal{F}g$ are bounded continuous functions, and hence $(\mathcal{F}f)(\mathcal{F}g)$ is a bounded continuous function as well. We cannot take its inverse Fourier transform (yet) directly, except under stronger assumptions (such as $f, g \in \mathcal{S}(\mathbb{R}^n)$), but we can ask whether $(\mathcal{F}f)(\mathcal{F}g)$ is the Fourier transform of a continuous function χ on $\mathbb{R}^n$ with $|x|^N \chi$ bounded for some $N > n$ (again, $\chi \in L^1(\mathbb{R}^n)$ would do). So we compute:

$$(\mathcal{F}f)(\xi)(\mathcal{F}g)(\xi) = \left(\int_{\mathbb{R}^n} e^{-ix\cdot\xi} f(x)\, dx\right)\left(\int_{\mathbb{R}^n} e^{-iy\cdot\xi} g(y)\, dy\right)$$
$$= \int_{\mathbb{R}^{2n}} e^{-ix\cdot\xi} e^{-iy\cdot\xi} f(x) g(y)\, dx\, dy;$$

note that the double integral is indeed absolutely convergent by our assumptions. We now change variables to make the exponential of the form $e^{-iz\cdot\xi}$; we thus let $z = x + y$, while keeping x, so $y = z - x$. Then we deduce

(8.10)
$$(\mathcal{F}f)(\xi)(\mathcal{F}g)(\xi) = \int_{\mathbb{R}^{2n}} e^{-iz\cdot\xi} f(x) g(z - x)\, dx\, dz$$
$$= \int_{\mathbb{R}^n} e^{-iz\cdot\xi}\left(\int_{\mathbb{R}^n} f(x) g(z - x)\, dx\right) dz = (\mathcal{F}(f * g))(\xi),$$

where we let

$$(f * g)(z) = \int_{\mathbb{R}^n} f(x) g(z - x)\, dx.$$

be the *convolution* of f and g. A change of variables shows that $(f * g)(z) = (g * f)(z)$, which is consistent with $(\mathcal{F}f)(\mathcal{F}g) = (\mathcal{F}g)(\mathcal{F}f)$. A simple calculation shows that if $f, g \in \mathcal{S}(\mathbb{R}^n)$, then $f * g \in \mathcal{S}(\mathbb{R}^n)$ as well. Again, this is consistent with (and indeed follows from, for here we can use the inverse Fourier transform already) $(\mathcal{F}f)(\mathcal{F}g) \in \mathcal{S}(\mathbb{R}^n)$.

If we write $\mathbb{R}^n \ni x = (x', x'') \in \mathbb{R}^m \times \mathbb{R}^k$ as above, we can talk about partial convolutions, and we still have the analogue of (8.10): we let

$$(f *_{x''} g)(x', x'') = \int_{\mathbb{R}^n} f(x', y'')g(x', x'' - y'') \, dy'',$$

and then

(8.11) $$(\mathcal{F}_{x''} f)(\mathcal{F}_{x''} g) = \mathcal{F}_{x''}(f *_{x''} g).$$

We now use this to rewrite the solution formula for the heat equation. By (8.9), (8.11) and the Fourier inversion formula, if we write

$$e^{-k|\xi|^2 t} = (\mathcal{F}_x f)(t, \xi)$$

for some $f \in \mathcal{C}^\infty((0, \infty)_t; \mathcal{S}(\mathbb{R}^n_x))$, then

$$u(t, x) = (f *_x \phi)(t, x) = \int f(t, x - y)\phi(y) \, dy.$$

But this is straightforward: we have computed the inverse Fourier transform of a Gaussian in (8.6), so with $b = kt$,

$$f(t, x) = (4\pi kt)^{-n/2} e^{-|x|^2/(4kt)},$$

and hence

$$u(t, x) = (4\pi kt)^{-n/2} \int e^{-|x-y|^2/(4kt)} \phi(y) \, dy,$$

yielding a more explicit solution formula for the heat equation. The function

$$K_t(x) = (4\pi kt)^{-n/2} e^{-|x|^2/(4kt)}, \ t > 0,$$

is called the *heat kernel*; in terms of this the solution formula for the heat equation is

$$u(t, x) = (K_t *_x \phi)(x).$$

4. Systems of PDE

The Fourier transform's use is not limited to single PDE. Consider the following system on $\mathbb{R}^2_{x,y}$:

(8.12)
$$u_x + v_y - u = f,$$
$$u_y - v_x - v = g,$$

with f, g given (Schwartz) functions. We Fourier transform both sides of both equations, writing the Fourier variables as (ξ, η), so the Fourier transform is

$$(\mathcal{F}f)(\xi, \eta) = \int_{\mathbb{R}^2} e^{-i(x\xi + y\eta)} f(x, y) \, dx \, dy.$$

This yields the system, with the hat denoting the Fourier transform,

$$i\xi \hat{u} + i\eta \hat{v} - \hat{u} = \hat{f},$$
$$i\eta \hat{u} - i\xi \hat{v} - \hat{v} = \hat{g},$$

or in matrix notation

$$\begin{pmatrix} i\xi - 1 & i\eta \\ i\eta & -i\xi - 1 \end{pmatrix} \begin{pmatrix} \hat{u} \\ \hat{v} \end{pmatrix} = \begin{pmatrix} \hat{f} \\ \hat{g} \end{pmatrix}.$$

Now, the determinant of the matrix on the left is

$$\det \begin{pmatrix} i\xi - 1 & i\eta \\ i\eta & -i\xi - 1 \end{pmatrix} = (i\xi - 1)(-i\xi - 1) - (i\eta)^2 = 1 + \xi^2 + \eta^2 \geq 1,$$

so the matrix is invertible, with its inverse given by

$$\begin{pmatrix} i\xi - 1 & i\eta \\ i\eta & -i\xi - 1 \end{pmatrix}^{-1} = \frac{1}{1 + \xi^2 + \eta^2} \begin{pmatrix} -i\xi - 1 & -i\eta \\ -i\eta & i\xi - 1 \end{pmatrix}.$$

Correspondingly,

$$\begin{pmatrix} \hat{u} \\ \hat{v} \end{pmatrix} = \frac{1}{1 + \xi^2 + \eta^2} \begin{pmatrix} -i\xi - 1 & -i\eta \\ -i\eta & i\xi - 1 \end{pmatrix} \begin{pmatrix} \hat{f} \\ \hat{g} \end{pmatrix},$$

and finally

$$\begin{pmatrix} u \\ v \end{pmatrix} = \mathcal{F}^{-1} \left[\frac{1}{1 + \xi^2 + \eta^2} \begin{pmatrix} -i\xi - 1 & -i\eta \\ -i\eta & i\xi - 1 \end{pmatrix} \begin{pmatrix} \hat{f} \\ \hat{g} \end{pmatrix} \right],$$

where we consider the inverse Fourier transform as acting independently on the two components of the vector (i.e. component by component).

This last expression can be rewritten somewhat, as

$$\begin{pmatrix} -i\xi - 1 & -i\eta \\ -i\eta & i\xi - 1 \end{pmatrix} \begin{pmatrix} \hat{f} \\ \hat{g} \end{pmatrix} = \begin{pmatrix} (-i\xi - 1)\hat{f} - i\eta \hat{g} \\ -i\eta \hat{f} + (i\xi - 1)\hat{g} \end{pmatrix}$$

as

$$\begin{pmatrix} \mathcal{F}(-f_x - f - g_y) \\ \mathcal{F}(-f_y + g_x - g) \end{pmatrix} = \mathcal{F} \begin{pmatrix} -f_x - f - g_y \\ -f_y + g_x - g \end{pmatrix},$$

where we again consider the Fourier transform as acting component by component on the vector. Then

$$\begin{pmatrix} u \\ v \end{pmatrix} = \mathcal{F}^{-1} \left[\frac{1}{1 + \xi^2 + \eta^2} \mathcal{F} \begin{pmatrix} -f_x - f - g_y \\ -f_y + g_x - g \end{pmatrix} \right]$$

$$= \begin{pmatrix} \mathcal{F}^{-1}(1 + \xi^2 + \eta^2)^{-1} \mathcal{F}(-f_x - f - g_y) \\ \mathcal{F}^{-1}(1 + \xi^2 + \eta^2)^{-1} \mathcal{F}(-f_y + g_x - g) \end{pmatrix}.$$

Thus, in this case solving the system amounts to first taking appropriate linear combinations of derivatives of f and g (with the combination being different for u and v), then performing a scalar Fourier transform, multiplying by $(1+\xi^2+\eta^2)^{-1}$, and then performing a scalar inverse Fourier transform. Note that multiplication by $(1+\xi^2+\eta^2)^{-1}$ on the Fourier transform side is exactly what showed up in our analysis of the Helmholtz equation $\Delta_{\mathbb{R}^2}+\lambda$ when $\lambda=-1$, up to an overall minus sign. It is instructive to see where this comes from directly. For this, write the PDE in a matrix form:

$$\begin{pmatrix} \partial_x - 1 & \partial_y \\ \partial_y & -\partial_x - 1 \end{pmatrix} \begin{pmatrix} u \\ v \end{pmatrix} = \begin{pmatrix} f \\ g \end{pmatrix}.$$

Now, it turns out that if we multiply the matrix of operators

$$\begin{pmatrix} \partial_x - 1 & \partial_y \\ \partial_y & -\partial_x - 1 \end{pmatrix}$$

by its transpose in the sense of Chapter 5 (so we are taking the transpose of the matrix, and then the transpose of its entries as differential operators), namely by

$$\begin{pmatrix} \partial_x - 1 & \partial_y \\ \partial_y & -\partial_x - 1 \end{pmatrix}^{\dagger} = \begin{pmatrix} -\partial_x - 1 & -\partial_y \\ -\partial_y & \partial_x - 1 \end{pmatrix},$$

then we get

(8.13)
$$\begin{pmatrix} -\partial_x - 1 & -\partial_y \\ -\partial_y & \partial_x - 1 \end{pmatrix} \begin{pmatrix} \partial_x - 1 & \partial_y \\ \partial_y & -\partial_x - 1 \end{pmatrix}$$
$$= \begin{pmatrix} -\partial_x^2 - \partial_y^2 + 1 & 0 \\ 0 & -\partial_x^2 - \partial_y^2 + 1 \end{pmatrix}.$$

This matrix is scalar in the sense that it is diagonal with the same diagonal entries, and so it can be written as

$$(-\partial_x^2 - \partial_y^2 - 1) \begin{pmatrix} 1 & 0 \\ 0 & 1 \end{pmatrix}.$$

Thus, solutions of the original system satisfy

$$(-\partial_x^2 - \partial_y^2 - 1) \begin{pmatrix} 1 & 0 \\ 0 & 1 \end{pmatrix} \begin{pmatrix} u \\ v \end{pmatrix} = \begin{pmatrix} \partial_x - 1 & \partial_y \\ \partial_y & -\partial_x - 1 \end{pmatrix} \begin{pmatrix} f \\ g \end{pmatrix},$$

i.e.

$$(-\partial_x^2 - \partial_y^2 - 1) \begin{pmatrix} u \\ v \end{pmatrix} = \begin{pmatrix} \partial_x - 1 & \partial_y \\ \partial_y & -\partial_x - 1 \end{pmatrix} \begin{pmatrix} f \\ g \end{pmatrix},$$

and now the Helmholtz equation's appearance is clear. While this may seem somewhat magical, and certainly arbitrary PDE are not so well behaved, such a special structure is often present in PDE of physical significance.

5. Integral transforms

As a final topic, we discuss the use of the Fourier transform for non-PDE purposes, concretely for integral transforms on $\mathbb{R}^n$, $n \geq 2$. An important question in many applications, in particular tomography, is the following: suppose we know the integral of a function f along straight lines, i.e. we know $\int_L f\,ds$, for every straight line L. Can we recover the function f? In practice, $f = f(x)$ might be for instance the absorption coefficient of a material (modelled by a bounded region $\Omega \subset \mathbb{R}^n$, so f has compact support) for a particular kind of wave, such as X-rays, and then for an incident wave along the line L, with incident amplitude a_0, $e^{-\int_L f\,ds}a_0$ would be the amplitude of the wave leaving the material, which is measured. Since we know a_0, this means that we also know $\int_L f\,ds$. Then we want to know $f(x)$ for all $x \in \Omega$ to find out the structure of the material. One example of this is the CT scan, in which the internal structure of the body is investigated using X-rays. Correspondingly, the map sending f to the collection of integrals $\int_L f\,ds$ is often called an *X-ray transform*.

To make progress on this problem let us parameterize the lines L explicitly. A line L can be specified by specifying its unit tangent vector $\omega \in \mathbb{R}^n$ (for which there are two choices) and the point y in which it intersects the plane Σ orthogonal to ω, so $y \cdot \omega = 0$. The line is then given by

$$(8.14) \qquad\qquad s \mapsto y + s\omega, \ s \in \mathbb{R}.$$

Correspondingly, the integral of a function f along the line is

$$(If)(y, \omega) = \int_L f\,ds = \int_{\mathbb{R}} f(y + s\omega)\,ds.$$

Thus, knowing $\int_L f\,ds$ for all L amounts to knowing $(If)(y, \omega)$ for all y, ω.

We start by considering the 2-dimensional setting, $n = 2$. We claim that we can compute the Fourier transform of f from If; then the Fourier inversion formula will give f. For $\xi \in \mathbb{R}^2$, let us write $\xi = \lambda\theta$, with θ a unit vector and $\lambda \in \mathbb{R}$. Then

$$(\mathcal{F}f)(\xi) = \int_{\mathbb{R}^2} e^{-i\lambda x \cdot \theta} f(x)\,dx.$$

Now let ω be a unit vector orthogonal to θ (there are two choices), and consider $\mathbb{R}^2$ as a union of lines parallel to ω; to be definite, say ω is the counterclockwise angle $\pi/2$ rotation of θ, so $\omega = \omega(\theta)$. Note for each of these lines L, which are of the form (8.14), $x \cdot \theta$ is constant as x varies along L since

$$x \cdot \theta = (y + s\omega) \cdot \theta = y \cdot \theta$$

as $\omega \cdot \theta = 0$. Thus, if we write the integral over $\mathbb{R}^2$ as an iterated integral, first integrating along these lines and then performing the integral as the lines vary, i.e. y varies, we obtain the integral over $\mathbb{R}^2$ and we will be able to pull out the exponential $e^{-i\lambda x \cdot \theta}$ from the inner integral. In this case, Σ is just the span of θ (since being orthogonal to ω means being parallel to θ), i.e. we can write $y = t\theta$, $t \in \mathbb{R}$.

Concretely,

$$\Sigma \times \mathbb{R} \ni (y, s) \mapsto y + s\omega \in \mathbb{R}^2$$

can be thought of as

$$\mathbb{R} \times \mathbb{R} \ni (t, s) \mapsto t\theta + s\omega \in \mathbb{R}^2.$$

This is an orthogonal change of variables (since θ, ω are orthogonal unit vectors), so by the change of variables formula for the integral (as this is a length preserving map, thus has Jacobian determinant of absolute value 1)

$$(\mathcal{F}f)(\xi) = \int_{\mathbb{R}} \int_{\mathbb{R}} e^{-i\lambda(t\theta + s\omega) \cdot \theta} f(t\theta + s\omega) \, ds \, dt$$

$$= \int_{\mathbb{R}} e^{-i\lambda t\theta \cdot \theta} \int_{\mathbb{R}} f(t\theta + s\omega) \, ds \, dt$$

$$= \int_{\mathbb{R}} e^{-i\lambda t} (If)(t\theta, \omega(\theta)) \, dt.$$

Thus, given If, we can find $\mathcal{F}f$ by taking the Fourier transform of If in the first variable, and then the Fourier inversion formula gives us f.

In the general $n > 2$-dimensional setting one way to proceed is to consider $\mathbb{R}_x^n$ as the union of 2-planes H, say given by $x_3, \ldots, x_n$ being constant, while x_1, x_2 vary:

$$H_{a_3, \ldots, a_n} = \{(x_1, \ldots, x_n) : x_3 = a_3, \ldots, x_n = a_n\}.$$

Then for each value of $a_3, \ldots, a_n$, $H_{a_3, \ldots, a_n}$ can be identified with $\mathbb{R}^2$ via the (x_1, x_2) coordinates, and the previous result can be applied. Note that this only uses lines in the planes $H_{a_3, \ldots, a_n}$, so we do not even need to know the integral of the function along lines that intersect these planes at an angle, e.g. lines in the direction $(0, \frac{1}{\sqrt{2}}, \frac{1}{\sqrt{2}})$. Correspondingly, the problem is *overdetermined*; we can forget some of the information we are given and still recover f! (In practice, this means that we do not need to send in X-rays from every direction.) See also Problem 8.11 for a related problem and a slightly different approach to finding f from If.

Additional material: A heat kernel proof of the Fourier inversion formula

In fact, the heat kernel provides an alternative way of showing the Fourier inversion formula, i.e. a proof of the inversion formula that uses all the basic results we proved, but not Lemma 8.2 or its consequences. Thus, we already know that $\mathcal{F}, \mathcal{F}^{-1} : \mathcal{S} \to \mathcal{S}$, and we have calculated the Fourier transform of the Gaussians. The point is that the collection of functions

$$K_t(x) = (4\pi k t)^{-n/2} e^{-|x|^2/(4kt)}, \ t > 0,$$

depending on the parameter t is well behaved: they have integral 1, are positive, and finally, for any $\delta > 0$, $\int_{\mathbb{R}^n \setminus B_\delta(0)} K_t(x)\, dx \to 0$ as $t \to 0$, which is easy to check. This has the following consequence:

Lemma 8.3. *If h is a bounded continuous function, then for every x, using $*$ simply to denote partial convolution, $(K_t * h)(x) \to h(x)$, and the convergence is uniform on sets A such that on $\tilde{A} = \{z \in \mathbb{R}^n : |z - x| \leq 1 \text{ for some } x \in A\}$ the function h is uniformly continuous (in particular, on compact subsets A).*

Proof. Let $A \subset \mathbb{R}^n$ be such that h is uniformly continuous on $\tilde{A}$ as above, and for $\epsilon > 0$ let $\delta' > 0$ be such that $x, x - y \in \tilde{A}$, $|y| < \delta'$ implies $|h(x - y) - h(x)| < \epsilon/2$. Let $\delta = \min(1, \delta')$. In particular, if $x \in A$ and $|y| < \delta$, then $x - y \in \tilde{A}$, so $|h(x - y) - h(x)| < \epsilon/2$.

Thus, using $\int K_t(y)\, dy = 1$, for $x \in A$,

$$(K_t * h)(x) - h(x) = \int K_t(y)(h(x - y) - h(x))\, dy$$

$$= \int_{B_\delta(0)} K_t(y)(h(x - y) - h(x))\, dy$$

$$+ \int_{\mathbb{R}^n \setminus B_\delta(0)} K_t(y)(h(x - y) - h(x))\, dy.$$

Now, the absolute value of the first integral is

$$\leq \int_{B_\delta(0)} |K_t(y)|\, |h(x - y) - h(x)|\, dy$$

$$\leq \frac{\epsilon}{2} \int_{B_\delta(0)} |K_t(y)|\, dy \leq \frac{\epsilon}{2} \int_{\mathbb{R}^n} |K_t(y)|\, dy \leq \frac{\epsilon}{2},$$

while that of the second integral is

$$\leq \int_{\mathbb{R}^n \setminus B_\delta(0)} |K_t(y)|(|h(x - y)| + |h(x)|)\, dy$$

$$\leq 2 \sup |h| \int_{\mathbb{R}^n \setminus B_\delta(0)} K_t(y)\, dy,$$

so it goes to 0 as $t \to 0$, and in particular there is $t_0 > 0$ such that this is $< \epsilon/2$ for $0 < t < t_0$. In summary, $\sup\{|(K_t * h)(x) - h(x)| : x \in A\} \leq \epsilon$ for $0 < t < t_0$, proving the uniform convergence on A. $\square$

In particular, if $h \in \mathcal{S}(\mathbb{R}^n)$, then h is uniformly continuous on $\mathbb{R}^n$. First, given $\epsilon > 0$, choose $R > 0$ such that $|h(x)| < \epsilon/2$ for $|x| > R$ (which is possible by the decay of h at infinity), so if $|x| > R + 1$, $|y| < 1$, then $|x - y| > R$ shows $|h(x - y) - h(x)| \leq |h(x - y)| + |h(y)| < \epsilon$. On the other hand, h is continuous, thus uniformly continuous, on the compact set $\{x : |x| \leq R+2\}$, so there is $\delta' > 0$ such that $|x|, |x-y| \leq R+2$ with $|y| < \delta'$ implies $|h(x-y)-h(x)| < \epsilon$. Now simply let $\delta = \min(\delta', 1)$; then $|x| \leq R+1$, $|y| < \delta$ implies $|x - y| < R + 2$ and $|h(x - y) - h(x)| < \epsilon$. In combination, these two estimates (for x with $|x| > R+1$ and for x with $|x| \leq R+1$) yield the uniform continuity on $\mathbb{R}^n$. Correspondingly, for Schwartz functions h, $K_t * h \to h$ uniformly on $\mathbb{R}^n$. (Notice that we actually only used the fact that h was continuous and $h(x) \to 0$ as $|x| \to \infty$.)

Now, as will be important in the next chapter, the Fourier transform satisfies the relation

$$(8.15) \qquad \int \hat{\phi}(\xi)\psi(\xi) \, d\xi = \int \phi(x)\hat{\psi}(x) \, dx, \qquad \phi, \psi \in \mathcal{S}.$$

(Of course, we could have denoted the variable of integration by x on both sides.) Indeed, explicitly writing out the Fourier transforms,

$$\int \left(\int e^{-ix\cdot\xi}\phi(x) \, dx \right) \psi(\xi) \, d\xi = \int_{\mathbb{R}^{2d}} e^{-ix\cdot\xi}\phi(x)\psi(\xi) \, dx \, d\xi$$

$$= \int \phi(x) \left(\int e^{-ix\cdot\xi}\psi(\xi) \, d\xi \right) dx.$$

We now apply this result with ψ replaced by the inverse Fourier transform of K_t, which is $\psi(\xi) = (2\pi)^{-n}e^{-k|\xi|^2 t}$ as we have already calculated the Fourier and inverse Fourier transform of Gaussians; this means that $\hat{\psi} = K_t$. Thus,

$$(2\pi)^{-n} \int \hat{\phi}(\xi)e^{-k|\xi|^2 t} \, d\xi = \int \phi(x)K_t(x) \, dx,$$

and the right-hand side converges to $\phi(0)$ as $t \to 0$ by our previous discussion (it is $K_t * \phi$ evaluated at 0). On the other hand, as $\hat{\phi} \in \mathcal{S}$ and $0 < e^{-k|\xi|^2 t} \leq 1$, and for each ξ, $e^{-k|\xi|^2 t} \to 1$ as $t \to 0$, the left-hand side converges to $(2\pi)^{-n} \int \hat{\phi}(\xi) \, d\xi$, which is the inverse Fourier transform of $\hat{\phi}$ evaluated at 0. This shows that the Fourier inversion formula holds at 0.

In order to prove the general Fourier inversion formula, use $K_t(x - a)$ in place of $K_t(x)$ in the argument above; then the inverse Fourier transform of $K_t(. - a)$ is $(2\pi)^{-n}e^{i\xi\cdot a}e^{-ik|\xi|^2 t}$ (cf. Problem 8.1; the analogue of (i) also holds for the inverse Fourier transform, with a sign change in the exponent), which is still bounded by $(2\pi)^{-n}$ in absolute value, but now converges to $(2\pi)^{-n}e^{i\xi\cdot a}$ pointwise, so we obtain

$$\mathcal{F}^{-1}\hat{\phi}(a) = \lim_{t\to 0} \int \phi(x)K_t(x-a) \, dx = \lim_{t\to 0} \int \phi(x)K_t(a-x) = \lim_{t\to 0}(K_t * \phi)(a),$$

and the proof is finished as above.

Problems

Problem 8.1. Suppose that f is (piecewise) continuous on $\mathbb{R}^n$ with $|x|^N f(x)$ bounded for some $N > n$ (or indeed simply that $f \in L^1(\mathbb{R}^n)$). Throughout this problem, $a \in \mathbb{R}^n$. Then:

(i) Let $f_a(x) = f(x - a)$. Show that $(\mathcal{F} f_a)(\xi) = e^{-ia \cdot \xi}(\mathcal{F} f)(\xi)$.

(ii) Let $g_a(x) = e^{ix \cdot a} f(x)$. Show that $(\mathcal{F} g_a)(\xi) = (\mathcal{F} f)(\xi - a)$.

(iii) Show that $(\mathcal{F}^{-1} f_a)(x) = e^{ia \cdot x}(\mathcal{F}^{-1} f)(x)$.

(iv) Show that $(\mathcal{F}^{-1} g_a)(x) = (\mathcal{F}^{-1} f)(x + a)$.

Problem 8.2. Use part (i) of Problem 8.1 to show that $(\mathcal{F}(\partial_j f))(\xi) = i\xi_j(\mathcal{F} f)(\xi)$ if f is C^1 and $|x|^N f$, $|x|^N \partial_j f$ are bounded for some $N > n$.

Problem 8.3. Find the Fourier transform on $\mathbb{R}$ of the following functions:

(i) $f(x) \doteq H(a - |x|)$, where $a > 0$ and H is the Heaviside step function, so $H(t) = 1$ if $t > 0$ and $H(t) = 0$ if $t < 0$.

(ii) $f(x) = H(x)e^{-ax}$, where $a > 0$.

(iii) $f(x) = |x|^n e^{-a|x|}$, where $a > 0$ and $n \geq 0$ integer.

(iv) $f(x) = (1 + x^2)^{-1}$. (*Hint:* Use the fact that if $f = \mathcal{F}^{-1} g$, then $g = \mathcal{F} f$ by the Fourier inversion formula. Rewrite $(1 + x^2)^{-1}$ as partial fractions (factor the denominator).)

Problem 8.4. Find the Fourier transform on $\mathbb{R}^3$ of the function $f(x) = |x|^n e^{-a|x|}$, where $a > 0$ and $n \geq -1$ integer. (*Hint:* Express the integral in the Fourier transform in polar coordinates.)

Problem 8.5.

(i) Compute the integral

$$\int_{\mathbb{R}} e^{-ax^2}\, dx$$

when $\operatorname{Re} a > 0$, a complex, by writing

$$\left(\int_{\mathbb{R}} e^{-ax^2}\, dx\right)^2 = \left(\int_{\mathbb{R}} e^{-ax^2}\, dx\right)\left(\int_{\mathbb{R}} e^{-ay^2}\, dy\right) = \int_{\mathbb{R}^2} e^{-a(x^2+y^2)}\, dx\, dy$$

and expressing the resulting integral in polar coordinates.

Note: The final expression involves the square root of a complex number. Make sure you specify which square root you are taking (choose the correct one!).

(ii) In the chapter we computed the Fourier transform of $e^{-a|x|^2}$ on $\mathbb{R}^n$ when $a > 0$. Extend the result to a complex with $\operatorname{Re} a > 0$.

Problem 8.6. Suppose $y \in \mathbb{R}^n$. Show that there really *is* a function $\chi \in C_c^\infty(\mathbb{R}^n)$ such that $\chi(x) = 1$ if $|x - y| < 4/3$ and $\chi(x) = 0$ if $|x - y| > 5/3$, to complete the proof of Lemma 8.2.

Hint: See Problem 5.2.

Problem 8.7. Show that if p is a polynomial and never vanishes, and $|p(\xi)| \geq C(1+|\xi|)^{-N}$ for some N and $C > 0$, then for $\psi \in \mathcal{S}(\mathbb{R}^n)$, $\rho(\xi) = \frac{\psi(\xi)}{p(\xi)}$ is also a Schwartz function. (*Hint:* The result is easy to see if you show that all derivatives of $\frac{1}{p(\xi)}$ have upper bounds such as $C'(1 + |\xi|)^{N'}$, with C', N' depending on the derivative. To see these, show that derivatives of $\frac{1}{p(\xi)}$ are the quotient of a polynomial and a power of p.)

Problem 8.8. Let $\mathbb{R}^n = \mathbb{R}^m \times \mathbb{R}^k$, and write $\mathbb{R}^n \ni x = (y, z) \in \mathbb{R}^m \times \mathbb{R}^k$. Suppose that $f \in C^1(\mathbb{R}^m \times \mathbb{R}^k)$ and $|z|^K f$, $|z|^K \partial_{x_j} f$ are bounded for all $j = 1, \ldots, n$, and $K > k$. Define the *partial Fourier transform* of f by

$$(\mathcal{F}_z f)(y, \zeta) = \int_{\mathbb{R}^k} e^{-iz \cdot \zeta} f(y, z) \, dz, \quad y \in \mathbb{R}^m, \; \zeta \in \mathbb{R}^k.$$

Show that

(i) $(\mathcal{F}_z D_{z_j} f)(y, \zeta) = \zeta_j (\mathcal{F}_z f)(y, \zeta)$.

(ii) $(\mathcal{F}_z D_{y_j} f)(y, \zeta) = (D_{y_j}(\mathcal{F}_z f))(y, \zeta)$.

Note: Under appropriate additional assumptions, as for the full Fourier transform, the formulae $\mathcal{F}_z(z_j f) = -D_{\zeta_j} \mathcal{F}_z f$, $\mathcal{F}_z(y_j f) = y_j \mathcal{F}_z f$, and the analogous formulae for

$$(\mathcal{F}_\zeta^{-1} \psi)(y, z) = (2\pi)^{-k} \int_{\mathbb{R}^k} e^{iz \cdot \zeta} \psi(y, \zeta) \, dz$$

also hold, but you do not need to prove this (but you should know these!).

Problem 8.9.

(i) Let $\hbar > 0$ be a fixed constant. Solve (for all $t \in \mathbb{R}$) the (mass 1) *Schrödinger equation*

$$i\hbar u_t = -\frac{\hbar^2}{2} \Delta_x u, \; u(x, 0) = \psi(x),$$

for a free particle in $\mathbb{R}_x^n$ using the Fourier transform. Here ψ is a given Schwartz function on $\mathbb{R}^n$. You may leave your solution as the inverse Fourier transform of a Schwartz function (you do not need to evaluate it explicitly). Note that the function u is called the *wave function* of the particle in quantum mechanics and $|u|^2$ is the probablity amplitude, so the probability of finding the particle in a region $\Omega \subset \mathbb{R}_x^n$ at time t is $\int_\Omega |u(x, t)|^2 \, dx$.

(ii) Evaluate the solution explicitly if $\psi(x) = (2\pi)^{-n/2}e^{-x^2/2}$ and show that $\int_{\mathbb{R}^n} |u(x,t)|^2\, dx = 1$ for all t, i.e. the total probability (the probability of being *somewhere* in space) remains 1 for all times.

(iii) Where (for what value of x) is $|u(x,t)|^2$ the largest for fixed values of t? Does the sharpness of the peak of $|u(x,t)|^2$ vary with t?

(iv) Answer questions (ii) and (iii) if $\psi(x) = (2\pi)^{-n/2}e^{ivx/\hbar}e^{-x^2/2}$, where $v \in \mathbb{R}^n$ is fixed.

Problem 8.10. Suppose that f is a Schwartz function on $\mathbb{R}^2$ and its integral along lines L is given by $s \mapsto y + s\omega$, $y \cdot \omega = 0$, is $e^{-|y|^2/2}$. Find f.

Problem 8.11. For $n \geq 2$ an important problem is to determine a function f from its integral along hyperplanes, i.e. $n-1$-dimensional flat surfaces. If $n = 2$ these are just straight lines, and this problem is the X-ray transform problem solved in the chapter. For $n = 3$ one integrates along 2-planes, which can be parameterized by their unit normal θ and the point $s\theta$ where they intersect the span of θ (i.e. are of the form $H = H_{s,\theta} = \{x \in \mathbb{R}^3 : x \cdot \theta = s\}$). The *Radon transform* of a function f is given by $(Rf)(s,\theta) = \int_{H_{s,\theta}} f(x)\, dS(x)$. Show that Rf determines the Fourier transform of f, and thus f, and give an explicit formula for f in terms of Rf.

Note: The Radon transform gives a different approach to inverting the X-ray transform. Indeed, the X-ray transform of a function f determines the integrals of the function along all hyperplanes, since one just needs to think of the hyperplane as a union of parallel lines. Then one can use the inversion result for the Radon transform to find the function f.

Chapter 9

The Fourier transform: Tempered distributions, the wave equation and Laplace's equation

1. Tempered distributions

We have previously constructed distributions by taking the set $\mathcal{C}_c^\infty(\mathbb{R}^n)$ as the set of 'very nice' functions and have defined distributions as continuous linear maps $u : \mathcal{C}_c^\infty(\mathbb{R}^n) \to \mathbb{C}$ (or into reals). While this was an appropriate class when studying just derivatives, we have seen that for the Fourier transform the set of very nice functions is that of Schwartz functions. Thus, we expect that the set $\mathcal{S}'(\mathbb{R}^n)$ of 'corresponding distributions' should consist of continuous linear maps $u : \mathcal{S}(\mathbb{R}^n) \to \mathbb{C}$. In order to make this into a definition, we need a notion of convergence on $\mathcal{S}(\mathbb{R}^n)$. Since $\mathcal{C}_c^\infty(\mathbb{R}^n) \subset \mathcal{S}(\mathbb{R}^n)$, it is not unreasonable to expect that if a sequence $\{\phi_j\}_{j=1}^\infty$ converges to some $\phi \in \mathcal{C}_c^\infty(\mathbb{R}^n)$ inside the space $\mathcal{C}_c^\infty(\mathbb{R}^n)$ (i.e. in the sense of $\mathcal{C}_c^\infty(\mathbb{R}^n)$-convergence), then it should also converge to ϕ in the sense of $\mathcal{S}$-convergence. We shall see that this is the case, which implies that every $u \in \mathcal{S}'$ lies in $\mathcal{D}'$ as well. For if $\phi_j \to \phi$ in $\mathcal{C}_c^\infty(\mathbb{R}^n)$, then $\phi_j \to \phi$ in $\mathcal{S}$, and hence $u(\phi_j) \to u(\phi)$ by the continuity of u as an element of $\mathcal{S}'$. Thus, $\mathcal{S}' \subset \mathcal{D}'$, i.e. $\mathcal{S}'$ is a special class of distributions.

In order to motivate the definition of $\mathcal{S}$-convergence, recall that $\mathcal{S} = \mathcal{S}(\mathbb{R}^n)$ is the set of functions $\phi \in C^\infty(\mathbb{R}^n)$ with the property that for any multiindices $\alpha, \beta \in \mathbb{N}^n$, $x^\alpha \partial^\beta \phi$ is bounded. Here we wrote $x^\alpha = x_1^{\alpha_1} x_2^{\alpha_2} \ldots x_n^{\alpha_n}$ and $\partial^\beta = \partial_{x_1}^{\beta_1} \ldots \partial_{x_n}^{\beta_n}$, with $\partial_{x_j} = \frac{\partial}{\partial x_j}$.

With this in mind, convergence of a sequence $\phi_k \in \mathcal{S}$, $k \in \mathbb{N}$, to some $\phi \in \mathcal{S}$, in $\mathcal{S}$ is defined as follows:

Definition 9.1. We say that ϕ_k converges to ϕ in $\mathcal{S}$ if for all multiindices α, β, $\sup |x^\alpha \partial^\beta (\phi_k - \phi)| \to 0$ as $k \to \infty$, i.e. if $x^\alpha \partial^\beta \phi_k$ converges to $x^\alpha \partial^\beta \phi$ uniformly.

Let's see an example:

Example 9.1. Let $\chi \in C_c^\infty(\mathbb{R}^n)$ be such that $\chi(x) = 1$ for $|x| < 1$ and $\chi(x) = 0$ for $|x| > 2$. For $\phi \in \mathcal{S}$, let $\phi_k(x) = \chi(x/k)\phi(x)$. Show that $\phi_k \to \phi$ in $\mathcal{S}$.

Note that $\phi_k(x) = \phi(x)$ for $|x| < k$, so the convergence claim is reasonable. In fact, if ϕ has compact support, say inside $\{x : |x| < R\}$, then $\phi_k(x) = \phi(x)$ of $k > R$, thus for all large k, so if ϕ happens to be compactly supported, the claim is very easy to see!

Solution. First consider $\beta = 0$. We have

$$x^\alpha(\phi_k(x) - \phi(x)) = x^\alpha(1 - \chi(x/k))\phi(x),$$

and if $1 - \chi(x/k)$ is non-zero, then necessarily $|x| \geq k$ (since otherwise $\chi(x/k) = 1$). Thus,

$$|x^\alpha(\phi_k(x) - \phi(x))| = (|x|^{-2}(1 - \chi(x/k)))\left(x^\alpha \left(\sum_{j=1}^n x_j^2\right)\phi(x)\right),$$

and the first factor is $\leq k^{-2}$ (as $|x| \geq k$ when $1 - \chi(x/k)$ is non-zero), while the second factor is a sum of powers of x times ϕ, and is thus bounded, with bound independent of k. Correspondingly, $\sup |x^\alpha(\phi_k(x) - \phi(x))| \to 0$ as $k \to \infty$.

Now let's add derivatives, i.e. consider general β. Then by Leibniz' rule (the product rule applied inductively)

$$x^\alpha \partial_x^\beta (\phi_k(x) - \phi(x)) = x^\alpha \partial^\beta (1 - \chi(x/k))\phi(x)$$
$$= x^\alpha (1 - \chi(x/k))\partial^\beta \phi(x)$$
$$- x^\alpha \sum_{\gamma < \beta} C_{\beta\gamma} k^{-(|\beta| - |\gamma|)} (\partial^{\beta - \gamma} \chi)(x/k)(\partial^\gamma \phi)(x),$$

where $C_{\beta\gamma}$ are combinatorial constants. Here, by Leibniz' rule, we have a sum of terms on the right-hand side, each of which is a derivative of

$1 - \chi(x/k)$ times a derivative of ϕ (times a constant times x^α), with γ derivatives falling on ϕ (so $\gamma \leq \beta$) and the rest, $\beta - \gamma$, falling on $1 - \chi(x/k)$. We separated out the term $\gamma = \beta$ as the first term; this is much like the case considered above with no derivatives, simply ϕ is replaced by $\partial^\beta \phi$. On the other hand, for the remaining terms $|\beta| - |\gamma| > 0$ gives decay in k, while

$$\sup_x |(\partial^{\beta-\gamma}\chi)(x/k)| = \sup_x |(\partial^{\beta-\gamma}\chi)(x)|$$

and $\sup |x^\alpha(\partial^\gamma \phi)(x)|$ are finite and k-independent, so the supremum of the remaining terms goes to 0 as $k \to \infty$.

Definition 9.2. A *tempered distribution* u is defined as a continuous linear functional on $\mathcal{S}$ (this is written as $u \in \mathcal{S}'$), i.e. as a map $u : \mathcal{S} \to \mathbb{C}$ which is linear: $u(a\phi + b\psi) = au(\phi) + bu(\psi)$ for all $a, b \in \mathbb{C}$, $\phi, \psi \in \mathcal{S}$, and which is continuous: if ϕ_k converges to ϕ in $\mathcal{S}$ then $\lim_{k\to\infty} u(\phi_k) = u(\phi)$ (this is the convergence of complex numbers).

Note that if $\phi \in \mathcal{S}$, then by Example 9.1 there is a sequence of compactly supported C^∞ functions ϕ_k converging to ϕ in $\mathcal{S}$, so if $u \in \mathcal{S}'$, then $u(\phi_k) \to u(\phi)$, i.e. $u \in \mathcal{S}'$ is determined by how it acts on $C_c^\infty(\mathbb{R}^n)$, since for any $\phi \in \mathcal{S}$, $u(\phi)$ is the limit of u applied to elements of $C_c^\infty(\mathbb{R}^n)$.

As mentioned already, any tempered distribution is a distribution, since $\phi \in C_c^\infty$ implies $\phi \in \mathcal{S}$, and convergence of a sequence in C_c^∞ implies that in $\mathcal{S}$ (recall that convergence of a sequence in C_c^∞ means that the supports stay inside a fixed compact set and the convergence of all derivatives is uniform). The converse is of course not true. That is, any continuous function f on $\mathbb{R}^n$ defines a distribution, but $\int_{\mathbb{R}^n} f(x)\phi(x)\,dx$ will not converge for all $\phi \in \mathcal{S}$ if f grows too fast at infinity; e.g. $f(x) = e^{|x|^2}$ does not define a tempered distribution. On the other hand, any continuous function f satisfying an estimate $|f(x)| \leq C(1 + |x|)^N$ for some N and C defines a tempered distribution $u = \iota_f$ via

$$u(\psi) = \iota_f(\psi) = \int_{\mathbb{R}^n} f(x)\psi(x)\,dx, \quad \psi \in \mathcal{S},$$

since

$$\left| \int_{\mathbb{R}^n} f(x)\psi(x)\,dx \right| \leq CM \int_{\mathbb{R}^n} (1 + |x|)^N (1 + |x|)^{-N-n-1}\,dx < \infty,$$
$$M = \sup \left((1 + |x|)^{N+n+1}|\psi| \right) < \infty.$$

This is the reason for the 'tempered' terminology: the growth of f is 'tempered' at infinity.

Another example of such a distribution $u \in \mathcal{S}'$ is the delta distribution: $u = \delta_a$, given by $\delta_a(\phi) = \phi(a)$ for $\phi \in \mathcal{S}$. A more extreme example is the following, on $\mathbb{R}$ for simplicity. Let $a > 0$, and let

$$u = \sum_{k \in \mathbb{Z}} \delta_{ak},$$

i.e. for $\phi \in \mathcal{S}(\mathbb{R})$, let

$$u(\phi) = \sum_{k \in \mathbb{Z}} \phi(ak).$$

Thus a delta distribution 'sits' at each scaled integer ak, $k \in \mathbb{Z}$. This sum converges, since $|\phi(x)| \leq C(1 + |x|^2)^{-1}$ as $\phi \in \mathcal{S}$, and $\sum_{k \in \mathbb{Z}} C(1 + a^2 k^2)^{-1}$ converges. It is also continuous since if $\phi_j \to \phi$ in $\mathcal{S}$, then $(1 + x^2)\phi_j \to (1 + x^2)\phi$ uniformly on $\mathbb{R}$. Hence

$$|u(\phi_j) - u(\phi)| \leq \sum_{k \in \mathbb{Z}} (1 + a^2 k^2)|\phi_j(ak) - \phi(ax)|(1 + a^2 k^2)^{-1}$$

$$\leq \sum_{k \in \mathbb{Z}} \sup_{x \in \mathbb{R}} \left((1 + x^2)|\phi_j(x) - \phi(x)|\right) (1 + a^2 k^2)^{-1}$$

$$\leq \sup_{x \in \mathbb{R}} \left((1 + x^2)|\phi_j(x) - \phi(x)|\right) \sum_{k \in \mathbb{Z}} (1 + a^2 k^2)^{-1},$$

and the last sum converges, so $|u(\phi_j) - u(\phi)| \to 0$ as $j \to \infty$.

2. The Fourier transform of tempered distributions

We defined the Fourier transform on $\mathcal{S}$ as

$$(\mathcal{F}\phi)(\xi) = \hat{\phi}(\xi) = \int_{\mathbb{R}^n} e^{-ix\cdot\xi} \phi(x)\, dx$$

and the inverse Fourier transform as

$$(\mathcal{F}^{-1}\psi)(x) = (2\pi)^{-n} \int_{\mathbb{R}^n} e^{ix\cdot\xi} \psi(\xi)\, d\xi.$$

The Fourier transform satisfies the relation

(9.1) $$\int \hat{\phi}(\xi)\psi(\xi)\, d\xi = \int \phi(x)\hat{\psi}(x)\, dx, \qquad \phi, \psi \in \mathcal{S}.$$

(Of course, we could have denoted the variable of integration by x on both sides.) Indeed, explicitly writing out the Fourier transforms,

$$\int \left(\int e^{-ix\cdot\xi}\phi(x)\, dx \right) \psi(\xi)\, d\xi = \int \int e^{-ix\cdot\xi}\phi(x)\psi(\xi)\, dx\, d\xi$$

$$= \int \phi(x) \left(\int e^{-ix\cdot\xi}\psi(\xi)\, d\xi \right) dx,$$

the middle integral converges absolutely (since ϕ, ψ decrease rapidly at infinity), and hence the order of integration can be changed. Of course, this argument does not really require $\phi, \psi \in \mathcal{S}$. It suffices if they decrease fast enough at infinity, e.g. $|\phi(x)| \leq C(1 + |x|)^{-s}$ for some $s > n$, and similarly for ψ.

In the language of distributional pairing this just says that the tempered distribtions ι_ϕ, resp. $\iota_{\hat{\phi}}$, defined by ϕ, resp. $\hat{\phi}$, satisfy

$$\iota_{\hat{\phi}}(\psi) = \iota_\phi(\hat{\psi}), \qquad \psi \in \mathcal{S}.$$

Motivated by this, we *define* the Fourier transform of an arbitrary tempered distribution $u \in \mathcal{S}'$ by

$$(\mathcal{F}u)(\psi) = u(\hat{\psi}), \qquad \psi \in \mathcal{S}.$$

It is easy to check that $\hat{u} = \mathcal{F}u$ is indeed a tempered distribution, and as observed above, this definition is consistent with the original one if u is a tempered distribution given by a Schwartz function ϕ (or one with enough decay at infinity). It is also easy to see that the Fourier transform, when thus extended to a map $\mathcal{F} : \mathcal{S}' \to \mathcal{S}'$, still has the standard properties, e.g. $\mathcal{F}(D_{x_j} u) = \xi_j \mathcal{F}u$. Indeed, by definition, for all $\psi \in \mathcal{S}$,

$$(\mathcal{F}(D_{x_j}u))(\psi) = (D_{x_j}u)(\mathcal{F}\psi) = -u(D_{x_j}\mathcal{F}\psi)$$
$$= u(\mathcal{F}(\xi_j\psi)) = (\mathcal{F}u)(\xi_j\psi) = (\xi_j\mathcal{F}u)(\psi),$$

finishing the proof.

The inverse Fourier transform of a tempered distribution is defined analogously,

$$\mathcal{F}^{-1}u(\psi) = u(\mathcal{F}^{-1}\psi),$$

and it satisfies

$$\mathcal{F}^{-1}\mathcal{F} = \mathrm{Id} = \mathcal{F}\mathcal{F}^{-1}$$

on tempered distributions as well. Again, this is an immediate consequence of the corresponding properties for $\mathcal{S}$, for

$$(\mathcal{F}^{-1}\mathcal{F}u)(\psi) = \mathcal{F}u(\mathcal{F}^{-1}\psi) = u(\mathcal{F}\mathcal{F}^{-1}\psi) = u(\psi).$$

As an example, we find the Fourier transform of the distribution $u = \iota_1$ given by the constant function 1. Namely, for all $\psi \in \mathcal{S}$,

$$\hat{u}(\psi) = u(\hat{\psi}) = \int_{\mathbb{R}^n} \hat{\psi}(x)\,dx = (2\pi)^n \mathcal{F}^{-1}(\hat{\psi})(0) = (2\pi)^n \psi(0) = (2\pi)^n \delta_0(\psi).$$

Here the first equality is from the definition of the Fourier transform of a tempered distribution, the second from the definition of u, the third by realizing that the integral of any function ϕ (in this case $\phi = \hat{\psi}$) is just $(2\pi)^n$ times its inverse Fourier transform evaluated at the origin (directly from the definition of $\mathcal{F}^{-1}$ as an integral), the fourth from $\mathcal{F}^{-1}\mathcal{F} = \mathrm{Id}$ on Schwartz functions, and the last from the definition of the delta distribution. Thus, $\mathcal{F}u = (2\pi)^n \delta_0$, which is often written as $\mathcal{F}1 = (2\pi)^n \delta_0$. Similarly, the Fourier transform of the tempered distribution u given by the function $f(x) = e^{ix \cdot a}$, where $a \in \mathbb{R}^n$ is a fixed constant, is given by $(2\pi)^n \delta_a$ since

$$\hat{u}(\psi) = u(\hat{\psi}) = \int_{\mathbb{R}^n} e^{ix \cdot a} \hat{\psi}(x)\, dx = (2\pi)^n \mathcal{F}^{-1}(\hat{\psi})(a)$$
$$= (2\pi)^n \psi(a) = (2\pi)^n \delta_a(\psi),$$

while its inverse Fourier transform is given by δ_{-a} since

$$\mathcal{F}^{-1}u(\psi) = u(\mathcal{F}^{-1}\psi) = \int_{\mathbb{R}^n} e^{ix \cdot a} \mathcal{F}^{-1}\psi(x)\, dx = \mathcal{F}(\mathcal{F}^{-1}\psi)(-a)$$
$$= \psi(-a) = \delta_{-a}(\psi).$$

We can also perform analogous calculations on δ_b, $b \in \mathbb{R}^n$:

$$\mathcal{F}\delta_b(\psi) = \delta_b(\mathcal{F}\psi) = (\mathcal{F}\psi)(b) = \int e^{-ix \cdot b} \psi(x)\, dx,$$

i.e. the Fourier transform of δ_b is the tempered distribution given by the function $f(x) = e^{-ix \cdot b}$. With $b = -a$, the previous calculations confirm what we knew anyway, namely that $\mathcal{F}\mathcal{F}^{-1}f = f$ (for this particular f).

3. The wave equation and the Fourier transform

We can now use tempered distributions to solve the wave equation on $\mathbb{R}^n$ – Laplace's equation can also be solved this way. This is done in the problems at the end of the chapter. Thus, consider the PDE

$$(\partial_t^2 - c^2 \Delta)u = 0, \ u(0, x) = \phi(x), \ u_t(0, x) = \psi(x).$$

Take the partial Fourier transform $\hat{u}$ of u in x to get

$$(\partial_t^2 + c^2 |\xi|^2)\hat{u} = 0, \ \hat{u}(0, \xi) = \mathcal{F}\phi(\xi), \ \hat{u}_t(0, \xi) = \mathcal{F}\psi(\xi).$$

For each $\xi \in \mathbb{R}^n$ this is an ODE that is easy to solve, with the result that

$$\hat{u}(t, \xi) = \cos(c|\xi|t)\mathcal{F}\phi(\xi) + \frac{\sin(c|\xi|t)}{c|\xi|}\mathcal{F}\psi(\xi).$$

Thus,

$$u = \mathcal{F}_\xi^{-1}\left(\cos(c|\xi|t)\mathcal{F}\phi(\xi) + \frac{\sin(c|\xi|t)}{c|\xi|}\mathcal{F}\psi(\xi)\right).$$

This can be rewritten in terms of convolutions, namely

$$u(t, x) = \mathcal{F}_\xi^{-1}(\cos(c|\xi|t)) *_x \phi + \mathcal{F}_\xi^{-1}\left(\frac{\sin(c|\xi|t)}{c|\xi|}\right) *_x \psi(\xi),$$

so it remains to evaluate the inverse Fourier transforms of these explicit functions. We only do this in $\mathbb{R}$ (i.e. $n = 1$) here; the case $n = 3$ is done in the problems at the end of the chapter.

Here we need to be a little careful, as we might be taking the convolution of two distributions in principle! However, any distribution can be convolved with elements of $C_c^\infty(\mathbb{R}^n)$. Indeed, if $f \in C(\mathbb{R}^n)$, $\phi \in C_c^\infty(\mathbb{R}^n)$, then

$$f * \phi(x) = \int f(y)\phi(x-y)\,dy = \iota_f(\phi_x),$$

where we write $\phi_x(y) = \phi(x-y)$. Note that $f * \phi \in C^\infty(\mathbb{R}^n)$, as differentiation under the integral sign shows (the derivatives fall on ϕ!). We make the consistent definition for $u \in \mathcal{D}'(\mathbb{R}^n)$ that

$$(u * \phi)(x) = u(\phi_x),$$

so $u * \phi \in C^\infty(\mathbb{R}^n)$ since

$$\partial_{x_j}(u * \phi) = u(\partial_{x_j}\phi_x),$$

in analogy with differentiation under the integral sign. As an example,

$$\delta_a * \phi(x) = \delta_a(\phi_x) = \phi(x-a).$$

With some work one can even make sense of convolving distributions, as long as one of them has compact support. However, we do not pursue this here, as we shall see directly that our formula even makes sense for distributions. We also note that for $f(x) = H(a - |x|)$, $a > 0$, where H is the Heaviside step function (so $H(s) = 1$ for $s \geq 0$, $H(s) = 0$ for $s < 0$),

$$f * \phi(x) = \int H(a - |y|)\phi(x-y)\,dx = \int_{-a}^a \phi(x-y)\,dy = \int_{x-a}^{x+a} \phi(s)\,ds,$$

where we write $s = x - y$.

Returning to the actual transforms (using that cos is even, sin is odd),

$$\mathcal{F}_\xi^{-1}(\cos(c\xi t)) = \frac{1}{2}(\mathcal{F}_\xi^{-1} e^{ict\xi} + \mathcal{F}_\xi^{-1} e^{-ict\xi}) = \frac{1}{2}(\delta_{-ct} + \delta_{ct}),$$

while (note that $\xi^{-1}\sin(\xi ct)$ is continuous, indeed C^∞, at $\xi = 0$!) from Problem 8.3,

$$\mathcal{F}_x H(ct - |x|) = \frac{2}{\xi}\sin(ct\xi),$$

so

$$\mathcal{F}_\xi^{-1}(c^{-1}\xi^{-1}\sin(c\xi t)) = \frac{1}{2c}H(ct - |x|).$$

In summary,

$$u(t,x) = \frac{1}{2}(\delta_{-ct} + \delta_{ct}) *_x \phi + \frac{1}{2c}H(ct - |x|) *_x \psi$$

$$= \frac{1}{2}(\phi(x - ct) + \phi(x + ct)) + \frac{1}{2c}\int_{x-ct}^{x+ct} \psi(y)\,dy,$$

so we recover d'Alembert's formula.

4. More on tempered distributions

In order to have more examples of tempered distributions, note that any distribution that 'vanishes at infinity' should be tempered. Of course, we need to make sense of this. We recall that the support of a continuous function f, denoted by $\operatorname{supp} f$, is the closure of the set $\{x : f(x) \neq 0\}$. Thus, $x \notin \operatorname{supp} f$ if and only if there exists a neighborhood U of x such that $y \in U$ implies $f(y) = 0$. Thus, $\operatorname{supp} f$ is closed by definition; so for continuous functions on $\mathbb{R}^n$, it is compact if and only if it is bounded.

The support of a distribution u is defined similarly. One says that $x \notin \operatorname{supp} u$ if there exists a neighborhood U of x such that on U, u is given by the zero function. That is, $x \notin \operatorname{supp} u$ if there exists U as above such that for all $\phi \in \mathcal{C}_c^\infty$ with $\operatorname{supp} \phi \subset U$, $u(\phi) = 0$. For example, if $u = \delta_a$ is the delta distribution at a, then $\operatorname{supp} u = \{a\}$, since $u(\phi) = \phi(a)$, so if $x \neq a$, taking U as a neighborhood of x that is disjoint from a, $u(\phi) = 0$ follows for all $\phi \in \mathcal{C}_c^\infty$ with $\operatorname{supp} \phi \subset U$. With this definition of the support one in fact concludes that if $u \in \mathcal{D}'$, $\phi \in \mathcal{C}_c^\infty$ and $\operatorname{supp} u \cap \operatorname{supp} \phi = \emptyset$, then $u(\phi) = 0$.

Note that if u is a distribution and $\operatorname{supp} u$ is compact, u, which is a priori a map $u : \mathcal{C}_c^\infty \to \mathbb{C}$, extends to a map $u : \mathcal{C}^\infty(\mathbb{R}^n) \to \mathbb{C}$, i.e. $u(\phi)$ is naturally defined if ϕ is just smooth, and does not have compact support. To see this, let $f \in \mathcal{C}_c^\infty$ be identically one in a neighborhood of $\operatorname{supp} u$, and for $\phi \in \mathcal{C}^\infty(\mathbb{R}^n)$ define $u(\phi) = u(f\phi)$, noting that $f\phi \in \mathcal{C}_c^\infty$. (The existence of such f follows from a partition of unity type construction; see Problem 5.2.) If $u = \iota_g$ is given by integration against a continuous function g of compact support, this just says that we defined for $\phi \in \mathcal{C}^\infty(\mathbb{R}^n)$

$$\iota_g(\phi) = \int_{\mathbb{R}^n} g(x) f(x) \phi(x)\, dx = \int_{\mathbb{R}^n} g(x) \phi(x)\, dx,$$

which is of course the standard definition if ϕ had compact support. Note that the second equality above holds since we are assuming that f is identically 1 on $\operatorname{supp} g$, i.e. wherever f is not 1, g necessarily vanishes. We should still check that the definition of the extension of u does not depend on the choice of f (which follows from the above calculation if u is given by a continuous function g). But this can be checked easily, for if f_0 is another function in $\mathcal{C}_c^\infty$ which is identically one on $\operatorname{supp} u$, then we need to make sure that $u(f\phi) = u(f_0\phi)$ for all $\phi \in \mathcal{C}^\infty(\mathbb{R}^n)$, i.e. that $u((f - f_0)\phi) = 0$ for all $\phi \in \mathcal{C}^\infty(\mathbb{R}^n)$. But $f = f_0 = 1$ on a neighborhood of $\operatorname{supp} u$, so $(f - f_0)\phi$ vanishes there, hence $u((f - f_0)\phi) = 0$ indeed.

Moreover, any distribution u of compact support, e.g. δ_a for $a \in \mathbb{R}^n$, is tempered. Indeed, $\psi \in \mathcal{S}$ certainly implies that $\psi \in \mathcal{C}^\infty(\mathbb{R}^n)$, so $u(\psi)$ is defined, and it is easy to check that this gives a tempered distribution.

In particular, $\delta_a(\psi) = \psi(a)$ is a tempered distribution, as we have already seen.

Note that the Fourier transform of a compactly supported distribution can be calculated directly. Indeed, $g_\xi(x) = e^{-ix\cdot\xi}$ is a $\mathcal{C}^\infty$ function (of x), and compactly supported distributions can be evaluated on these. Thus, we can define $\mathcal{F}u$ as the tempered distribution given by the function $\xi \mapsto u(g_\xi)$. For example, if $u = \delta_b$, then $\mathcal{F}u$ is given by the function $\delta_b(g_\xi) = g_\xi(b) = e^{i\xi\cdot b}$ in accordance with our previous calculation. Of course, if u is given by a continuous function f of compact support, then $u(g_\xi) = \int f(x)g_\xi(x)\,dx = \int e^{-ix\cdot\xi}f(x)\,dx = (\mathcal{F}f)(\xi)$; indeed, this motivated the definition of $\mathcal{F}u$. This definition is also consistent with the general one for tempered distributions, as we have seen in the particular example of delta distributions. Indeed, in general,

$$\int u(g_\xi)\phi(\xi)\,d\xi = u\left(\int g_\xi\,\phi(\xi)\,d\xi\right) = u(\mathcal{F}\phi) = (\mathcal{F}u)(\phi).$$

The fact that for compactly supported distributions u, $\mathcal{F}u$ is given by $\xi \mapsto u(g_\xi)$, shows directly that for such u, $\mathcal{F}u$ is given by a $\mathcal{C}^\infty$ function, $u(g_\xi) = u(e^{-ix\cdot\xi})$, and differentiating this with respect to ξ simply differentiates g_ξ, i.e. simply gives another exponential (times a linear function), which is still $\mathcal{C}^\infty$.

Problems

Problem 9.1.

(i) Show that for $\phi \in C^0(\mathbb{R}^n)$ with $(1 + |x|)^N\phi(x)$ bounded for some $N > n$,

$$\overline{\mathcal{F}\phi(\xi)} = (2\pi)^n(\mathcal{F}^{-1}\overline{\phi})(\xi),$$

where the bar denotes complex conjugation.

(ii) Show the *Parseval/Plancherel formula*, namely that for $\phi, \psi \in \mathcal{S}(\mathbb{R}^n)$,

$$\int_{\mathbb{R}^n} \phi(x)\,\overline{\psi(x)}\,dx = (2\pi)^{-n}\int_{\mathbb{R}^n}(\mathcal{F}\phi)(\xi)\,\overline{(\mathcal{F}\psi)(\xi)}\,d\xi,$$

and hence conclude that

$$\int_{\mathbb{R}^n} |(\mathcal{F}\phi)(\xi)|^2\,d\xi = (2\pi)^n\int_{\mathbb{R}^n} |\phi(x)|^2\,dx.$$

Hint: Use part (i) to write $\overline{\mathcal{F}\psi(\xi)}$ as $\mathcal{F}^{-1}\chi(\xi)$ for some Schwartz function χ, and then use (9.1).

(iii) Use this result to show that for solutions (with initial condition given by ψ an arbitrary Schwartz function) of the Schrödinger equation, as in Problem 8.9, if $\int_{\mathbb{R}^n} |\psi(x)|^2 \, dx = 1$, then

$$\int_{\mathbb{R}^n} |u(x,t)|^2 \, dx = 1$$

for all t, i.e. 'total probability is conserved'.

Problem 9.2.

(i) On $\mathbb{R}^3$, find the Fourier transform of the function $g(x) = |x|^{-1}$. (*Hint:* To do this efficiently, consider $g(x)$ as the limit of $g_a(x) = e^{-a|x|}|x|^{-1}$, and use your result from Problem 8.4.)

(ii) Solve $\Delta u = f$ on $\mathbb{R}^3$, where $f \in S(\mathbb{R}^3)$, writing your answer as a convolution.

Problem 9.3. Compute the Fourier transform of the tempered distribution on $\mathbb{R}^n$ given by the function $f(x) = e^{ib|x|^2}$ and the inverse Fourier transform of the tempered distribution given by the function $g(\xi) = e^{ic|x|^2}$, where $b, c \in \mathbb{R}$. Make sure you specify the choice of the square root of any non-positive number you are taking.

Hint: Write f as the limit of $e^{(ib-\epsilon)|x|^2}$, $\epsilon > 0$, as $\epsilon \to 0$; cf. Problem 8.5.

Problem 9.4. Using Problem 9.3, find an explicit formula for the solution of the Schrödinger equation from Problem 8.9, writing your solution in a form that does not involve the Fourier transform.

Problem 9.5. Find the Fourier transform on $\mathbb{R}^3$ of the distribution $u = \delta_{|x|-R}$, i.e. for $\psi \in S(\mathbb{R}^3)$,

$$u(\psi) = R^2 \int_{\mathbb{S}^2} \psi(R\omega) \, dS(\omega),$$

or in spherical coordinates,

$$u(\psi) = R^2 \int_0^\pi \int_0^{2\pi} \psi(R\cos\theta, R\sin\theta\cos\phi, R\sin\theta\sin\phi) \sin\theta \, d\phi \, d\theta.$$

(*Hint:* Use the fact that u is compactly supported, so you can evaluate the Fourier transform directly by applying it to $e^{-ix\cdot\xi}$, and use spherical coordinates centered around ξ.)

Problem 9.6. Write the solution of the wave equation on $\mathbb{R}^3_x \times \mathbb{R}_t$,

$$u_{tt} = c^2 \Delta_x u, \quad u(x,0) = \phi(x), \quad u_t(x,0) = \psi(x),$$

in a form that does not involve the Fourier transform by using convolutions.

Hint: Use the previous problem to deal with the ψ-term. To deal with the ϕ term, note that

$$\cos(c|\xi|t) = \frac{\partial}{\partial t}\left(\frac{\sin(c|\xi|t)}{c|\xi|}\right).$$

Problem 9.7. In quantum mechanics, the Pauli matrices are

$$\sigma_x = \begin{pmatrix} 0 & 1 \\ 1 & 0 \end{pmatrix}, \ \sigma_y = \begin{pmatrix} 0 & -i \\ i & 0 \end{pmatrix}, \ \sigma_z = \begin{pmatrix} 1 & 0 \\ 0 & -1 \end{pmatrix}.$$

The corresponding modified momentum operator for spin systems is

$$P = -i\hbar\sigma_x\partial_x - i\hbar\sigma_y\partial_y - i\hbar\sigma_z\partial_z.$$

(i) Solve the following equation on $\mathbb{R}^3_{x,y,z} \times \mathbb{R}_t$,

$$i\hbar\begin{pmatrix} u_t \\ v_t \end{pmatrix} = P\begin{pmatrix} u \\ v \end{pmatrix}$$

with initial condition

$$\begin{pmatrix} u(x,y,z,0) \\ v(x,y,z,0) \end{pmatrix} = \begin{pmatrix} \phi(x,y,z) \\ \psi(x,y,z) \end{pmatrix},$$

ϕ, ψ given Schwartz functions, using the Fourier transform. Write your answer in a form that does not involve the Fourier transform by using convolutions.

   Hint: Solve the ODE on the Fourier transform side by exponentiating a matrix (cf. Chapter 3), and rewrite it by separating the even and odd order terms in the Taylor series of the exponential, comparing it with Problem 9.6.

(ii) Solve the equation differently by showing that (u, v) solves the PDE

$$-\hbar^2\begin{pmatrix} u_{tt} \\ v_{tt} \end{pmatrix} = P^2\begin{pmatrix} u \\ v \end{pmatrix},$$

   where

$$P^2\begin{pmatrix} u \\ v \end{pmatrix} = P\left[P\begin{pmatrix} u \\ v \end{pmatrix}\right],$$

   with appropriate initial conditions, and compute P^2 to see that it is a scalar operator in the sense of (8.13).

Problem 9.8. Show that if $u \in \mathcal{S}'(\mathbb{R}^n)$, then there is an integer $m \geq 0$ and $C > 0$ such that for all $\phi \in \mathcal{S}(\mathbb{R}^n)$,

$$|u(\phi)| \leq C\|\phi\|_m,$$

where

$$\|\phi\|_m = \sum_{|\alpha|\leq m, |\beta|\leq m} \sup_{x\in\mathbb{R}^n} |x^\alpha\partial_x^\beta\phi|.$$

Hints: This relies on the continuity of u as a map $u : \mathcal{S} \to \mathbb{C}$. So suppose for the sake of contradiction that no such m and C exist; in particular, for an integer $j > 0$, $m = j$ and $C = j$ do not work, i.e. there exists $\phi_j \in \mathcal{S}$ such that

$$|u(\phi_j)| > j\|\phi_j\|_j.$$

Note that ϕ_j cannot be 0 (for then $u(\phi_j)$ would vanish by linearity). Let $\psi_j = \frac{1}{j\|\phi_j\|_j}\phi_j$, so $\psi_j \in \mathcal{S}$, $\|\psi_j\|_j = \frac{1}{j}$ and

$$|u(\psi_j)| > j\|\psi_j\|_j = 1.$$

Now show that $\psi_j \to 0$ in $\mathcal{S}$ as $j \to \infty$, and use this to get a contradiction with the continuity of u.

Problem 9.9. Suppose that $u \in \mathcal{S}'(\mathbb{R}^n)$ and u has compact support. (This means that there is a function $f \in \mathcal{C}_c^\infty(\mathbb{R}^n)$ such that $u = fu$; namely, one would take f identically 1 on a neighborhood of the support of u.) Show that the $\mathcal{C}^\infty$ function $\mathcal{F}u$ satisfies

$$|(\mathcal{F}u)(\xi)| \le C(1 + |\xi|)^m$$

for some C and m.

Hint: Use the result of the previous problem and the fact that $(\mathcal{F}u)(\xi) = u(\phi_\xi)$, $\phi_\xi(x) = f(x)e^{-ix\cdot\xi}$.

Problem 9.10. Suppose that $u, v \in \mathcal{S}'(\mathbb{R}^n)$ and v has compact support. Give a definition of $u * v \in \mathcal{S}'(\mathbb{R}^n)$ that is consistent with the definition if one of the two distributions is in $\mathcal{S}(\mathbb{R}^n)$.

Problem 9.11. Show that every $u \in \mathcal{S}'(\mathbb{R}^n)$ can be approximated by elements of $\mathcal{S}(\mathbb{R}^n)$, i.e. show that there exist $f_j \in \mathcal{S}(\mathbb{R}^n)$ such that $\iota_{f_j} \to u$ in $\mathcal{S}'(\mathbb{R}^n)$. Here we recall that $u_j \to u$ in $\mathcal{S}'(\mathbb{R}^n)$ means that $u_j(\phi) \to u(\phi)$ for all $\phi \in \mathcal{S}(\mathbb{R}^n)$.

Hints: It suffices to show that there is f_j and m such that $|\iota_{f_j}(\phi) - u(\phi)| \le j^{-1}\|\phi\|_m$ for all $\phi \in \mathcal{S}$; see Problem 9.8 for the notation. So first consider $v_j = \chi_j u$, where $\chi_j(x) = \chi(x/j)$, and $\chi \in \mathcal{C}_c^\infty(\mathbb{R}^n)$ is identically 1 for $|x| < 1$ and identically 0 for $|x| > 2$. Thus, v_j are compactly supported distributions, and show that $v_j \to u$ in $\mathcal{S}'(\mathbb{R}^n)$ in the strong sense that for some $\tilde{m}$, $|v_j(\phi) - u(\phi)| \le \tilde{C}j^{-1}\|\phi\|_{\tilde{m}}$. Now to get the f_j, we approximate the v_j as follows: $\mathcal{F}v_j$ is $\mathcal{C}^\infty$, but does not decay at infinity. So let $g_{jk}(\xi) = \chi_k(\xi)(\mathcal{F}v_j)(\xi)$, χ_j as above. Show that $g_{jk} \in \mathcal{S}(\mathbb{R}^n)$ and that $\iota_{g_{jk}} \to \mathcal{F}v_j$ in $\mathcal{S}'(\mathbb{R}^n)$ as $k \to \infty$. Now consider $\mathcal{F}^{-1}g_{jk}$.

Problem 9.12. Suppose that $P = \sum_{|\alpha| \le m} a_\alpha D^\alpha$ on $\mathbb{R}^n$. A *fundamental solution* for P is a distribution E such that $PE = \delta_0$. E is also called a *Green's function* with pole at 0.

(i) Show that if E is a fundamental solution for P, then for $f \in C_c^\infty(\mathbb{R}^n)$, $u = E * f$ solves $Pu = f$.

(ii) Show that the same holds even if $f \in C_c^0(\mathbb{R}^n)$ (or indeed $f \in \mathcal{D}'(\mathbb{R}^n)$ with compact support). You may assume that $E \in \mathcal{S}'(\mathbb{R}^n)$ to use your results from Problem 9.10, if you wish (though this is not strictly necessary).

(iii) Show that the distribution E given by the function $\frac{-1}{4\pi|x|}$ is a fundamental solution for Δ in $\mathbb{R}^3$ in two different ways: using the Fourier transform, and directly.

   Hint: For the direct calculation, to find ΔE, recall that

$$\Delta E(\phi) = E(\Delta\phi)$$

and write the right-hand side as $-\lim_{\epsilon \to 0} \int_{|x| \geq \epsilon} \frac{1}{4\pi|x|} \Delta\phi(x)\,dx$, and use the divergence theorem (integrate by parts in polar coordinates).

Problem 9.13. The goal of this problem is to show that if $u \in \mathcal{D}'(\mathbb{R}^3)$ and $\Delta u = f$ satisfies $x_0 \notin \operatorname{singsupp} f$, i.e. f is C^∞ near x_0, then u is C^∞ near x_0. This is called elliptic regularity: Δ is elliptic, and for an elliptic operator P, if Pu is C^∞ near some x_0, then so is u.

We achieve this as follows:

(i) First suppose that u is a C^2 function. Let $\phi \in C_c^\infty(\mathbb{R}^3)$ be identically 1 near x_0 such that f is C^∞ on $\operatorname{supp}\phi$. Then show that $\Delta(\phi u) = \phi \Delta u + v$, where v is a compactly supported distribution that vanishes near x_0. Now as $w = \phi u$ is compactly supported,

$$w(x) = -\int_{\mathbb{R}^3} \frac{1}{4\pi|x-y|} \Delta_y(\phi(y)u(y))\,dy.$$

(ii) Expand $\Delta_y(\phi(y)u(y))$ as above. To analyze

$$\int_{\mathbb{R}^3} \frac{1}{4\pi|x-y|} v(y)\,dy$$

for x near x_0, note that if x is near x_0 and $y \in \operatorname{supp} v$, then $x \neq y$, so $|x-y|^{-1}$ is C^∞. On the other hand, $\phi\Delta u$ is C^∞ by assumption. Write the corresponding part of the convolution as

$$\int_{\mathbb{R}^3} \frac{1}{4\pi|y|} \phi(x-y)(\Delta u)(x-y)\,dy,$$

and deduce that it is C^∞.

(iii) Suppose now that $u \in \mathcal{D}'$. Proceed as above, writing

$$w = -\frac{1}{4\pi|x|} * (\Delta(\phi u)),$$

convolution in the sense of distributions (so w is merely a distribution), and show that both parts are C^∞ near x_0. You do not have to be very careful in writing up this part; there are some technicalities, but the point is to get the main idea.

Problem 9.14. We work out a version of *elliptic regularity* for general elliptic P here, e.g. $P = \Delta^2$. We start with a definition:

$$P = \sum_{|\alpha| \leq m} a_\alpha D^\alpha$$

is *elliptic* if its full symbol, $p(\xi) = \sum_{|\alpha| \leq m} a_\alpha \xi^\alpha$, satisfies the following: there are $c > 0$ and $R > 0$ such that

$$|\xi| \geq R \Rightarrow |p(\xi)| \geq c|\xi|^m.$$

Thus, p is a polynomial such that $|p|$ has comparable upper and lower bounds near infinity: $c|\xi|^m \leq |p(\xi)| \leq C|\xi|^m$.

(i) Show that $\Delta + \lambda$ and $\Delta^2 + \lambda$ are elliptic for any $\lambda \in \mathbb{C}$.

(ii) Show that if P is elliptic, $u \in \mathcal{S}'(\mathbb{R}^n)$ and $Pu = f \in \mathcal{S}(\mathbb{R}^n)$, then $u \in C^\infty(\mathbb{R}^n)$.

   Hint: Write $p\mathcal{F}u = \mathcal{F}f \in \mathcal{S}(\mathbb{R}^n)$. Let $\chi \in C_c^\infty(\mathbb{R}^n)$ be such that $\chi(\xi) = 1$ if $|\xi| \leq R$ and $\chi(\xi) = 0$ if $|\xi| \geq 2R$. Now decompose $\mathcal{F}u = \chi \mathcal{F}u + (1 - \chi)\mathcal{F}u$. For the first part just use the fact that $\chi \mathcal{F}u$ is compactly supported and for the second part use $p\mathcal{F}u = \mathcal{F}f \in \mathcal{S}(\mathbb{R}^n)$.

Note: The full result is as above: if $u \in \mathcal{D}'(\mathbb{R}^n)$ and Pu is C^∞ near x_0, then u is C^∞ at x_0. This requires a little more work, but you don't need to show this. One way to think about this is that the approximate Green's function $G = \mathcal{F}^{-1}((1 - \chi)p^{-1})$ satisfies that G is C^∞ away from 0, much like $\frac{-1}{4\pi|x|}$ was in the previous problem.

Chapter 10

PDE and boundaries

We have used the Fourier transform and other tools (factoring the PDE) to solve PDEs on $\mathbb{R}^n$. We now study how we can use these results to solve problems on the half space, or indeed on intervals, cubes, etc.

1. The wave equation on a half space

As shown in the problems in Chapter 7, the solution of the wave equation on $\mathbb{R}^n$ (so u is a function on $\mathbb{R}^n_x \times \mathbb{R}_t$) is even, resp. odd, in x_n if the initial conditions and the inhomogeneity are even, resp. odd, in x_n. That is, write $x = (x', x_n)$ where $x' = (x_1, \ldots, x_{n-1})$. The wave equation is

$$u_{tt} - c^2 \Delta u = f, \ u(x,0) = \phi(x), \ u_t(x,0) = \psi(x).$$

If

$$f(x', x_n, t) = \pm f(x', -x_n, t), \ \phi(x', x_n) = \pm \phi(x', -x_n), \ \psi(x', x_n)$$
$$= \pm \psi(x', -x_n)$$

for all x and t, i.e. if f, ϕ, ψ are all even $(+)$, resp. odd $(-)$, functions of x_n (so every $\pm$ is the same, all $+$ or all $-$, i.e. always the upper or always the lower sign is used), then u is an even, resp. odd, function of x_n as well, i.e.

$$u(x', x_n, t) = \pm u(x', -x_n, t).$$

Recall that this was based on considering $u(x', x_n, t) \mp u(x', -x_n, t)$ and showing that it solves the homogeneous wave equation with 0 initial conditions. The problems in Chapter 7 also showed that if u is continuous and is an odd function of x_n, then $u(x', 0, t) = 0$ for all x' and t, while if u is a C^1 and is an even function of x_n, then $\partial_{x_n} u(x', 0, t) = 0$ for all x' and t.

These observations reduce the solution of the wave equation in $x_n > 0$ with either Dirichlet or Neumann boundary condition to solving the PDE on all of $\mathbb{R}^n$. For the sake of definiteness, suppose we want to solve the Dirichlet problem:

$$u_{tt} - c^2 \Delta u = f, \ x_n \geq 0,$$

$$u(x', 0, t) = 0 \text{ (DBC)},$$

$$u(x, 0) = \phi(x), \ u_t(x, 0) = \psi(x), \ x_n \geq 0 \text{ (IC)}.$$

Here f, ϕ and ψ are given functions, defined in $x_n \geq 0$ only. To solve the PDE, we consider the *odd extensions* of f, ϕ, ψ in x_n, i.e. *define*

$$f_{\text{odd}}(x', x_n, t) = \begin{cases} f(x', x_n, t), & x_n \geq 0, \\ -f(x', -x_n, t), & x_n < 0, \end{cases}$$

and analogously

$$\phi_{\text{odd}}(x', x_n) = \begin{cases} \phi(x', x_n), & x_n \geq 0, \\ -\phi(x', -x_n), & x_n < 0, \end{cases}$$

with a similar definition for ψ. The resulting function is odd and continuous, provided that $f(x', 0, t) = 0 = \phi(x', 0) = \psi(x', 0)$, i.e. if the data are compatible with the boundary condition. Indeed, for $x_n < 0$ then

$$\phi_{\text{odd}}(x', x_n) = -\phi(x', -x_n) = -\phi_{\text{odd}}(x', -x_n),$$

and similarly in all other cases.

Now let v be the solution of the wave equation on $\mathbb{R}^n$ with these odd data:

$$v_{tt} - c^2 \Delta v = f_{\text{odd}},$$

$$v(x, 0) = \phi_{\text{odd}}(x), \ v_t(x, 0) = \psi_{\text{odd}}(x) \text{ (IC)}.$$

As we have seen, v is an odd function of x_n, hence $v(x', 0, t) = 0$ for all x' and t. Now simply let u be the restriction of v to $x_n \geq 0$, so

$$u(x', x_n, t) = v(x', x_n, t), \ x_n \geq 0.$$

Then u solves the PDE and satisfies the initial conditions as well as the Dirichlet boundary conditions, so we have solved our problem!

Concretely, if $n = 1$, we obtain the following result for the solution of the homogeneous wave equation with Dirichlet BC:

$$v(x, t) = \frac{1}{2}(\phi_{\text{odd}}(x + ct) + \phi_{\text{odd}}(x - ct)) + \frac{1}{2c} \int_{x-ct}^{x+ct} \psi_{\text{odd}}(\sigma) \, d\sigma.$$

Then u is the restriction of v to $x \geq 0$, so it remains to work out these formulae in terms of ϕ and ψ themselves. If $x \geq 0$, $t \geq 0$ and $x \geq ct$, then

$x - ct \geq 0$, so we simply have

$$u(x,t) = \frac{1}{2}(\phi(x + ct) + \phi(x - ct)) + \frac{1}{2c} \int_{x-ct}^{x+ct} \psi(\sigma) \, d\sigma, \ x \geq ct \geq 0,$$

i.e. the standard solution formula, which was to be expected in view of the propagation speed of waves: if $x > ct$, the effects of the boundary cannot be felt yet. On the other hand, if $x \geq 0$, $t \geq 0$ and $x < ct$, then $x - ct < 0$ (but $x + ct \geq 0$ still!) and then

$$\begin{aligned}
u(x,t) &= \frac{1}{2}(\phi(x + ct) - \phi(ct - x)) - \frac{1}{2c} \int_{x-ct}^{0} \psi(-\sigma) \, d\sigma + \frac{1}{2c} \int_{0}^{x+ct} \psi(\sigma) \, d\sigma \\
&= \frac{1}{2}(\phi(x + ct) - \phi(ct - x)) + \frac{1}{2c} \int_{ct-x}^{0} \psi(\sigma) \, d\sigma + \frac{1}{2c} \int_{0}^{x+ct} \psi(\sigma) \, d\sigma \\
&= \frac{1}{2}(\phi(x + ct) - \phi(ct - x)) + \frac{1}{2c} \int_{ct-x}^{x+ct} \psi(\sigma) \, d\sigma.
\end{aligned}$$

We can solve the Neumann problem similarly:

$$\begin{aligned}
&u_{tt} - c^2 \Delta u = f, \ x_n \geq 0, \\
&(\partial_{x_n} u)(x', 0, t) = 0 \ \text{(NBC)}, \\
&u(x, 0) = \phi(x), \ u_t(x, 0) = \psi(x), \ x_n \geq 0 \ \text{(IC)}.
\end{aligned}$$

Again, f, ϕ and ψ are given functions, defined in $x_n \geq 0$ only. To solve the PDE, we consider the *even extensions* of f, ϕ, ψ in x_n, i.e. *define*

$$f_{\text{even}}(x', x_n, t) = \begin{cases} f(x', x_n, t), & x_n \geq 0, \\ f(x', -x_n, t), & x_n < 0, \end{cases}$$

and analogously

$$\phi_{\text{even}}(x', x_n) = \begin{cases} \phi(x', x_n), & x_n \geq 0, \\ \phi(x', -x_n), & x_n < 0, \end{cases}$$

with a similar definition for ψ. The resulting function is even and C^1, provided that $f_{x_n}(x', 0, t) = 0 = \phi_{x_n}(x', 0) = \psi_{x_n}(x', 0)$, which again means that the data are compatible with the boundary condition. We again check, e.g. for f this time, that for $x_n < 0$,

$$f_{\text{even}}(x', x_n, t) = f(x', -x_n, t) = f_{\text{even}}(x', -x_n, t),$$

and similarly in all other cases.

We now let v be the solution of the wave equation on $\mathbb{R}^n$ with these even data:

$$\begin{aligned}
&v_{tt} - c^2 \Delta v = f_{\text{even}}, \\
&v(x, 0) = \phi_{\text{even}}(x), \ v_t(x, 0) = \psi_{\text{even}}(x) \ \text{(IC)}.
\end{aligned}$$

Now v is an even function of x_n, and hence $v_{x_n}(x', 0, t) = 0$ for all x' and t. We finally let u be the restriction of v to $x_n \geq 0$, so

$$u(x', x_n, t) = v(x', x_n, t), \ x_n \geq 0.$$

Then u solves the PDE and satisfies the initial conditions as well as the Neumann boundary conditions, so we have solved our problem!

If $n = 1$, a calculation analogous to the DBC gives for the homogeneous PDE in $x - ct < 0$, $x \geq 0$, $t \geq 0$ (the usual formula still holds if $x - ct \geq 0$, so we skip writing it here):

$$u(x, t) = \frac{1}{2}(\phi(x + ct) + \phi(ct - x)) + \frac{1}{2c}\int_{x-ct}^{0} \psi(-\sigma)\, d\sigma + \frac{1}{2c}\int_{0}^{x+ct} \psi(\sigma)\, d\sigma$$

$$= \frac{1}{2}(\phi(x + ct) + \phi(ct - x)) + \frac{1}{c}\int_{0}^{ct-x} \psi(\sigma)\, d\sigma + \frac{1}{2c}\int_{ct-x}^{x+ct} \psi(\sigma)\, d\sigma.$$

The main reason our method worked is that for any function (or indeed distribution) u, letting $u_-(x', x_n, t) = u(x', -x_n, t)$, we have

$$((\partial_t^2 - c^2\Delta)u_-)(x', x_n, t) = ((\partial_t^2 - c^2\Delta)u)(x', -x_n, t).$$

Since

$$(\partial_{x'}^\alpha \partial_t^k u_-)(x', x_n, t) = (\partial_{x'}^\alpha \partial_t^k u)(x', -x_n, t)$$

for all $\alpha \in \mathbb{N}^{n-1}$, $k \in \mathbb{N}$, as these differentiations are not in the x_n variable, this boils down to

$$(\partial_{x_n}^k u_-)(x', x_n, t) = (-1)^k(\partial_{x_n}^k u)(x', -x_n, t),$$

and only even number (in this case $k = 2$) x_n-derivatives of u enter into the wave operator. Thus, for instance, the wave equation for the bi-Laplacian, $(\partial_t^2 + c^2\Delta^2)u = 0$ with appropriate boundary conditions (namely either the function and its second derivative vanishing, in which case we could use odd extensions, or the first and third derivatives vanishing, in which case we could use even extensions), could also be treated by this method.

2. The heat equation on a half space

The heat equation is completely analogous. Thus, to solve, for instance,

$$u_t - k\Delta u = f, \ x_n \geq 0,$$
$$u(x', 0, t) = 0 \ (\text{DBC}),$$
$$u(x, 0) = \phi(x), \ x_n \geq 0 \ (\text{IC}),$$

where f and ϕ are given functions, defined in $x_n \geq 0$ only, we consider the *odd extensions* f_{odd} and ϕ_{odd} of f and ϕ in x_n, and solve

$$v_t - k\Delta v = f_{\text{odd}},$$
$$v(x, 0) = \phi_{\text{odd}}(x) \ (\text{IC}).$$

Once we check that v is an odd function of x_n, hence $v(x', 0, t) = 0$ for all x' and t, we are done as before: by letting u be the restriction of v to $x_n \geq 0$, so

$$u(x', x_n, t) = v(x', x_n, t), \quad x_n \geq 0,$$

then u solves the PDE and satisfies the initial conditions as well as the Dirichlet boundary conditions.

It remains to check that v is indeed an odd function of x_n. This can be done as for the wave equation. Namely, consider $w(x', x_n, t) = v(x', x_n, t) + v(x', -x_n, t)$. Since v solves the inhomogeneous heat equation with odd f, w solves the homogeneous heat equation:

$$\begin{aligned}
(w_t &- k\Delta w)(x, t) \\
&= v_t(x', x_n, t) + v_t(x', -x_n, t) - k(\Delta v)(x', x_n, t) - k(\Delta v)(x', -x_n, t) \\
&= f_{\text{odd}}(x', x_n, t) + f_{\text{odd}}(x', -x_n, t) = 0,
\end{aligned}$$

since f_{odd} is odd. In addition, w has vanishing initial condition:

$$w(x, 0) = v(x', x_n, 0) + v(x', -x_n, 0) = \phi_{\text{odd}}(x', x_n) + \phi_{\text{odd}}(x', -x_n) = 0,$$

as ϕ_{odd} is odd. Assuming, for instance, that $v(x, t) \to 0$ as $|x| \to \infty$, hence the same holds for w, the maximum principle, as in the problems of Chapter 7, shows that w vanishes, so v is indeed odd.

We would still need to check the needed decay of v (assuming decay on f and ϕ), but this follows from our solution formula because of the exponential decay of the Gaussian. In fact, as an alternative, we can see *directly* from the solution formula, at this point for $f = 0$, that the v we constructed is odd. Indeed, the solution formula is

$$v(x, t) = (4\pi k t)^{-n/2} \int_{\mathbb{R}^n} e^{-|x-y|^2/(4kt)} \phi_{\text{odd}}(y) \, dy,$$

so

$$\begin{aligned}
v(x', -x_n, t) &= (4\pi k t)^{-n/2} \int_{\mathbb{R}^n} e^{-|x'-y'|^2/(4kt)} e^{-(-x_n-y_n)^2/(4kt)} \phi_{\text{odd}}(y', y_n) \, dy \\
&= (4\pi k t)^{-n/2} \int_{\mathbb{R}^n} e^{-|x'-y'|^2/(4kt)} e^{-(-x_n+y_n)^2/(4kt)} \phi_{\text{odd}}(y', -y_n) \, dy \\
&= -(4\pi k t)^{-n/2} \int_{\mathbb{R}^n} e^{-|x'-y'|^2/(4kt)} e^{-(x_n-y_n)^2/(4kt)} \phi_{\text{odd}}(y', y_n) \, dy \\
&= -v(x', x_n, t),
\end{aligned}$$

where the second equality is a change of variables from y_n to $-y_n$, and the third equality uses the fact that ϕ_{odd} is odd.

We obtain an explicit solution formula this way (for $x \geq 0$, $t \geq 0$):

$$u(x,t) = -(4\pi kt)^{-n/2} \int_{y_n \leq 0} e^{-|x-y|^2/(4kt)} \phi(y', -y_n) \, dy$$

$$+ (4\pi kt)^{-n/2} \int_{y_n \geq 0} e^{-|x-y|^2/(4kt)} \phi(y) \, dy$$

$$= -(4\pi kt)^{-n/2} \int_{y_n \geq 0} e^{-|x'-y'|^2/(4kt)} e^{-(x_n+y_n)^2/(4kt)} \phi(y) \, dy$$

$$+ (4\pi kt)^{-n/2} \int_{y_n \geq 0} e^{-|x-y|^2/(4kt)} \phi(y) \, dy$$

$$= (4\pi kt)^{-n/2}$$

$$\int_{y_n \geq 0} e^{-|x'-y'|^2/(4kt)} \left(e^{-(x_n-y_n)^2/(4kt)} - e^{-(x_n+y_n)^2/(4kt)} \right) \phi(y) \, dy$$

$$= \int_{y_n \geq 0} G(x,y,t)\phi(y) \, dy,$$

where

$$G(x,y,t) = (4\pi kt)^{-n/2} e^{-|x'-y'|^2/(4kt)} \left(e^{-(x_n-y_n)^2/(4kt)} - e^{-(x_n+y_n)^2/(4kt)} \right)$$

and where the second equality is a change of variable from y_n to $-y_n$. Note that this is still an integral over the space, $y_n \geq 0$ in this case, but the Gaussian has been amended by subtracting its 'reflection' around the $x_n = 0$-plane, $e^{-|x'-y'|^2/(4kt)} e^{-(x_n+y_n)^2/(4kt)}$. It is also instructive to notice that we can see that the boundary condition is indeed satisfied directly from

$$G(x', 0, y, t) = (4\pi kt)^{-n/2} e^{-|x'-y'|^2/(4kt)} \left(e^{-(-y_n)^2/(4kt)} - e^{-y_n^2/(4kt)} \right) = 0$$

for all x', y, t.

Laplace's equation works similarly. Thus, to solve

$$\Delta u = f, \ x_n \geq 0,$$

$$u(x', 0) = 0 \ (\text{DBC}),$$

where f is defined for $x_n \geq 0$ only, we take the odd extension f_{odd} of f to $\mathbb{R}^n$, solve

$$\Delta v = f_{\text{odd}}$$

on $\mathbb{R}^n$, and let u be the restriction of v to $x_n \geq 0$. If f decays at infinity, so does f_{odd}, and by the maximum principle, $\Delta v = f_{\text{odd}}$ has at most one solution that decays at infinity. Explicitly, for $n = 3$, as shown in the problems at the end of Chapter 9, one has

$$v(x) = -\frac{1}{4\pi} \int_{\mathbb{R}^3} |x - y|^{-1} f_{\text{odd}}(y) \, dy,$$

so, letting $x_- = (x', -x_3)$ be the reflection of x around the x_3-plane, one has

$$u(x', x_n) = -\frac{1}{4\pi} \int_{y_3 \geq 0} |x - y|^{-1} f(y) \, dy + \frac{1}{4\pi} \int_{y_3 \leq 0} |x - y|^{-1} f(y', -y_3) \, dy$$

$$= -\frac{1}{4\pi} \int_{y_3 \geq 0} |x - y|^{-1} f(y) \, dy$$

$$+ \frac{1}{4\pi} \int_{y_3 \geq 0} (|x' - y'|^2 + (x_3 + y_3)^2)^{-1/2} f(y', y_3) \, dy$$

$$= -\frac{1}{4\pi} \int_{y_3 \geq 0} |x - y|^{-1} f(y) \, dy$$

$$+ \frac{1}{4\pi} \int_{y_3 \geq 0} (|x' - y'|^2 + (-x_3 - y_3)^2)^{-1/2} f(y', y_3) \, dy$$

$$= -\frac{1}{4\pi} \int_{y_3 \geq 0} (|x - y|^{-1} - |x_- - y|^{-1}) f(y) \, dy$$

$$= \int_{y_3 \geq 0} G(x, y) f(y) \, dy,$$

where

$$G(x, y) = -\frac{1}{4\pi} (|x - y|^{-1} - |x_- - y|^{-1})$$

and where the second equality is a change of variable from y_3 to $-y_3$, while the third is rewriting

$$(|x' - y'|^2 + (x_3 + y_3)^2)^{-1/2} = (|x' - y'|^2 + (-x_3 - y_3)^2)^{-1/2} = |x_- - y|^{-1}.$$

Again, note that for all x' and all y, $G(x', 0, y) = 0$ since $x_- = x$ if $x_3 = 0$, so indeed the boundary condition is satisfied. In physics, $\Delta u = f$ finds the electrostatic potential u associated to a charge distribution f. Correspondingly, considering f_{odd} means that we place imaginary charges in $x_3 < 0$, with opposite signs to those in $x_3 > 0$, and solve the problem in the whole space. Thus, this method is also called the *method of images*.

3. More complex geometries

These results can easily be generalized to quadrants, etc. For instance, to solve the heat equation in the quadrant $x_{n-1}, x_n \geq 0$, write $x = (x', x_{n-1}, x_n)$, $x' = (x_1, \ldots, x_{n-2})$. Impose, for instance, the Dirichlet boundary condition at $x_n = 0$ and the Neumann at $x_{n-1} = 0$, so the PDE is

$$u_t - k\Delta u = f, \quad x_{n-1}, x_n \geq 0,$$

$$(10.1) \qquad u(x', x_{n-1}, 0, t) = 0 \text{ (DBC)}, \quad u(x', 0, x_n, t) = 0 \text{ (NBC)},$$

$$u(x, 0) = \phi(x), \quad x_{n-1}, x_n \geq 0 \text{ (IC)},$$

where f and ϕ are defined if both x_{n-1} and x_n are ≥ 0. The method is then to extend ϕ and f to all of $\mathbb{R}^n_x$ in such a manner that the extensions ϕ_{ext}, resp. f_{ext}, are odd in x_n and even in x_{n-1}. This can be achieved by first extending ϕ (and similarly f) to be even in x_{n-1}, defined now for only $x_n \geq 0$:

$$\phi_{\text{even}}(x', x_{n-1}, x_n) = \begin{cases} \phi(x', x_{n-1}, x_n), & x_{n-1} \geq 0, \\ \phi(x', -x_{n-1}, x_n), & x_{n-1} < 0, \end{cases}$$

and then extending ϕ_{even} to be defined on all of $\mathbb{R}^n$ by making it odd in x_n:

$$\phi_{\text{ext}}(x', x_{n-1}, x_n) = \begin{cases} \phi_{\text{even}}(x', x_{n-1}, x_n), & x_n \geq 0, \\ -\phi_{\text{even}}(x', x_{n-1}, -x_n), & x_n < 0. \end{cases}$$

Of course, we could have done the extensions in the opposite order. Then the extensions ϕ_{ext} and f_{ext} are odd in x_n and even in x_{n-1}, as desired.

We again let v be the solution of the PDE with the extended data on all of $\mathbb{R}^n$:

$$v_t - k\Delta v = f_{\text{ext}},$$
$$v(x, 0) = \phi_{\text{ext}}(x) \ (\text{IC}).$$

It is now even in x_{n-1} and odd in x_n, so letting $u(x, t) = v(x, t)$ if $x_{n-1}, x_n \geq 0$, we have solved the PDE. See Problem 10.1 for an explicit formula.

4. Boundaries and properties of solutions

When solving PDEs in half spaces, quadrants, etc., it is sometimes easier to read off properties of the solution of the PDE directly from the solution v on $\mathbb{R}^n$, rather than trying to work out an explicit formula in the region. For instance, the solution of the homogeneous heat equation on $\mathbb{R}^n \times (0, \infty)$ is C^∞ in $t > 0$, even with merely bounded continuous, or L^1, or tempered distributional, initial data. Hence the same remains true for the heat equation in a half space with either Dirichlet or Neumann initial data.

On the other hand, for the wave equation, singularities propagate along characteristics. If $n = 1$, with either Dirichlet or Neumann boundary conditions, if the original initial condition is C^∞ at some x with $x > 0$, the extension will be C^∞ at x and $-x$; conversely, if the original initial condition is not C^∞ at some x with $x > 0$, the extension will not be C^∞ at x or $-x$. The situation at $x = 0$ is more delicate; e.g. the function $\phi(x) = x^4$, $x \geq 0$, is a perfectly nice C^∞ function, but its odd extension is not: $\phi_{\text{odd}}(x) = -x^4$ for $x < 0$, so $\lim_{x \to 0+} \phi^{(4)}(x) = 24$, while $\lim_{x \to 0-} \phi^{(4)}(x) = -24$, so its 4th derivative is not continuous. This may be avoided, e.g., by assuming that all derivatives of ϕ vanish at 0, or even more strongly, ϕ vanishes identically near 0. Under either of these assumptions, the singularities of the extensions of ϕ lie exactly at x and $-x$ as x runs through the points on the positive half-line at which ϕ is singular. Correspondingly, the solution of the wave equation on $\mathbb{R}_x \times \mathbb{R}_t$ will be singular on the characteristics through the two points $(x, 0)$ and $(-x, 0)$ for each such x. This gives that singularities of solutions *reflect* from the boundary with the usual (equal angle) law of reflection: fixing $x_0 > 0$ at which ϕ is singular, in $t > 0$ the characteristic through $(x_0, 0)$ along which x is decreasing, $x = x_0 - ct$ (or $x + ct = x_0$) hits $x = 0$ at $t = x_0/c$, which is the same place where the characteristic from $(-x_0, 0)$ along which x is increasing, $x = -x_0 + ct$, hits $x = 0$. Thus, restricted to $x \geq 0$, the singularity seems to reflect from the boundary, since we do not 'see' the images that arose by our even or odd extensions; see Figure 10.1.

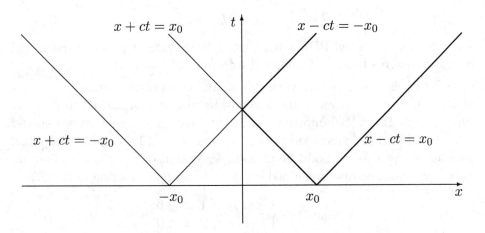

Figure 10.1. Singularities of the initial data at x_0 propagating along reflected characteristics. Note that the region $x < 0$ is 'non-physical', namely it is the image region, so the characteristics shown there are thin.

5. PDE on intervals and cubes

While we conveniently had all our boundaries on a coordinate plane, this is by no means necessary. For instance, consider the PDE

$$u_t = k\Delta_x u, \ x_n \geq L,$$
$$u(x', 0, t) = 0,$$
$$u(x, 0) = \phi(x), \ x_n \geq L.$$

By introducing the new variable, $y = (y', y_n)$ with $y_n = x_n - L$ and $y' = x'$, for the function $\tilde{u}(y', y_n, t) = u(y', y_n + L, t)$, the PDE becomes

$$\tilde{u}_t = k\Delta_y \tilde{u}, \ y_n \geq 0,$$

$$\tilde{u}(y', 0, t) = 0,$$

$$\tilde{u}(y', y_n, 0) = \phi(y', y_n + L), \ y_n \geq 0,$$

which we can solve as above. Then we let $u(x', x_n, t) = \tilde{u}(x', x_n - L, t)$ to get back the solution of the original equation.

Rather than going through this process, we could have obtained the the same result by extending u *directly* to an *odd function about the plane* $x_n = L$. The reflection of a point (x', x_n) about this plane is $(x', L - (x_n - L)) = (x', 2L - x_n)$, so the extension we would have considered is

$$\phi_{\text{ext}}(x', x_n) = \begin{cases} \phi(x', x_n), & x_n \geq L, \\ -\phi(x', 2L - x_n), & x_n < L. \end{cases}$$

We can then solve the PDE on $\mathbb{R}^n$ using these extended initial data, and restrict ourselves to $x_n \geq L$ to get the desired solution.

We can also work on intervals, cubes, etc., using further iterated versions of these methods. As an example, consider the heat equation on $[0, \ell]_x \times (0, \infty)_t$ with Dirichlet boundary condition. We have seen that we should take the odd extension around both $x = 0$ and $x = \ell$. This can be simplified as follows: we take the odd 2ℓ-periodic extension of ϕ. That is, first we extend ϕ, which at first is defined on $[0, \ell]$, to an odd function on $[-\ell, \ell]$:

$$\phi_{\text{odd}}(x) = \begin{cases} \phi(x), & x \geq 0, \\ -\phi(-x), & x < 0. \end{cases}$$

Next, we extend ϕ_{odd} to a 2ℓ-periodic function on $\mathbb{R}$, i.e. to a function ϕ_{ext}, i.e. to a function satisfying

$$\phi_{\text{ext}}(x + 2\ell) = \phi_{\text{ext}}(x), \ x \in \mathbb{R}.$$

This can be done, since any $x \in \mathbb{R}$ can be translated by an integer multiple of 2ℓ so that the result lies in $[-\ell, \ell]$, and the result is a unique point except if we started at an odd multiple of ℓ (in which case both $\pm\ell$ are acceptable results). Moreover, ϕ_{ext} is still odd, for if $x \in \mathbb{R}$ and $x - 2k\ell \in [-\ell, \ell]$, then $-x + 2k\ell = -(x - 2k\ell) \in [-\ell, \ell]$, so the oddness of the 2ℓ-periodic extension reduces to the oddness of ϕ_{odd}. This way we obtain an extension to all of $\mathbb{R}$ which is *odd and 2ℓ-periodic*, and it satisfies

$$\phi_{\text{ext}}(2\ell - x) = \phi_{\text{ext}}(-x) = -\phi_{\text{ext}}(x),$$

where the first equality follows from being 2ℓ-periodic and the second from being odd, so ϕ_{ext} is also odd about $x = \ell$.

Now we solve the heat equation on $\mathbb{R}_x \times (0, \infty)$:

$$v_t = kv_{xx}, \ x \in \mathbb{R},$$
$$v(x, 0) = \phi_{\text{ext}}(x),$$

and let u be the restriction of v to $[0, \ell]_x \times [0, \infty)_t$. Since ϕ_{ext} is odd both about $x = 0$ and $x = \ell$, we deduce that v is also such, and thus $v(0, t) = 0 = v(\ell, t)$; hence u satisfies the Dirichlet boundary condition. One can again work out the formula explicitly, but now one will end up with an infinite sum corresponding to the 2ℓ-periodic nature of the extension.

Solving the wave equation again works similarly. At the qualitative level, solutions on these intervals, cubes, etc., behave just like the solutions on half spaces, quadrants, etc., so in particular for the heat equation we get smoothness of the solutions, and for the wave equation we get reflected singularities following characteristics.

Problems

Problem 10.1. Write the solution of (10.1) as explicitly as possible in terms of f and ϕ.

Problem 10.2. Solve the Schrödinger equation in the half space $x_n \geq 0$ on $\mathbb{R}^n \times \mathbb{R}_t$, with $\mathbb{R}^n = \mathbb{R}^{n-1}_{x'} \times \mathbb{R}_{x_n}$:

$$i\hbar u_t = -\frac{\hbar^2}{2}\Delta_x u, \ x_n \geq 0,$$
$$u(x, 0) = \psi(x), \ x_n \geq 0,$$
$$u(x', 0, t) = 0, \ x' \in \mathbb{R}^{n-1}, \ t \in \mathbb{R},$$

with $\hbar > 0$ fixed. This corresponds to a particle meeting an infinite potential barrier (so the wave function must vanish in $x_n \leq 0$). Write your answer as explicitly in terms of ψ as possible. (See Problem 9.4.)

Problem 10.3. On $\mathbb{R}^3$, write $x = (x', x_3)$, so $x' = (x_1, x_2)$. Suppose f is a compactly supported C^1 function in $x_3 \geq 0$ in $\mathbb{R}^3$ vanishing near $x_3 = 0$. Find the solution u of

$$\Delta u = f, \ x_3 \geq 0,$$
$$\partial_{x_3} u(x', 0) = 0$$

which goes to 0 at infinity. Write your solution as explicitly in terms of f as possible.

Problem 10.4. Suppose f is a compactly supported C^1 function in $\{(x_1, x_2, x_3) \in \mathbb{R}^3 : x_2 \geq 0, \ x_3 \geq 0\}$ vanishing at $x_2 = 0$ and at $x_3 = 0$.

Find the solution u of

$$\Delta u = f, \ x_2, x_3 \geq 0,$$
$$u(x_1, x_2, 0) = 0, \ u(x_1, 0, x_3) = 0,$$

which goes to 0 at infinity. Write your solution as explicitly in terms of f as possible.

Problem 10.5. Although some special conditions are needed, one can also solve PDE in domains with angles that are not right angles, for instance for conic sectors in $\mathbb{R}^2$ with vertex angle $\pi/3$:

Let Ω be the region $\{(x, y) \in \mathbb{R}^2 : \ 0 < y < \sqrt{3}x\}$. Solve the heat equation on $\Omega \times (0, \infty)_t$ with Dirichlet boundary conditions:

$$u_t = k\Delta u \text{ on } \Omega \times (0, \infty)_t,$$
$$u|_{\partial\Omega \times (0,\infty)} = 0,$$
$$u|_{t=0} = \phi,$$

where ϕ is a given function. Write your solution as explicitly in terms of ϕ as possible.

Problem 10.6. Using the method of reflection, in the chapter we solved the wave equation with Dirichlet boundary conditions on the interval $[0, \ell]_x$:

$$u_{tt} - c^2 u_{xx} = 0, \ u(x, 0) = \phi(x), \ u_t(x, 0) = \psi(x), \ u_x(0, t) = 0 = u_x(\ell, t).$$

If $\psi = 0$ and ϕ is C^∞ except at a point $x_0 \in (0, \ell)$, where do you know for sure that u is C^∞? What if ϕ vanishes near $x = 0, \ell$? Draw a picture.

Problem 10.7.

(i) Using the method of reflection, solve the wave equation with Neumann boundary conditions on the interval $[0, \ell]_x$:

$$u_{tt} - c^2 u_{xx} = 0, \ u(x, 0) = \phi(x), \ u_t(x, 0) = \psi(x), \ u_x(0, t) = 0 = u_x(\ell, t).$$

You do not need to write an explicit formula containing only ϕ and ψ; the appropriate extension of ϕ and ψ to $\mathbb{R}$ may appear in the formula.

(ii) If $\psi = 0$ and ϕ is C^∞ except at a point $x_0 \in (0, \ell)$, and vanishes near $x = 0, \ell$, where do you know for sure that u is C^∞? What do you need to assume about the appropriate extension of ϕ to get the same conclusion without the vanishing assumption at $x = 0, \ell$?

Chapter 11

Duhamel's principle

1. The inhomogeneous heat equation

Although we have solved only the homogeneous heat equation on $\mathbb{R}^n$, the same method employed there also solves the inhomogeneous PDE. As an application of these methods, let's solve the heat equation on $(0, \infty)_t \times \mathbb{R}^n_x$:

$$(11.1) \qquad u_t - k\Delta u = f, \ u(0, x) = \phi(x),$$

with $f \in \mathcal{C}^\infty([0, \infty)_t; \mathcal{S}(\mathbb{R}^n_x))$, $\phi \in \mathcal{S}(\mathbb{R}^n)$ (say) given. Recall here that the notation $\mathcal{C}^\infty([0, \infty)_t; \mathcal{S}(\mathbb{R}^n_x))$ was discussed in Chapter 8; see (8.8). Namely, taking the partial Fourier transform in x and writing $\mathcal{F}_x u(t, \xi) = \hat{u}(t, \xi)$ gives

$$\frac{\partial \hat{u}}{\partial t}(t, \xi) + k|\xi|^2 \hat{u}(t, \xi) = \hat{f}(t, \xi), \ \hat{u}(0, \xi) = (\mathcal{F}\phi)(\xi).$$

We again solve the ODE for each fixed ξ. To do so, we multiply through by $e^{k|\xi|^2 t}$,

$$e^{k|\xi|^2 t} \frac{\partial \hat{u}}{\partial t}(t, \xi) + e^{k|\xi|^2 t} k|\xi|^2 \hat{u}(t, \xi) = e^{k|\xi|^2 t} \hat{f}(t, \xi),$$

and realize that the left-hand side is

$$\frac{d}{dt} \left(e^{k|\xi|^2 t} \hat{u}(t, \xi) \right).$$

Thus, integrating from $t = 0$ and using the fundamental theorem of calculus yields

$$e^{k|\xi|^2 t} \hat{u}(t, \xi) - \hat{u}(0, \xi) = \int_0^t e^{k|\xi|^2 s} \hat{f}(s, \xi) \, ds,$$

so

$$\hat{u}(t, \xi) = e^{-k|\xi|^2 t} (\mathcal{F}\phi)(\xi) + \int_0^t e^{-k|\xi|^2 (t-s)} \hat{f}(s, \xi) \, ds.$$

Finally,
(11.2)

$$u(t,x) = \mathcal{F}_\xi^{-1}\left(e^{-k|\xi|^2 t}(\mathcal{F}\phi)(\xi) + \int_0^t e^{-k|\xi|^2(t-s)}\hat{f}(s,\xi)\,ds\right)$$

$$= \mathcal{F}_\xi^{-1}\left(e^{-k|\xi|^2 t}(\mathcal{F}\phi)(\xi)\right) + \int_0^t \mathcal{F}_\xi^{-1}\left(e^{-k|\xi|^2(t-s)}\hat{f}(s,\xi)\right)ds$$

$$= (4\pi kt)^{-n/2}\int_{\mathbb{R}^n} e^{-|x-y|^2/4kt}\phi(y)\,dy$$

$$+ \int_0^t (4\pi k(t-s))^{-n/2}\int_{\mathbb{R}^n} e^{-|x-y|^2/4k(t-s)}f(s,y)\,dy\,ds.$$

Note that the parts of the solution formula corresponding to the initial condition and the forcing term are very similar. In fact, let the solution operator of the homogeneous PDE,

$$u_t - k\Delta u = 0, \quad u(0,x) = \phi(x),$$

be denoted by $S(t)$, so

$$(S(t)\phi)(x) = u(t,x).$$

Then

$$(4\pi k(t-s))^{-n/2}\int_{\mathbb{R}^n} e^{-|x-y|^2/4k(t-s)}f(s,y)\,dy = (S(t-s)f_s)(x),$$

where we let $f_s(x) = f(s,x)$ be the restriction of f to the time s slice. Thus, the solution formula for the inhomogeneous PDE, (11.1), takes the form

(11.3) $$u(t,x) = (S(t)\phi)(x) + \int_0^t (S(t-s)f_s)(x)\,ds.$$

The fact that we can write the solution of the inhomogeneous PDE in terms of the solution of the Cauchy problem for the homogeneous PDE is called *Duhamel's principle*.

Changing the initial condition to one at $t = t_0$, namely

$$u_t - k\Delta u = 0, \quad u(t_0,x) = \phi(x),$$

but with $S(t)$ still as above (solution operator with initial condition at $t = 0$), one similarly has, for $t \geq t_0$,

(11.4) $$u(t,x) = (S(t-t_0)\phi)(x) + \int_{t_0}^t (S(t-s)f_s)(x)\,ds,$$

as can be seen by time translation. Indeed, $v(t,x) = u(t+t_0,x)$ solves the heat equation

$$v_t(t,x) - k(\Delta v)(t,x) = f(t+t_0,x)$$

with initial condition $v(0,x) = \phi(x)$, so by (11.3)

$$v(t,x) = (S(t)\phi)(x) + \int_0^t (S(t-s)f_{s+t_0})(x)\, ds,$$

and then $u(t,x) = v(t-t_0, x)$ is given by substituting $t - t_0$ in place of t into v:

$$u(t,x) = (S(t-t_0)\phi)(x) + \int_0^{t-t_0} (S(t-t_0-s)f_{s+t_0})(x)\, ds.$$

Also changing the variable of integration, namely replacing s by $s' = s + t_0$ in the integral (so s' runs from t_0 to $(t - t_0) + t_0 = t$), gives (11.4) (after renaming s' as s).

Physically one may think of (11.3) as follows. The expression $S(t-s)f_s$ is the solution of the heat equation at time t with initial condition $f(s,x)$ imposed at time s. Thus, we think of the forcing as a superposition (namely, integral) of initial conditions given at times s between 0 (when the actual initial condition is imposed) and time t (when the solution is evaluated). Conversely, one could say that the initial condition amounts to a delta-distributional forcing, $\phi(x)\delta_0(t)$, provided that f vanished for time ≤ 0, and we impose $u(T,x) = 0$ for some $T < 0$ (say, $T = -1$). The delta distribution should be considered as being paired against the characteristic function $[-T,t]$, giving $S(t)\phi$ as a result (at least if $t > 0$). See Problem 11.3.

In fact, this calculation did not depend significantly on particular features of the Laplacian on $\mathbb{R}^n$. That is, suppose that we have another differential operator L on $\mathbb{R}^n$ and consider the PDE

(11.5) $$\partial_t u - Lu = f, \ u(0,x) = \phi(x).$$

Suppose that we can solve the homogeneous problem, i.e. that

$$\partial_t u - Lu = 0, \ u(0,x) = \phi(x),$$

and let $S(t)$ denote the solution operator, so that $(S(t)\phi)(x) = u(t,x)$. Hence

$$(S(0)\phi)(x) = \phi(x),$$

as the initial condition holds, and

$$(\partial_t - L)S(t)\phi = 0,$$

since the PDE holds. We claim that under these assumptions the solution of (11.5) is given by Duhamel's formula,

$$u(t,x) = (S(t)\phi)(x) + \int_0^t (S(t-s)f_s)(x)\, ds,$$

just as above. To see this, first note that the initial condition is certainly satisfied, for the integral vanishes if $t = 0$. Next, as $S(t)$ is the solution operator for the homogeneous PDE, $\partial_t - L$ applied to the first term of Duhamel's formula vanishes. Thus,

$$(\partial_t u - Lu)(t,x) = (\partial_t - L)\int_0^t (S(t-s)f_s)(x)\,ds$$

$$= (S(0)f_t)(x) + \int_0^t (\partial_t - L)(S(t-s)f_s)(x)\,ds$$

$$= f(t,x),$$

where the $S(0)f_t$ term arose by differentiating the upper limit of the integral, while the other term by differentiating under the integral sign, and we used the fact that $S(0)f_t = f_t$ and

$$(\partial_t - L)(S(t-s)f_s) = ((\partial_\tau - L)S(\tau)f_s)|_{\tau=t-s} = 0.$$

We can also deal with inhomogeneous boundary conditions. For instance, we can solve

(11.6)
$$u_t - k\Delta u = f, \quad x_n \geq 0,$$
$$u(x',0,t) = h(x',t) \text{ (DBC)},$$
$$u(x,0) = \phi(x) \text{ (IC)},$$

where f and ϕ are given functions for $x_n \geq 0$, and h is a given function of (x',t). (Now we write t as the last variable!) To solve this, we reduce it to a PDE with homogeneous boundary conditions. Thus, let F be any function on $\mathbb{R}_x^n \times [0,\infty)_t$ such that F satisfies the boundary condition, i.e. $F(x',0,t) = h(x',t)$. For instance, we may take $F(x',0,t) = h(x',t)$. If u solves (11.6), then $v = u - F$ solves

(11.7)
$$v_t - k\Delta v = f - (F_t - k\Delta F), \quad x_n \geq 0,$$
$$v(x',0,t) = 0 \text{ (DBC)},$$
$$v(x,0) = \phi(x) - F(x,0) \text{ (IC)},$$

i.e. a PDE with homogeneous boundary condition, but different initial condition and forcing term. Here we used

$$v_t - k\Delta v = u_t - F_t - k\Delta u + k\Delta F = f - (F_t - k\Delta F).$$

Conversely, if we solve (11.7), then $u = v + F$ will solve (11.6). But we know how to solve (11.7): either use an appropriate (odd) extension of the data to $\mathbb{R}^n$ and solve the PDE there, or use Duhamel's principle and the solution of the homogeneous PDE (with a vanishing right-hand side) on the half space; see the problems at the end of the chapter. Thus, (11.6) can be solved as well.

Similarly, for inhomogeneous Neumann boundary conditions,

$$u_t - k\Delta u = f, \ x_n \geq 0,$$

(11.8) $\qquad (\partial_{x_n} u)(x', 0, t) = h(x', t) \ \text{(NBC)},$

$$u(x, 0) = \phi(x) \ \text{(IC)},$$

we take some F such that F satisfies the boundary condition, i.e.

$$(\partial_{x_n} F)(x', 0, t) = h(x', t).$$

For instance, we can take

$$F(x', x_n, t) = x_n h(x', t).$$

Proceeding as above, it suffices to solve

$$v_t - k\Delta v = f - (F_t - k\Delta F), \ x_n \geq 0,$$

(11.9) $\qquad (\partial_{x_n} v)(x', 0, t) = 0 \ \text{(NBC)},$

$$v(x, 0) = \phi(x) - F(x, 0) \ \text{(IC)},$$

and let $u = v + F$. Since (11.9) has homogeneous boundary conditions, it can be solved as above (either take even extensions or use Duhamel's principle on the half space).

2. The inhomogeneous wave equation

One can also derive a version of Duhamel's principle for the wave equation, or indeed equations of the form

(11.10) $\qquad (\partial_t^2 - L)u = f, \ u(x, 0) = \phi(x), \ u_t(x, 0) = \psi(x).$

Let $S(t)$ be the solution operator corresponding to $f = 0$, $\phi = 0$, i.e. for functions ψ, $u(t, x) = (S(t)\psi)(x)$ solves

$$(\partial_t^2 - L)u = 0, \ u(x, 0) = 0, \ u_t(x, 0) = \psi(x),$$

so

$$(\partial_t^2 - L)(S(t)\psi) = 0, \ S(0)\psi(x) = 0, \ (\partial_t S(t)\psi)(x)|_{t=0} = \psi(x).$$

Then the solution for (11.10) is

(11.11) $\qquad u(x, t) = \partial_t(S(t)\phi)(x) + (S(t)\psi)(x) + \int_0^t (S(t-s)f_s)(x) \, ds.$

One easily checks that this indeed solves (11.10). First, the last term vanishes when $t = 0$ (since the integral is from 0 to 0 then) and its t-derivative at $t = 0$ is

$$S(0)f_0 + \int_0^0 (\partial_t S(t-s)f_s)(x) \, ds = 0,$$

so it does not contribute to the initial conditions. We have $S(0)\psi = 0$ and $\partial_t(S(t)\psi)|_{t=0} = \psi$. Moreover, $\partial_t(S(t)\phi)|_{t=0} = \phi$ while

$$\partial_t(\partial_t(S(t)\phi)) = \partial_t^2(S(t)\phi) = L(S(t)\phi),$$

so

$$\partial_t(\partial_t(S(t)\phi))|_{t=0} = L(S(0)\phi) = 0$$

since $S(0)\phi = 0$. Adding up, we see that all the initial conditions are satisfied. We still need to check that the PDE holds. Certainly $S(t)\psi$ satisfies the homogeneous PDE, and the same follows for $\partial_t(S(t)\phi)$ since

$$(\partial_t^2 - L)\partial_t(S(t)\phi) = \partial_t((\partial_t^2 - L)S(t)\phi) = \partial_t 0 = 0$$

as ∂_t commutes with $\partial_t^2 - L$. It remains to check that the last term of (11.11) solves the inhomogeneous PDE. First,

$$\partial_t\left(\int_0^t (S(t-s)f_s)(x)\,ds\right) = S(0)f_t + \int_0^t \partial_t(S(t-s)f_s)(x)\,ds$$

$$= \int_0^t \partial_t(S(t-s)f_s)(x)\,ds.$$

Thus,

$$\partial_t^2\left(\int_0^t (S(t-s)f_s)(x)\,ds\right) = (\partial_t S)(0)f_t(x) + \int_0^t \partial_t^2 S(t-s)f_s(x)\,ds$$

$$= f(t,x) + \int_0^t LS(t-s)f_s(x)\,ds$$

$$= f(t,x) + L\left(\int_0^t S(t-s)f_s(x)\,ds\right),$$

so

$$(\partial_t^2 - L)\left(\int_0^t (S(t-s)f_s)(x)\,ds\right) = f(t,x),$$

as desired. Combining these calculations shows that (11.11) solves the inhomogeneous PDE.

Of course, (11.11) here showed up out of the blue. We could have derived this by taking a Fourier transform and solving the inhomogeneous second order ODE

$$(\partial_t^2 + c^2|\xi|^2)\hat{u}(t,\xi) = \hat{f}(t,\xi), \quad \hat{u}(0,\xi) = \mathcal{F}\phi(\xi), \quad \hat{u}_t(0,\xi) = \mathcal{F}\psi(\xi).$$

One can solve this by factoring the differential operator on the left-hand side as $(\partial_t + ic|\xi|)(\partial_t - ic|\xi|)$ and by solving two ODEs. Rather than doing this directly, one could convert the wave equation into a system; this is done in the problems at the end of the chapter.

We write out explicitly the solution of the inhomogeneous wave equation on $\mathbb{R}$:

(11.12) $\qquad (\partial_t^2 - c^2 \partial_x^2)u = f, \ u(x,0) = \phi(x), \ u_t(x,0) = \psi(x).$

Then

$$(S(t)\psi)(x) = \frac{1}{2c} \int_{x-ct}^{x+ct} \psi(\sigma) \, d\sigma,$$

so by (11.11), the solution of the inhomogeneous PDE is

$$u(x,t) = \partial_t \left(\frac{1}{2c} \int_{x-ct}^{x+ct} \phi(\sigma) \, d\sigma \right) + \frac{1}{2c} \int_{x-ct}^{x+ct} \psi(\sigma) \, d\sigma$$
$$+ \int_0^t \frac{1}{2c} \int_{x-c(t-s)}^{x+c(t-s)} f(\sigma, s) \, d\sigma \, ds$$

(11.13)

$$= \frac{1}{2}(\phi(x+ct) + \phi(x-ct)) + \frac{1}{2c} \int_{x-ct}^{x+ct} \psi(\sigma) \, d\sigma$$
$$+ \frac{1}{2c} \int_0^t \int_{x-c(t-s)}^{x+c(t-s)} f(\sigma, s) \, d\sigma \, ds.$$

The first two terms give d'Alembert's formula for the homogeneous wave equation. The last term is a constant times the integral of f over the region

$$D_{x,t}^- \cap \{(\sigma, s) : \ s \geq 0\} = \{(\sigma, s) : \ 0 \leq s \leq t, \ x - c(t-s) \leq \sigma \leq x + c(t-s)\}$$
$$= \{(\sigma, s) : \ 0 \leq s \leq t, \ |x - \sigma| \leq c(t-s)\},$$

i.e. the backward characteristic triangle from (x,t), truncated at the x-axis (i.e. $t = 0$); see Figure 11.1. Thus, $u(x,t)$ depends on f in the part of $D_{x,t}^-$ after the initial data are imposed, which is another reason why we defined the domain of dependence for the wave equation to be the whole region $D_{x,t}^-$ (apart from the issue that initial conditions could be imposed at other, non-zero, times $\leq t$).

We can again solve the wave equation on the half space (and similarly on quadrants, intervals, etc.) with inhomogeneous boundary conditions by using the same method as for the heat equation. Thus, consider

(11.14)
$$(\partial_t^2 - c^2 \Delta)u = f, \ x_n \geq 0,$$
$$u(x', 0, t) = h(x', t) \ \text{(DBC)},$$
$$u(x, 0) = \phi(x), \ u_t(x, 0) = \psi(x) \ \text{(IC)},$$

where f, ϕ, ψ are given functions in $x_n \geq 0$ and h is a given function of x' and t. Namely, let F be any function in $x_n \geq 0$ such that it satisfies the boundary condition, i.e. $F(x', 0, t) = h(x', t)$. Then let v be the solution of

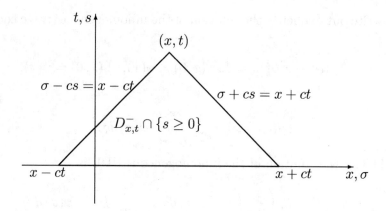

Figure 11.1. The backward characteristic triangle over which we need
to integrate in Duhamel's formula for the wave equation.

the PDE with homogeneous boundary condition (which we can thus solve):

$$(\partial_t^2 - c^2\Delta)v = f - (\partial_t^2 - c^2\Delta)F, \ x_n \geq 0,$$

$$v(x',0,t) = 0 \ \text{(DBC)},$$

$$v(x,0) = \phi(x) - F(x,0), \ u_t(x,0) = \psi(x) - F_t(x,0) \ \text{(IC)}.$$

Then $u = v + F$ solves (11.14).

One possible choice is of course $F(x', x_n, t) = h(x', t)$, as above. However, there can be better choices. In particular, suppose that $n = 1$. Then any function of the form $\tilde{g}(x - ct)$, or equivalently of the form $g\left(t - \frac{x}{c}\right)$, solves the wave equation. Now let

(11.15)
$$g(s) = \begin{cases} h(s), & x > 0, \\ 0, & x < 0. \end{cases}$$

Then

$$F(x,t) = g\left(t - \frac{x}{c}\right)$$

solves the homogeneous wave equation, $F(0,t) = h(t)$ for $t \geq 0$ and $F(x,0) = F_t(x,0) = 0$ for $x > 0$. Thus, the PDE for v is the 'same' as the PDE for u, with the only change being that the boundary condition is now homogeneous:

$$(\partial_t^2 - c^2\partial_x^2)v = f, \ x \geq 0,$$

$$v(x',0,t) = 0 \ \text{(DBC)},$$

$$v(x,0) = \phi(x), \ u_t(x,0) = \psi(x) \ \text{(IC)}.$$

In particular, if $f = 0$, $\phi = \psi = 0$, then the solution of the original PDE, (11.14), (for $n = 1$) is simply

$$u(x,t) = g\left(t - \frac{x}{c}\right),$$

with g as in (11.15). Note that u is a distributional solution of the PDE even if g is discontinuous. The latter can happen even if h is C^∞, namely if it does not vanish at 0. In general, in view of (11.15), the smoothness of u depends not only on the smoothness of h, but on its vanishing at $t = 0$, which should be thought of as a matching condition of the boundary condition with the initial condition (which vanish!).

Problems

Problem 11.1. Solve the inhomogeneous heat equation on the half-line for Dirichlet boundary conditions,

$$u_t - k u_{xx} = f, \ u(x,0) = \phi(x), \ u(0,t) = 0,$$

in two different ways:

(i) Using Duhamel's principle along with the solution formula for the homogeneous equation derived in text (i.e. with $f = 0$) on the half-line.

(ii) Using the appropriate extension of f and ϕ to the whole real line and solving the inhomogeneous PDE on the real line.

Problem 11.2. Derive Duhamel's principle for the wave equation on $\mathbb{R}$,

$$u_{tt} - c^2 \partial_x^2 u = f, \ u(x,0) = \phi(x), \ u_t(x,0) = \psi(x),$$

by setting up a first order system for $U = \begin{bmatrix} u \\ v \end{bmatrix}$, $v = u_t$, namely

$$u_t - v = 0, \ u(x,0) = \phi(x),$$
$$v_t - c^2 \partial_x^2 u = f, \ v(x,0) = \psi(x).$$

Thus, one has

$$\partial_t U - AU = \begin{bmatrix} 0 \\ f \end{bmatrix}, \ U(0,x) = \begin{bmatrix} \phi(x) \\ \psi(x) \end{bmatrix},$$

where

$$A = \begin{bmatrix} 0 & \mathrm{Id} \\ c^2 \partial_x^2 & 0 \end{bmatrix}.$$

This is now a first order equation in time, so Duhamel's principle for first order equations is applicable and gives the solution of the inhomogeneous equation as

$$U(x,t) = \mathcal{S}(t) \begin{bmatrix} \phi \\ \psi \end{bmatrix}(x) + \int_0^t \mathcal{S}(t-s) \begin{bmatrix} 0 \\ f_s \end{bmatrix}(x)\, ds,$$

where $\mathcal{S}$ is the solution operator for the homogeneous problem $\partial_t U - AU = 0$. You need to work this out explicitly, in particular what $\mathcal{S}$ is, to derive the solution of the wave equation.

Problem 11.3. Not only can one use the solution formula for the homogeneous PDE (with inhomogeneous initial conditions) to solve an inhomogeneous one, but if one knows how to solve the inhomogenenous PDE with *homogeneous* initial conditions, one can also solve the inhomogenenous PDE with *inhomogeneous* initial conditions, in particular the homogenenous PDE with *inhomogeneous* initial conditions. This is useful for theoretical considerations, e.g. for those in Chapter 17, as forcing terms are often the most convenient to work with.

(i) For $T > 0$, use the solution formula for the inhomogeneous heat equation, (11.2), applied with 0 initial data at $t = 0$, for $f = \phi(x)\delta_T(t)$, where δ_T is the delta distribution at T, and show that this gives exactly the solution of the heat equation

$$u_t = k\Delta u, \ t > T,$$

with $u(T, x) = \phi(x)$.

(ii) For $T > 0$, use the solution formula for the inhomogeneous wave equation, (11.13), applied with 0 initial data at $t = 0$, for $f = \psi(x)\delta_T(t)$, where δ_T is the delta distribution at T, and show that this gives exactly the solution of the wave equation

$$u_{tt} = c^2\Delta u, \ t > T,$$

with $u(x, T) = 0$, $u_t(x, T) = \psi(x)$.

(iii) Express the solution of the wave equation

$$u_{tt} = c^2\Delta u, \ t > T,$$

with $u(x, T) = \phi(x)$, $u_t(x, T) = 0$ in terms of the solution of the wave equation with 0 initial data at $t = 0$ with a differentiated delta distributional forcing at $t = T$.

Chapter 12

Separation of variables

1. The general method

Separation of variables is a method to solve certain PDEs which have a 'warped product' structure. The general idea is the following: suppose we have a linear PDE $Lu = 0$ on a space $M_x \times N_y$. We look for solutions $u(x, y) = X(x)Y(y)$. In general, there are no non-trivial solutions (the identically 0 function being trivial), but in special cases we might be able to find some. We cannot expect even then that all solutions of the PDE are of this form. However, if we have a family

$$u_n(x, y) = X_n(x)Y_n(y), \ n \in \mathcal{I},$$

of separated solutions, where $\mathcal{I}$ is some index set (e.g. the positive integers), then, this being a linear PDE,

$$u(x, y) = \sum_{n \in \mathcal{I}} c_n u_n(x, y) = \sum_{n \in \mathcal{I}} c_n X_n(x) Y_n(y)$$

solves the PDE as well in a distributional sense for any constants $c_n \in \mathbb{C}$, $n \in \mathcal{I}$, provided the sum converges in distributions, and we may be able to choose the constants so that this in fact gives an arbitrary solution of the PDE.

We emphasize that our endeavor, in general, is very unreasonable. Thus, we may make assumptions as we see fit — we need to justify our results *after* we derive them.

As an example, consider the wave equation

$$u_{tt} - c^2 \Delta_x u = 0$$

on $M_x \times \mathbb{R}_t$, where M is the space; for instance, M is $\mathbb{R}^n$ or a cube $[a, b]^n$ or a ball $\mathbb{B}^n = \{x \in \mathbb{R}^n : |x| \leq 1\}$. A separated solution is one of the form $u(x, t) = X(x)T(t)$. Substituting into the PDE yields

$$X(x)T''(t) - c^2 T(t)(\Delta_x X)(x) = 0.$$

Rearranging, and assuming T and X do not vanish, we obtain

$$\frac{T''(t)}{c^2 T(t)} = \frac{\Delta_x X(x)}{X(x)}.$$

Now, the left-hand side is a function independent of x and the right-hand side is a function independent of t, so they are both equal to a constant, $-\lambda$. Namely pick your favorite value of x_0 and t_0, and then for any x and t,

$$\mathrm{RHS}(x) = \mathrm{LHS}(t_0) = \mathrm{RHS}(x_0) = \mathrm{LHS}(t),$$

so the constant in question is $\mathrm{LHS}(t_0)$. Thus, we get two ODEs:

$$T''(t) = -\lambda c^2 T(t),$$
$$(\Delta_x X)(x) = -\lambda X(x).$$

Now typically one has additional conditions. For instance, one has boundary conditions at ∂M:

$$u|_{\partial M \times \mathbb{R}} = 0 \text{ (DBC) or}$$
$$\frac{\partial u}{\partial n}\Big|_{\partial M \times \mathbb{R}} = 0 \text{ (NBC)}.$$

Then $X(x)T(t)$ has to satisfy these conditions for all $x \in \partial M$ and all $t \in \mathbb{R}$. Taking some t for which $T(t)$ does not vanish, we deduce that the analogous boundary condition is satisfied, namely

$$X|_{\partial M} = 0 \text{ (DBC) or}$$
$$\frac{\partial X}{\partial n}\Big|_{\partial M} = 0 \text{ (NBC)}.$$

We also have initial conditions such as

$$u(x, 0) = \phi(x), \ u_t(x, 0) = \psi(x),$$

but as these are not homogeneous, we do not impose these at this point, and hope that we will have sufficient flexibility from the c_n to match these.

We start by solving the ODE for T, which is easy:

$$T(t) = A\cos(\sqrt{\lambda}ct) + B\sin(\sqrt{\lambda}ct), \ \lambda \neq 0,$$

and

$$T(t) = A + Bt$$

if $\lambda = 0$. (We could have used complex exponentials instead. If λ is not positive, the trigonometric functions should be thought of as given by the corresponding complex exponentials, so e.g. $\cos(\sqrt{\lambda}ct) = \frac{1}{2}(e^{i\sqrt{\lambda}ct} + e^{-i\sqrt{\lambda}ct})$.)

Now, in general the spatial equation

$$-\Delta X = \lambda X,$$

(12.1)
$$X|_{\partial M} = 0 \ \text{(DBC) or}$$

$$\frac{\partial X}{\partial n}|_{\partial M} = 0 \ \text{(NBC)}$$

is impossible to solve explicitly. However, we point out that it is an *eigenvalue equation* for Δ: the statement is that X is an eigenfunction of $-\Delta$ with eigenvalue λ, in the strong sense that it also satisfies the boundary condition. If we let $X_n(x)$, $n \in \mathbb{N}$, be the eigenfunctions of $-\Delta$ with this boundary condition, with corresponding eigenvalue λ_n, then the conclusion is that

$$u(x,t) = \sum_{n \in \mathbb{N}} A_n \cos(\sqrt{\lambda_n} ct) X_n(x) + B_n \sin(\sqrt{\lambda_n} ct) X_n(x)$$

is the general separated solution. Note that matching the initial conditions requires

$$\phi(x) = \sum_{n \in \mathbb{N}} A_n X_n(x), \ \psi(x) = \sum_{n \in \mathbb{N}} \sqrt{\lambda_n} c B_n X_n(x),$$

i.e. writing the given functions ϕ and ψ as an infinite linear combination of the eigenfunctions X_n—of course, in addition to finding some A_n and B_n which *should* work, we need to actually prove that these series indeed converge to the desired limit. Due to the decomposition of u into eigenfunctions of the spatial operator, X, these methods for solving the PDE are also called *spectral methods*.

2. Interval geometries

In a simple situation, such as when $M = [0, \ell]$, we can find the eigenfunctions of $-\Delta$ explicitly. Namely, for Dirichlet boundary condition, the equation is

$$-\frac{d^2 X}{dx^2} = \lambda X, \ X(0) = 0 = X(\ell)$$

and the solution of the ODE (without the boundary conditions) is

(12.2)
$$X(x) = C \cos(\sqrt{\lambda} x) + D \sin(\sqrt{\lambda} x), \ \lambda \neq 0,$$
$$X(x) = C + Dx, \ \lambda = 0.$$

Evaluating at $x = 0$ and enforcing $X(0) = 0$ yields $C = 0$ in either case. Evaluating at ℓ yields $D = 0$ if $\lambda = 0$, so $\lambda = 0$ is of no interest (we are only interested in non-trivial solutions). If $\lambda \neq 0$, then evaluation at $x = \ell$ yields that either $D = 0$, which again would give a trivial solution, or $\sin(\sqrt{\lambda} \ell) = 0$, which we assume is the case. But the zeros of the sine are at $n\pi$, $n \in \mathbb{Z}$, so

$$\sqrt{\lambda} \ell = n\pi,$$

and hence
$$\lambda = \left(\frac{n\pi}{\ell}\right)^2.$$

Note that n and $-n$ give essentially the same X_n (up to an overall minus sign, which is irrelevant, as we are allowing an arbitrary coefficient D), while $n = 0$ yields the trivial solution, so we conclude that
$$X_n(x) = \sin\left(\frac{n\pi x}{\ell}\right), \ \lambda_n = \left(\frac{n\pi}{\ell}\right)^2, \ n > 0, \ n \in \mathbb{Z}.$$

Returning to the full problem, we deduce that
$$u(x,t) = \sum_{n=1}^{\infty} A_n \cos(\sqrt{\lambda_n}ct)\sin\left(\frac{n\pi x}{\ell}\right) + B_n \sin(\sqrt{\lambda_n}ct)\sin\left(\frac{n\pi x}{\ell}\right).$$

It remains to determine the coefficients A_n and B_n, which is a subject of the next chapter.

Next, still on $[0, \ell]$, we consider the Neumann boundary conditions
$$-\frac{d^2 X}{dx^2} = \lambda X, \ X'(0) = 0 = X'(\ell)$$

Then (12.2) still holds. Substituting in $X'(0) = 0$ yields $D = 0$ whether $\lambda = 0$ or not. If $\lambda = 0$, the constant C satisfies the boundary condition at ℓ, so 1 is an eigenfunction with eigenvalue 0. If $\lambda \neq 0$, we obtain the requirement (as we want a non-trivial solution)
$$\sqrt{\lambda}\sin(\sqrt{\lambda}\ell) = 0,$$

and hence
$$\sqrt{\lambda}\ell = n\pi, \ n \in \mathbb{Z},$$

as above. Since $\lambda \neq 0$, we in fact have $n \neq 0$ in this case. Again, n and $-n$ give essentially the same eigenfunctions (in fact, exactly!), so our eigenfunctions are
$$X_n(x) = \cos\left(\frac{n\pi x}{\ell}\right), \ \lambda_n = \left(\frac{n\pi}{\ell}\right)^2, \ n > 0, \ n \in \mathbb{Z},$$
$$X_0(x) = 1, \ \lambda_0 = 0.$$

Note that $n = 0$ can be considered as simply a special case of the general formula in this particular situation, and thus we may write our answer as
$$X_n(x) = \cos\left(\frac{n\pi x}{\ell}\right), \ \lambda_n = \left(\frac{n\pi}{\ell}\right)^2, \ n \geq 0, \ n \in \mathbb{Z},$$

and hence the general separated solution as
$$u(x,t) = A_0 + B_0 t + \sum_{n=1}^{\infty} A_n \cos(\sqrt{\lambda_n}ct)\cos\left(\frac{n\pi x}{\ell}\right) + B_n \sin(\sqrt{\lambda_n}ct)\cos\left(\frac{n\pi x}{\ell}\right).$$

Next, consider the heat equation on a space M, i.e.

$$u_t - k\Delta_x u = 0$$

with either Dirichlet or Neumann boundary conditions. For separated solutions $u(x,t) = X(x)T(t)$ we obtain

$$X(x)T'(t) - kT(t)\Delta_x X = 0,$$

so

$$\frac{T'}{kT} = \frac{\Delta_x X}{X}.$$

Again, both sides must be equal to a constant, $-\lambda$. The ODE for T,

$$T' = -\lambda k T,$$

is then easy to solve:

$$T(t) = Ae^{-\lambda kt}.$$

The PDE for X still has boundary conditions, and is

(12.3)
$$-\Delta X = \lambda X,$$
$$X|_{\partial M} = 0 \text{ (DBC) or}$$
$$\frac{\partial X}{\partial n}|_{\partial M} = 0 \text{ (NBC)},$$

the same as (12.1). The solution is thus also the same. For instance, for Dirichlet boundary conditions on $[0, \ell]$,

$$X_n(x) = \sin\left(\frac{n\pi x}{\ell}\right), \quad \lambda_n = \left(\frac{n\pi}{\ell}\right)^2, \quad n > 0, \ n \in \mathbb{Z};$$

hence the general separated solution is

$$u(x,t) = \sum_{n=1}^{\infty} A_n e^{-k\lambda_n t} \sin\left(\frac{n\pi x}{\ell}\right).$$

Similarly, for Neumann boundary conditions we obtain

$$u(x,t) = \sum_{n=0}^{\infty} A_n e^{-k\lambda_n t} \cos\left(\frac{n\pi x}{\ell}\right).$$

3. Circular geometries

Similar techniques also work for some other problems with less product-type behavior. Thus, consider Laplace's equation on the *disk* of radius R,

$$D = \mathbb{B}_{r_0}^2 = \{x \in \mathbb{R}^2 : |x| < r_0\},$$

namely

$$\Delta u = 0, \ x \in D,$$
$$u|_{\partial D} = h,$$

with h a given function on ∂D. We again ignore the inhomogeneous conditions, so we are left with the PDE. Now, we need to think of the disk as a product space. This is not the case in Cartesian coordinates. However, in polar coordinates, we can identify D with

$$[0, r_0)_r \times \mathbb{S}^1_\theta,$$

with $\mathbb{S}^1_\theta$ the unit circle, on which the angular variable is denoted by θ. The Laplacian in polar coordinates is

$$\Delta = \partial_r^2 + r^{-1}\partial_r + r^{-2}\partial_\theta^2.$$

Note that polar coordinates are singular at $r = 0$ (the whole circle at $r = 0$ is squashed into a single point), and Δ has coefficients which are not smooth at $r = 0$. We again consider separated solutions,

$$u(r, \theta) = R(r)\Theta(\theta).$$

Substituting into the PDE yields

$$R''\Theta + r^{-1}R'\Theta + r^{-2}\Theta''R = 0.$$

Dividing by ΘR and multiplying by r^2,

$$-\frac{r^2 R'' + rR'}{R} = \frac{\Theta''}{\Theta},$$

and as before both sides must be equal to a constant, $-\lambda$.

Thus, the Θ ODE is

$$\Theta'' = -\lambda\Theta,$$

where Θ is a function on $\mathbb{S}^1$. A function on $\mathbb{S}^1$ can be thought of as a function on $[0, 2\pi]$ such that the values at 0 and 2π agree (since they represent the same point on $\mathbb{S}^1$). However, a better way of thinking of it as a 2π-periodic function on $\mathbb{R}$ is that the points $\theta + 2n\pi$, $n \in \mathbb{Z}$, correspond to the same point in the circle. Explicitly, if the circle is considered the unit circle in $\mathbb{R}^2$, as usual, the map $\mathbb{R} \to \mathbb{S}^1$ is

$$\mathbb{R} \ni \theta \mapsto (\cos\theta, \sin\theta) \in \mathbb{S}^1,$$

making the 2π-periodicity clear. Thus, we want solutions of the Θ ODE, considered as an ODE on the real line, which are 2π-periodic. Now, without the periodicity condition, the solution of the ODE is, as before (when this was the ODE for T),

$$\Theta(\theta) = A\cos(\sqrt{\lambda}\theta) + B\sin(\sqrt{\lambda}\theta), \ \lambda \neq 0,$$

and

$$\Theta(\theta) = A + B\theta$$

if $\lambda = 0$. Adding the 2π periodicity condition we need that $\sqrt{\lambda} = n \in \mathbb{Z}$ if $\lambda \neq 0$, and $B = 0$ if $\lambda = 0$. If $\lambda \neq 0$, then this gives $\lambda = n^2$. As positive and negative values of n give rise to the same eigenfunctions, we can restrict ourselves to $n > 0$ (note $n \neq 0$ if $\lambda \neq 0$). Thus, in summary, the solutions are

$$\Theta_n(\theta) = A_n \cos(n\theta) + B_n \sin(n\theta), \ n > 0, \ n \in \mathbb{Z},$$
$$\Theta_0(\theta) = A_0.$$

It remains to deal with the R ODE which is

$$r^2 R'' + r R' - \lambda R = 0.$$

Note that the coefficient of the highest derivative, R^2, vanishes at $r = 0$, so this ODE is degenerate, or singular, there. However, note that each derivative comes with a factor of r, thus R' (one derivative falling on R) has a factor of r in the coefficient and R'' (two derivatives falling on R) has a factor of r^2. A slightly better way to rewrite it is

$$r(rR')' - \lambda R = 0,$$

then explicitly you can see that each time you differentiate, you multiply by r. Such an ODE is called a regular singular ODE (in general, one could multiply by a smooth function times r each time one differentiates) and can (usually) be solved in a power series (in general with non-integer powers) around $r = 0$, with possible logarithmic terms. Our ODE is a particularly simple regular singular ODE. This can be seen by changing variables $s = \log r$ and writing $S(r) = \log r$, so $\frac{dS}{dr} = r^{-1}$. Then for a function f on $\mathbb{R}$,

$$r \frac{d(f \circ S)}{dr} = r \left(\frac{df}{ds} \circ S \right) \frac{dS}{dr} = \frac{df}{ds} \circ S,$$

or informally,

$$r \frac{df}{dr} = r \frac{df}{ds} \frac{ds}{dr} = \frac{df}{ds},$$

so the ODE for $F(t) = R(e^s)$ becomes

$$\frac{d^2}{ds^2} F - \lambda F = 0,$$

so

$$F(t) = A e^{\sqrt{\lambda} s} + B e^{-\sqrt{\lambda} s}, \ \lambda \neq 0,$$
$$F(t) = A + Bs, \ \lambda = 0.$$

Using $s = \log r$, we get for the original function R,

$$R(r) = A r^{\sqrt{\lambda}} + B r^{-\sqrt{\lambda}}, \ \lambda \neq 0,$$
$$R(r) = A + B \log r, \ \lambda = 0.$$

Recalling that $\sqrt{\lambda}$ is a non-negative integer from the Θ ODE, we conclude that

$$R(r) = A r^n + B r^{-n}, \ n > 0, \ n \in \mathbb{Z},$$
$$R(r) = A + B \log r, \ n = 0.$$

Now, there is nothing special about the origin: it is just a point in the interior of the disk, and we know by elliptic regularity that the solution of Laplace's equation should be C^∞ in the disk and so, in particular, it should be bounded. Hence we throw out the exponents that would yield unbounded terms to get

$$R_n(r) = r^n, \; n > 0, \; n \in \mathbb{Z},$$
$$R_0(r) = 1, \; n = 0,$$

and the $n = 0$ case could be simply included in the $n > 0$ one by letting $n \geq 0$ there. Note that the 'badly behaved' solutions r^{-n} and $\log r$ arose because we used badly behaved 'coordinates' on D: recall that these themselves were singular at the origin.

Combining this with our results for Θ, we obtain the general separated solution

$$u(r, \theta) = A_0 + \sum_{n=1}^{\infty} r^n (A_n \cos(n\theta) + B_n \sin(n\theta)).$$

Again, the question is whether we can determine the coefficients to match the boundary conditions. Explicitly, the boundary condition is

$$h(\theta) = A_0 + \sum_{n=1}^{\infty} R^n (A_n \cos(n\theta) + B_n \sin(n\theta)),$$

so we need to write the given function h as an infinite linear combination of the sines and cosines, and discuss convergence of the result. This is the topic of the next chapters.

Problems

Problem 12.1.

(i) Consider the following eigenvalue problem on $[0, \ell]$:
$$-X'' = \lambda X, \; X(0) = 0, \; X'(\ell) = 0.$$
Find all eigenvalues and eigenfunctions.

(ii) Using separation of variables, find the general 'separated' solution of the wave equation
$$u_{tt} = c^2 u_{xx}, \; u(0, t) = 0, \; u_x(\ell, t) = 0.$$

(iii) Solve the wave equation with initial conditions
$$u(x, 0) = \sin(3\pi x/(2\ell)) - 2\sin(5\pi x/(2\ell)), \; u_t(x, 0) = 0.$$

(iv) Using separation of variables, find the general 'separated' solution of the *heat equation*
$$u_t = k u_{xx}, \; u(0, t) = 0, \; u_x(\ell, t) = 0;$$
here $k > 0$ is constant.

Problem 12.2. Consider Laplace's equation in a *circular sector* S_{α,r_0} which in polar coordinates is given by

$$S_{\alpha,r_0} = \{(r,\theta) : \ 0 < r < r_0, \ 0 < \theta < \alpha\}.$$

We impose a homogeneous Dirichlet boundary condition on the straight sides,

$$u(r,0) = 0 = u(r,\alpha),$$

and an inhomogeneous boundary condition

$$u(r_0,\theta) = h(\theta)$$

on the circular side.

    (i) Using separation of variables find the general 'separated' solution of Laplace's equation.

    (ii) Solve Laplace's equation if $h(\theta) = \sin(2\pi\theta/\alpha)$.

Problem 12.3. Consider Schrödinger's equation in a potential well $[0,a]$:

$$i\hbar u_t = -\hbar^2 u_{xx}, \ (x,t) \in (0,a) \times \mathbb{R},$$

where $a > 0$, and $\hbar > 0$ is constant (called Planck's constant) with Dirichlet boundary condition

$$u(0,t) = 0 = u(a,t)$$

for all t, and initial condition

$$u(x,0) = \phi(x),$$

with ϕ given. Physically u is the wave function of a particle, and the PDE describes its evolution as time progresses.

    (i) Find the general 'separated' solution of the equation.

    (ii) Solve the equation if $\phi(x) = 3\sin(2\pi x/a)$.

Problem 12.4. Separation of variables can be iterated, so one might first separate time and space variables, as we did at the beginning, and then separate various spatial coordinates. As an example:

Use separation of variables for the wave equation

$$u_{tt} = c^2(u_{xx} + u_{yy}), (x,y,t) \in (0,a) \times (0,b) \times \mathbb{R}$$

on the *rectangle* $[0,a]_x \times [0,b]_y$, where $a,b,c > 0$, and impose Dirichlet boundary conditions

$$u(0,y,t) = 0, \ u(a,y,t) = 0, \ u(x,0,t) = 0, \ u(x,b,t) = 0$$

for all x,y,t.

(i) Show that the general 'separated' solution is of the form

$$u(x, y, t) = \sum_{n=1}^{\infty} \sum_{m=1}^{\infty} A_{nm} \cos(\sqrt{\lambda_{nm}} ct) \sin\left(\frac{n\pi x}{a}\right) \sin\left(\frac{m\pi y}{b}\right)$$

$$+ \sum_{n=1}^{\infty} \sum_{m=1}^{\infty} B_{nm} \sin(\sqrt{\lambda_{nm}} ct) \sin\left(\frac{n\pi x}{a}\right) \sin\left(\frac{m\pi y}{b}\right),$$

and find λ_{nm}.

(ii) Find the solution of the wave equation on the rectangle with Dirichlet boundary condition if we also impose initial conditions

$$u(x, y, 0) = 0, \quad u_t(x, y, 0) = \sin(2\pi x/a) \sin(3\pi y/b).$$

Problem 12.5. Use separation of variables for the heat equation

$$u_t = k(u_{xx} + u_{yy}), (x, y, t) \in (0, a) \times (0, b) \times (0, \infty)$$

on the rectangle $[0, a]_x \times [0, b]_y$, where $a, b, k > 0$, and impose Neumann boundary conditions

$$u_x(0, y, t) = 0, \quad u_x(a, y, t) = 0, \quad u_y(x, 0, t) = 0, \quad u_y(x, b, t) = 0$$

for all x, y, t.

(i) Find the general 'separated' solution.

(ii) Find the solution of the heat equation on the rectangle with Neumann boundary condition if we also impose initial conditions

$$u(x, y, 0) = 4\cos(\pi x/a)\cos(3\pi y/b) - 5\cos(7\pi x/a)\cos(2\pi y/b).$$

Chapter 13

Inner product spaces, symmetric operators, orthogonality

1. The basics of inner product spaces

When discussing separation of variables, we noted that at the last step we need to express the inhomogeneous initial or boundary data as a super-position of functions arising in the process of separation of variables. For instance, for the Dirichlet problem $\Delta u = 0$, $u|_{\partial D} = h$, we had to express h as

$$(13.1) \qquad h(\theta) = A_0 + \sum_{n=1}^{\infty} R^n (A_n \cos(n\theta) + B_n \sin(n\theta)),$$

i.e. we had to find constants A_n and B_n so that this expression holds. The basic framework for this is inner product spaces, which we now discuss.

Definition 13.1. An *inner product* on a complex vector space V is a map

$$\langle .,. \rangle : V \times V \to \mathbb{C}$$

such that

    (i) $\langle .,. \rangle$ is linear in the first slot:

$$\langle c_1 v_1 + c_2 v_2, w \rangle = c_1 \langle v_1, w \rangle + c_2 \langle v_2, w \rangle, \ c_1, c_2 \in \mathbb{C}, \ v_1, v_2, w \in V,$$

    (ii) $\langle .,. \rangle$ is *Hermitian symmetric*:

$$\langle v, w \rangle = \overline{\langle w, v \rangle},$$

with the bar denoting complex conjugate,

(iii) $\langle .,. \rangle$ is *positive definite*:

$$v \in V \Rightarrow \langle v,v \rangle \geq 0, \text{ and } \langle v,v \rangle = 0 \Leftrightarrow v = 0.$$

A vector space with an inner product is also called an inner product space.

While one should write $(V, \langle .,. \rangle)$ to specify the inner product space, one typically says merely that V is an inner product space when the inner product is understood.

For real vector spaces, one makes essentially the same definition, except that, as the complex conjugate does not make sense, one simply has symmetry:

$$V \text{ real vector space } \Rightarrow \langle v,w \rangle = \langle w,v \rangle, \ v,w \in V.$$

We also introduce the notation for the *norm* associated to this inner product:

$$\|v\| = \langle v,v \rangle^{1/2},$$

where the square root is the unique non-negative square root of a non-negative number (see (iii)). Thus,

$$\langle v,v \rangle = \|v\|^2.$$

We check below that this is indeed a norm in the sense discussed in the Introduction, with the main point to be proved being the triangle inequality.

We note some examples:

(i) $V = \mathbb{R}^n$, with inner product

$$\langle x,y \rangle = \sum_{j=1}^{n} x_j y_j,$$

where $x = (x_1, \ldots, x_n)$, $y = (y_1, \ldots, y_n)$. Thus $\|x\|^2 = \sum_{j=1}^{n} x_j^2$.

(ii) $V = \mathbb{C}^n$, with inner product

$$\langle x,y \rangle = \sum_{j=1}^{n} x_j \overline{y_j},$$

where $x = (x_1, \ldots, x_n)$, $y = (y_1, \ldots, y_n)$. Thus $\|x\|^2 = \sum_{j=1}^{n} |x_j|^2$, which explains why we need Hermitian symmetry for complex vector spaces, namely to satisfy property (iii) of the inner product, positive definiteness.

(iii) $V = \mathbb{R}^n$, with inner product

$$\langle x,y \rangle = \sum_{j=1}^{n} a_j x_j y_j,$$

where $x = (x_1, \ldots, x_n)$, $y = (y_1, \ldots, y_n)$, $a_j > 0$ for all j. Thus $\|x\|^2 = \sum_{j=1}^{n} a_j x_j^2$.

(iv) $V = \ell^2$, the set of square summable (complex-valued) sequences $\{x_j\}_{j=1}^{\infty}$, i.e. sequences with $\sum_{j=1}^{\infty} |x_j|^2 < \infty$, with the inner product

(13.2)
$$\langle \{x_j\}_{j=1}^{\infty}, \{y_j\}_{j=1}^{\infty} \rangle = \sum_{j=1}^{\infty} x_j \overline{y_j}.$$

Note that this is the infinite dimensional version of example (ii). The convergence of the series on the right-hand side of (13.2) follows from

$$|x_j \overline{y_j}| = |x_j||y_j| \leq \frac{|x_j|^2 + |y_j|^2}{2}$$

and the assumed convergence of $\sum_{j=1}^{\infty} |x_j|^2$ and $\sum_{j=1}^{\infty} |y_j|^2$, since the latter implies that $\sum_{j=1}^{\infty} x_j \overline{y_j}$ converges absolutely (i.e. $\sum_{j=1}^{\infty} |x_j \overline{y_j}|$ converges), and thus it converges. Now the squared norm is

$$\|\{x_j\}_{j=1}^{\infty}\|^2 = \|\{x_j\}_{j=1}^{\infty}\|_{\ell^2}^2 = \sum_{j=1}^{\infty} |x_j|^2.$$

(v) $V = \ell^2(\mathbb{Z})$, the set of square summable (complex-valued) bi-infinite sequences $\{x_j\}_{j=-\infty}^{\infty}$, i.e. sequences with $\sum_{j=-\infty}^{\infty} |x_j|^2 < \infty$, with the inner product

(13.3)
$$\langle \{x_j\}_{j=-\infty}^{\infty}, \{y_j\}_{j=-\infty}^{\infty} \rangle = \sum_{j=-\infty}^{\infty} x_j \overline{y_j}.$$

This is closely related to the previous example.

(vi) $V = C^0(\overline{\Omega})$ (complex-valued continuous functions on the closure of Ω), where Ω is a bounded domain in $\mathbb{R}^n$, with inner product

$$\langle f, g \rangle = \int_{\Omega} f(x) \overline{g(x)} \, dx.$$

Thus,

$$\|f\|^2 = \int_{\Omega} |f(x)|^2 \, dx.$$

We often write $\|f\|_{L^2} = \|f\|_{L^2(\Omega)}$ for this norm.

(vii) $V = C^0(\overline{\Omega})$, with Ω as above, with inner product

$$\langle f, g \rangle = \int_{\Omega} f(x) \overline{g(x)} \, a(x) \, dx,$$

where $a \in C^0(\overline{\Omega})$, $a > 0$ is fixed. Thus,

$$\|f\|^2 = \int_{\Omega} |f(x)|^2 \, a(x) \, dx.$$

We may write $\|f\|_{L^2(\Omega, a(x) \, dx)}$ for this norm.

(viii) Let $N > n/2$ and let $V = \{f \in C^0(\mathbb{R}^n) : (1 + |x|)^N f \text{ is bounded}\}$, with the inner product

$$\langle f, g \rangle = \int_{\mathbb{R}^n} f(x) \, \overline{g(x)} \, dx.$$

Note that the integral converges under this decay assumption. Thus,

$$\|f\|^2 = \int_{\mathbb{R}^n} |f(x)|^2 \, dx.$$

We often write $\|f\|_{L^2} = \|f\|_{L^2(\mathbb{R}^n)}$ for this norm.

Note that examples (iv)-(v) and (vi)-(viii) are related: a sequence is a function defined on the positive integers, so it is a function on a discrete space, and the sum is the discrete version of the integral.

A few properties on inner products should be observed immediately. First, the inner product is conjugate-linear in the second variable (which simply means linear if the vector space is real):

$$\langle v, c_1 w_1 + c_2 w_2 \rangle = \overline{\langle c_1 w_1 + c_2 w_2, v \rangle} = \overline{c_1 \langle w_1, v \rangle + c_2 \langle w_2, v \rangle}$$

$$= \overline{c_1} \overline{\langle w_1, v \rangle} + \overline{c_2} \overline{\langle w_2, v \rangle} = \overline{c_1} \langle v, w_1 \rangle + \overline{c_2} \langle v, w_2 \rangle.$$

A map $V \times V \to \mathbb{C}$ which is linear in the first variable and conjugate-linear in the second variable is called *sesquilinear*.

Remark 13.2. Our convention is that the inner product is linear in the first variable and conjugate linear in the second. This is the standard convention in mathematics. The standard convention in physics reverses this, i.e. the inner products in physics are linear in the second variable and conjugate linear in the first. Both of these conventions are perfectly reasonable; the important thing is to *know* which convention one is using and not to confuse the conventions in the middle of a calculation. In order to move between the two conventions, simply reverse the order of the two factors, i.e. write $\langle v, w \rangle = \langle w, v \rangle_{\text{physics}}$, where the right-hand side is the physics inner product, often written as $\langle w | v \rangle$.

Second, by (ii), if $v \in V$, then $\langle v, v \rangle = \overline{\langle v, v \rangle}$, so $\langle v, v \rangle$ is real. Thus, (iii) is the statement that this real number is non-negative, and it is actually positive if $v \neq 0$. Also, by linearity, denoting the 0 vector in V by 0_V (usually we simply denote this by 0, but here this care clarifies the calculation),

$$\langle 0_V, v \rangle = \langle 0 \cdot 0_V, v \rangle = 0 \langle 0_V, v \rangle = 0, \quad v \in V,$$

and by Hermitian symmetry, then

$$\langle v, 0_V \rangle = \overline{\langle 0_V, v \rangle} = 0.$$

We also note a useful property of $\|.\|$:

$$\|cv\|^2 = \langle cv, cv \rangle = c\langle v, cv \rangle = c\bar{c}\langle v, v \rangle = |c|^2\|v\|^2, \ c \in \mathbb{C}, \ v \in V,$$

so

$$\|cv\| = |c| \, \|v\|.$$

This property of $\|.\|$ is called absolute homogeneity (of degree 1).

One concept that is tremendously useful in inner product spaces is orthogonality:

Definition 13.3. Suppose V is an inner product space. For $v, w \in V$ we say that v is orthogonal to w if $\langle v, w \rangle = 0$.

Note that $\langle v, w \rangle = 0$ if and only of $\langle w, v \rangle = 0$, so v is orthogonal to w if and only if w is orthogonal to v. Therefore we often say simply that v and w are orthogonal.

As an illustration of its use, let's prove *Pythagoras' theorem*:

Lemma 13.4. *Suppose V is an inner product space, $v, w \in V$ and v and w are orthogonal. Then*

$$\|v + w\|^2 = \|v\|^2 + \|w\|^2 = \|v - w\|^2.$$

Proof. Since $v - w = v + (-w)$, the statement about $v - w$ follows from the statement for $v + w$ and $\| - w\| = \|w\|$. Now,

$$\langle v + w, v + w \rangle = \langle v, v + w \rangle + \langle w, v + w \rangle = \langle v, v \rangle + \langle v, w \rangle + \langle w, v \rangle + \langle w, w \rangle$$
$$= \langle v, v \rangle + \langle w, w \rangle$$

by the orthogonality of v and w, proving the result. □

One use of orthogonality is the following:

Lemma 13.5. *Suppose $v, w \in V$, $w \neq 0$. Then there exist unique $v_\|, v_\perp \in V$ such that $v = v_\| + v_\perp$, $v_\| = cw$ for some $c \in \mathbb{C}$ and $\langle v_\perp, w \rangle = 0$; see Figure* 13.1.

Proof. If $v = v_\| + v_\perp$, then taking the inner product with w and using $v_\| = cw$ we deduce

$$\langle v, w \rangle = \langle cw, w \rangle + \langle v_\perp, w \rangle = c\|w\|^2,$$

so as $w \neq 0$,

$$c = \frac{\langle v, w \rangle}{\|w\|^2}.$$

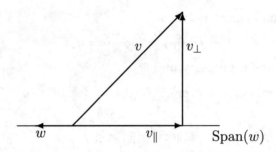

Figure 13.1. The decomposition $v = v_\| + v_\perp$.

Thus, $v_\| = cw$ and $v_\perp = v - cw$, giving uniqueness.

On the other hand, if we *let*

$$c = \frac{\langle v, w \rangle}{\|w\|^2}, \quad v_\| = cw, \quad v_\perp = v - cw,$$

then $v_\perp + v_\| = v$ and $v_\| = cw$ are satisfied, so we merely need to check $\langle v_\perp, w \rangle = 0$. But

$$\langle v_\perp, w \rangle = \langle v, w \rangle - c\langle w, w \rangle = \langle v, w \rangle - \frac{\langle v, w \rangle}{\|w\|^2}\|w\|^2 = 0,$$

so the desired vectors $v_\perp$ and $v_\|$ indeed exist. $\square$

One calls $v_\|$ the *orthogonal projection* of v to the span of w.

The inner product can be used to define a notion of convergence:

Definition 13.6. We say that a sequence v_j converges to v in V if $\|v_j - v\| \to 0$ as $j \to \infty$.

In order to make this a useful tool, we need to be able to estimate the inner product using the norm. This is achieved by the *Cauchy-Schwarz inequality*, which we already mentioned when discussing energy estimates:

Lemma 13.7 (Cauchy-Schwarz). *In an inner product space V,*

$$|\langle v, w \rangle| \le \|v\|\|w\|, \quad v, w \in V.$$

Remark 13.8. In the case of function spaces as in example (vi), Cauchy-Schwarz says explicitly that

$$\left| \int_\Omega f(x)\overline{g(x)} \, dx \right| \le \sqrt{\int_\Omega |f(x)|^2 \, dx} \sqrt{\int_\Omega |g(x)|^2 \, dx}.$$

In the case of the sequence space as in example (iv), it says explicitly that

$$\left| \sum_{j=1}^\infty x_j \overline{y_j} \right| \le \sqrt{\sum_{j=1}^\infty |x_j|^2} \sqrt{\sum_{j=1}^\infty |y_j|^2}.$$

Proof. If $w = 0$, then both sides vanish, so we may assume $w \neq 0$. Write $v = v_\parallel + v_\perp$ as in Lemma 13.5, so

$$v_\parallel = cw, \quad c = \frac{\langle v, w \rangle}{\|w\|^2}.$$

Then by Pythagoras' theorem, using $\langle v_\parallel, v_\perp \rangle = c \langle w, v_\perp \rangle = 0$,

$$\|v\|^2 = \|v_\parallel\|^2 + \|v_\perp\|^2 \geq \|v_\parallel\|^2 = |c|^2 \|w\|^2 = \frac{|\langle v, w \rangle|^2}{\|w\|^2}.$$

Multiplying through by $\|w\|^2$ and taking the non-negative square root completes the proof of the lemma. $\qquad\square$

A useful consequence of the Cauchy-Schwarz inequality is the triangle inequality for the norm:

Lemma 13.9. *In an inner product space V,*

$$\|v + w\| \leq \|v\| + \|w\|.$$

Proof. One only needs to prove the equivalent estimate where one takes the square of both sides:

$$\|v + w\|^2 \leq \|v\|^2 + 2\|v\| \, \|w\| + \|w\|^2.$$

But

$$\|v + w\|^2 = \langle v + w, v + w \rangle = \|v\|^2 + \langle v, w \rangle + \langle w, v \rangle + \|w\|^2$$
$$\leq \|v\|^2 + 2\|v\| \, \|w\| + \|w\|^2,$$

where the last inequality is the Cauchy-Schwarz inequality. $\qquad\square$

Recall from Chapter 1 that in general, if one has a vector space V, one defines the notion of a norm on it as follows:

Definition 13.10. Suppose V is a vector space. A norm on V is a map

$$\|.\| : V \to \mathbb{R}$$

such that

    (i) (positive definiteness) $\|v\| \geq 0$ for all $v \in V$, and $v = 0$ if and only if $\|v\| = 0$,

(ii) (absolute homogeneity) $\|cv\| = |c|\,\|v\|$, $v \in V$, and c a scalar (so $c \in \mathbb{R}$ or $c \in \mathbb{C}$, depending on whether V is real or complex),

(iii) (triangle inequality) $\|v + w\| \le \|v\| + \|w\|$ for all $v, w \in V$.

Thus, Lemma 13.9 shows that the map $\|.\| : V \to \mathbb{R}$ we defined on an inner product space is indeed a norm in this sense, i.e. our use of the word norm was justified.

Some examples of a normed vector space where the norm does not come from an inner product are:

(i) $V = C^0(\overline{\Omega})$ (complex-valued continuous functions on the closure of Ω), where Ω is a bounded domain in $\mathbb{R}^n$, and

$$\|f\|_{C^0} = \sup_{x \in \overline{\Omega}} |f(x)|;$$

(ii) $V = C^0(\overline{\Omega})$ (complex-valued continuous functions on the closure of Ω), where Ω is a bounded domain in $\mathbb{R}^n$, with norm

$$\|f\|_{L^1} = \int_{\Omega} |f(x)|\,dx.$$

We recall from the Introduction that in a normed vector space the vector space operations are (jointly) continuous; see Lemma 1.2.

Lemma 13.11. *If V is a normed vector space, $v_j \to v$, $w_j \to w$ in V and $c_j \to c$ in the scalars, then $c_j v_j \to cv$ and $v_j + w_j \to v + w$ in V.*

We can now show that the inner product is jointly continuous with respect to the notion of convergence we discussed:

Lemma 13.12. *Suppose that V is an inner product space. If $v_j, w_j \in V$, and $v_j \to v$, $w_j \to w$ in V, then $\langle v_j, w_j \rangle \to \langle v, w \rangle$.*

Proof. We have
$$\langle v_j, w_j \rangle - \langle v, w \rangle = \langle v_j, w_j \rangle - \langle v_j, w \rangle + \langle v_j, w \rangle - \langle v, w \rangle$$
$$= \langle v_j, w_j - w \rangle + \langle v_j - v, w \rangle.$$
Thus,
$$|\langle v_j, w_j \rangle - \langle v, w \rangle| \le |\langle v_j, w_j - w \rangle| + |\langle v_j - v, w \rangle|$$
$$\le \|v_j\|\,\|w_j - w\| + \|v_j - v\|\,\|w\| \to 0,$$
since $\{\|v_j\|\}_{j=1}^{\infty}$ is bounded. $\qquad\square$

Suppose now that we have a sequence of orthogonal non-zero vectors $\{x_j\}_{j=1}^{\infty}$ in an inner product space V, and suppose that

(13.4)
$$v = \sum_{j=1}^{\infty} c_j x_j,$$

i.e. this sum converges to v, which, recall, means that with $v_N = \sum_{j=1}^{N} c_j x_j$, $v_N \to v$ as $N \to \infty$. Then, by the continuity of the inner product, for any w,

$$\langle v, w \rangle = \sum_{j=1}^{\infty} c_j \langle x_j, w \rangle.$$

Applying this with $w = x_k$, all inner products on the right-hand side but the one with $j = k$ vanish, and we deduce that

$$\langle v, x_k \rangle = c_k \|x_k\|^2;$$

hence

(13.5)
$$c_k = \frac{\langle v, x_k \rangle}{\|x_k\|^2}.$$

As these orthogonal collections of vectors are so useful, we make a definition.

Definition 13.13. An *orthogonal set S* of vectors in an inner product space V is a subset of V consisting of non-zero vectors such that if $x, x' \in S$ and $x \neq x'$, then x is orthogonal to x'. The series (13.4) is called the corresponding *generalized Fourier series*.

2. Symmetric operators

Now, one could easily check by an explicit computation that the functions we have considered, in terms of which we would like to express our initial or boundary data, are orthogonal to each other. Thus, *if* we can write our data as an infinite linear combination of these functions, then the coefficients can be determined easily. For instance, if we want to write

$$h(\theta) = a_0 + \sum_{n=1}^{\infty} a_n \cos(n\theta) + b_n \sin(n\theta),$$

where we consider the inner product on functions on $\mathbb{S}^1$ to be given by

$$\langle f, g \rangle = \int_0^{2\pi} f(\theta) \overline{g(\theta)} \, d\theta,$$

then we must have

$$a_0 = \frac{\langle h, 1 \rangle}{\|1\|^2} = \frac{\int_0^{2\pi} h(\theta) \, d\theta}{\int_0^{2\pi} 1 \, d\theta} = \frac{1}{2\pi} \int_0^{2\pi} h(\theta) \, d\theta,$$

$$a_n = \frac{\int_0^{2\pi} h(\theta) \cos(n\theta) \, d\theta}{\int_0^{2\pi} \cos^2(n\theta) \, d\theta} = \frac{1}{\pi} \int_0^{2\pi} h(\theta) \cos(n\theta) \, d\theta,$$

$$b_n = \frac{\int_0^{2\pi} h(\theta) \sin(n\theta) \, d\theta}{\int_0^{2\pi} \sin^2(n\theta) \, d\theta} = \frac{1}{\pi} \int_0^{2\pi} h(\theta) \sin(n\theta) \, d\theta.$$

Here we used the computation $\cos(2n\theta) = 2\cos^2(n\theta) - 1$, so $\cos^2(n\theta) = \frac{1}{2}(\cos(2n\theta) + 1)$, and thus

$$\int_0^{2\pi} \cos^2(n\theta)\, d\theta = \frac{1}{2} \int_0^{2\pi} (\cos(2n\theta) + 1)\, d\theta = \frac{1}{4n} \sin(2n\theta)\big|_0^{2\pi} + \frac{1}{2}\theta\big|_0^{2\pi} = \pi,$$

with a similar result with cosine replaced by sine. In particular, returning to (13.1), we deduce that

$$A_0 = \frac{1}{2\pi} \int_0^{2\pi} h(\theta)\, d\theta,$$

$$A_n = \frac{1}{\pi R^n} \int_0^{2\pi} h(\theta)\, \cos(n\theta)\, d\theta,$$

$$B_n = \frac{1}{\pi R^n} \int_0^{2\pi} h(\theta)\, \sin(n\theta)\, d\theta.$$

We of course need to discuss whether we can indeed write h in this form, but before this we should give a conceptual reason why the functions we considered are automatically orthogonal to each other. For this purpose we need to consider symmetric operators on V. Indeed, the operators A we want to consider are not defined on all of V (or if we define A on V, it will not map V to itself), so we have to enlarge our framework.

So we consider operators defined on a linear subspace D of V. In order to make this definition well behaved, we need to assume that D is *dense* in V, i.e. if $v \in V$, then there exists $v_j \in D$ such that $v_j \to v$.

Definition 13.14. A linear operator $A : D \to V$ is called *symmetric* if

$$\langle Av, w \rangle = \langle v, Aw \rangle$$

for all $v, w \in D$.

Recall that an *eigenvector* of A is an element $v \neq 0$ of D such that $Av = \lambda v$ for some $\lambda \in \mathbb{C}$; λ is then an *eigenvalue* of A.

Lemma 13.15. *Suppose $A : D \to V$ is symmetric. Then all eigenvalues of A are real, and eigenvectors corresponding to distinct eigenvalues are orthogonal to each other.*

Proof. Suppose that λ is an eigenvalue of A, so for some $v \in D$, $v \neq 0$, $Av = \lambda v$. Then

$$\lambda \|v\|^2 = \langle \lambda v, v \rangle = \langle Av, v \rangle = \langle v, Av \rangle = \langle v, \lambda v \rangle = \overline{\lambda} \|v\|^2,$$

so dividing through by $\|v\|^2$ we deduce that $\lambda = \overline{\lambda}$, so λ is real.

Now suppose that λ and μ are eigenvalues of A (so they are both real), and $Av = \lambda v$, $Aw = \mu w$. Then

$$\lambda \langle v, w \rangle = \langle \lambda v, w \rangle = \langle Av, w \rangle = \langle v, Aw \rangle = \langle v, \mu w \rangle = \mu \langle v, w \rangle.$$

If $\lambda \neq \mu$, then rearranging and dividing by $\lambda - \mu$ gives $\langle v, w \rangle = 0$, as claimed. $\qquad\square$

Now, the operator $A = -\frac{d^2}{d\theta^2}$ defined for functions $f \in C^2(\mathbb{S}^1)$ (which, recall, can be identified with 2π-periodic functions on $\mathbb{R}$) is symmetric, since

$$\langle Af, g \rangle = \int_0^{2\pi} -f''(\theta)\overline{g(\theta)}\, d\theta = -f'(\theta)\overline{g(\theta)}\big|_0^{2\pi} + \int_0^{2\pi} f'(\theta)\overline{g'(\theta)}\, d\theta$$

$$= f(\theta)\overline{g'(\theta)}\big|_0^{2\pi} - \int_0^{2\pi} f(\theta)\overline{g''(\theta)}\, d\theta = \langle f, Ag \rangle,$$

where the boundary terms vanish due to periodicity. A similar computation for the operator $A = -\frac{d^2}{dx^2}$ defined for functions $f \in C^2([0, \ell])$ gives

$$\langle Af, g \rangle = \int_0^\ell -f''(x)\overline{g(x)}\, dx = -f'(x)\overline{g(x)}\big|_0^\ell + \int_0^\ell f'(x)\overline{g'(x)}\, dx$$

$$= -f'(x)\overline{g(x)}\big|_0^\ell + f(x)\overline{g'(x)}\big|_0^\ell - \int_0^\ell f(x)\overline{g''(x)}\, dx$$

$$= \left(f(x)\overline{g'(x)} - f'(x)\overline{g(x)} \right)\big|_0^\ell + \langle f, Ag \rangle.$$

Thus, the operator A is symmetric provided we restrict it to a domain D so that the boundary terms vanish. For instance,

(i) $A = -\frac{d^2}{dx^2}$ defined for functions $f \in C^2([0, \ell])$ such that $f(0) = 0 = f(\ell)$ (Dirichlet boundary condition) is symmetric,

(ii) $A = -\frac{d^2}{dx^2}$ defined for functions $f \in C^2([0, \ell])$ such that $f'(0) = 0 = f'(\ell)$ (Neumann boundary condition) is symmetric,

(iii) $A = -\frac{d^2}{dx^2}$ defined for functions $f \in C^2([0, \ell])$ such that $f'(0) - af(0) = 0$ and $f'(\ell) - bf(\ell) = 0$ for some given $a, b \in \mathbb{R}$ (Robin boundary condition) is symmetric.

As another example, the operator $A = -i\frac{d}{dx}$ defined on C^1 functions which are 2ℓ-periodic, mapping into $C^0([0, 2\ell])$ with the standard inner product, is symmetric since

$$\langle Af, g \rangle = \int_0^{2\ell} -if'(x)\overline{g(x)}\, dx = -if(x)\overline{g(x)}\big|_0^{2\ell} + i\int_0^{2\ell} f(x)\overline{g'(x)}\, dx = \langle f, Ag \rangle.$$

As a consequence, we deduce:

Corollary 13.16. *The following sets of functions are all orthogonal to each other in the respective vector spaces:*

(i) *In* $V = C^0([0, \ell])$, *with the standard inner product,*

$$f_n(x) = \sin\left(\frac{n\pi x}{\ell} \right), \quad n \geq 1,\ n \in \mathbb{Z};$$

these are the Fourier sine basis functions *on* $[0, \ell]$.

(ii) *In $V = C^0([0, \ell])$, with the standard inner product,*

$$f_0(x) = 1, \ f_n(x) = \cos\left(\frac{n\pi x}{\ell}\right), \ n \geq 1, \ n \in \mathbb{Z};$$

these are the Fourier cosine basis *functions on $[0, \ell]$.*

(iii) *In $V = C^0([0, 2\ell])$, with the standard inner product,*

$$f_n(x) = e^{in\pi x/\ell}, \ n \in \mathbb{Z}.$$

(iv) *In $V = C^0([0, 2\ell])$, with the standard inner product,*

$$f_0(x) = 1, \ f_n(x) = \cos\left(\frac{n\pi x}{\ell}\right), \ n \geq 1, \ n \in \mathbb{Z},$$

$$g_n(x) = \sin\left(\frac{n\pi x}{\ell}\right), \ n \geq 1, \ n \in \mathbb{Z};$$

these are the full Fourier basis *functions on $[0, 2\ell]$.*

Proof. These are all consequences of being eigenfunctions of symmetric operators; namely:

(i) $A = -\frac{d^2}{dx^2}$, with Dirichlet boundary condition, the eigenvalue of f_n is $\lambda_n = \frac{n^2\pi^2}{\ell^2}$,

(ii) $A = -\frac{d^2}{dx^2}$, with Neumann boundary condition, the eigenvalue of f_n is $\lambda_n = \frac{n^2\pi^2}{\ell^2}$,

(iii) $A = -i\frac{d}{dx}$, on 2ℓ-periodic C^1 functions, the eigenvalue of f_n is $\lambda_n = \frac{n\pi}{\ell}$,

(iv) $A = -\frac{d^2}{dx^2}$, on 2ℓ-periodic C^2 functions, the eigenvalue of f_n is $\lambda_n = \frac{n^2\pi^2}{\ell^2}$ which is also the eigenvalue of g_n.

Thus, in all cases but the last the functions are eigenfunctions of a symmetric operator with distinct eigenvalues, and hence are orthogonal. In the last case, we have all of the claimed orthogonality by the same argument, except the orthogonality of f_n to g_n. This can be easily seen, however, as the cosines are even around $x = \ell$ while the sines are odd around $x = \ell$, so the product is odd, and hence its integral over the interval $[0, 2\ell]$, which is symmetric around ℓ, vanishes. $\qquad\square$

Definition 13.17. The generalized Fourier series corresponding to

(i) the Fourier sine basis is called the *Fourier sine series*,

(ii) the Fourier cosine basis is called the *Fourier cosine series*,

(iii) the full Fourier basis is called the *full Fourier series*.

We also have the following general symmetry result for the Laplacian:

Proposition 13.18. *Let Ω be a bounded domain with smooth boundary, $V = C^0(\bar\Omega)$. The Laplacian Δ, defined on any one of the domains*

    (i) *(Dirichlet)* $D = \{f \in C^2(\overline{\Omega}) : f|_{\partial\Omega} = 0\}$,

    (ii) *(Neumann)* $D = \{f \in C^2(\overline{\Omega}) : \frac{\partial f}{\partial n}|_{\partial\Omega} = 0\}$,

    (iii) *(Robin)* $D = \{f \in C^2(\overline{\Omega}) : \left(\frac{\partial f}{\partial n} - af\right)|_{\partial\Omega} = 0\}$,

is symmetric. Here, in case (iii)*, a is a given real-valued continuous function on* $\partial\Omega$.

Proof. We recall Green's identity,

$$\int_{\Omega} (f\Delta g - (\Delta f)g)\, dx = \int_{\partial\Omega} \left(f\frac{\partial g}{\partial n} - \frac{\partial f}{\partial n}g\right) dS(x).$$

Replacing g by $\bar{g}$, we have

$$\int_{\Omega} (f\overline{\Delta g} - (\Delta f)\bar{g})\, dx = \int_{\partial\Omega} \left(f\frac{\overline{\partial g}}{\partial n} - \frac{\partial f}{\partial n}\bar{g}\right) dS(x).$$

Under each of the conditions listed above, the right-hand side vanishes. Thus,

$$\int_{\Omega} f\overline{\Delta g}\, dx = \int_{\Omega} (\Delta f)\bar{g}\, dx,$$

so Δ is symmetric.

    For the sake of completeness, we recall how Green's identity is proved. Consider the vector field $f\nabla g$. By the divergence theorem,

$$\int_{\Omega} \operatorname{div}(f\nabla g)\, dx = \int_{\partial\Omega} f\hat{n} \cdot \nabla g\, dS(x) = \int_{\partial\Omega} f\frac{\partial g}{\partial n}\, dS(x),$$

and similarly

$$\int_{\Omega} \operatorname{div}(g\nabla f)\, dx = \int_{\partial\Omega} g\frac{\partial f}{\partial n}\, dS(x).$$

Subtracting these two yields

$$\int_{\partial\Omega} \left(f\frac{\partial g}{\partial n} - \frac{\partial f}{\partial n}g\right) dS(x) = \int_{\Omega} (\operatorname{div}(f\nabla g) - \operatorname{div}(g\nabla f))\, dx$$

$$= \int_{\Omega} (f\Delta g + \nabla f \cdot \nabla g - g\Delta f - \nabla g \cdot \nabla f)\, dx = \int_{\Omega} (f\Delta g - (\Delta f)g)\, dx,$$

as claimed. $\square$

3. Completeness of orthogonal sets and of the inner product space

We still need to discuss whether 'any' v in V can be written as a stated linear combination. This is in general not the case. For instance, if $V = \mathbb{R}^2$, and $x_1 = (1,0)$, then $\{x_1\}$ is an orthogonal set of non-zero vectors, but not every element of V can be written as a linear combination of $\{x_1\}$ (i.e. is not, in general, a multiple of x_1). We need another vector, such as $x_2 = (0,1)$.

We thus make the following definition.

Definition 13.19. Suppose V is an inner product space. We say that an *orthogonal set* $\{x_n\}_{n=1}^{\infty}$ is *complete* if for every $v \in V$ there exist scalars c_n such that $v = \sum_{n=1}^{\infty} c_n x_n$.

Note that, as discussed before, if the c_n exist, they are automatically unique; they are determined by (13.5). In fact, we can always *define* a sequence by the formula (13.5): given $v \in V$, let

$$(13.6) \qquad v_n = \sum_{k=1}^{n} c_k x_k, \ c_k = \frac{\langle v, x_k \rangle}{\|x_k\|^2};$$

the question is whether $v_n \to v$. The coefficients c_k in (13.6) may be called the *generalized Fourier coefficients* of v. Note that v_n has the useful property that $v - v_n$ is orthogonal to x_j for $1 \le j \le n$. Indeed,

$$\langle v - v_n, x_j \rangle = \langle v, x_j \rangle - \sum_{k=1}^{n} c_k \langle x_k, x_j \rangle = c_j - c_j = 0, \ j \le n,$$

hence $v - v_n$ is also orthogonal to any vector of the form $\sum_{k=1}^{n} b_k x_k$, i.e. to the linear span of $\{x_1, \ldots, x_n\}$, in particular to v_n itself. Thus, by Pythagoras' theorem, we have

$$\|v\|^2 = \|v - v_n\|^2 + \|v_n\|^2 \ge \|v_n\|^2 = \sum_{j,k=1}^{n} c_j \overline{c_k} \langle x_j, x_k \rangle = \sum_{k=1}^{n} |c_k|^2 \|x_k\|^2.$$

This is *Bessel's inequality*. Thus, we deduce that for any $v \in V$, the generalized Fourier coefficients satisfy

$$\sum_{k=1}^{n} |c_k|^2 \|x_k\|^2 \le \|v\|^2.$$

Hence, as these partial sums form a bounded monotone increasing (which, recall, means non-decreasing) sequence, $\sum_{k=1}^{\infty} |c_k|^2 \|x_k\|^2$ converges, and as $\|v\|^2$ is an upper bound for the partial sums, the limit is $\le \|v\|^2$. If this limit is $\|v\|^2$, then we deduce that $\|v - v_n\| \to 0$ as $n \to \infty$, so with the converse being clear we conclude the following:

Lemma 13.20. *Suppose V is an inner product space. An orthogonal set $\{x_n\}_{n=1}^{\infty}$ is complete if and only if for every $v \in V$ the generalized Fourier coefficients satisfy*

$$\sum_{k=1}^{\infty} |c_k|^2 \|x_k\|^2 = \|v\|^2.$$

This may be called *Bessel's equality.*

Now, let's compute $\|v - \sum_{k=1}^{n} a_k x_k\|^2$, where $a_k \in \mathbb{C}$. This is best done by writing $v = v_\| + v_\perp$, where

$$v_\| = \sum_{k=1}^{n} c_k x_k,$$

the c_k being the generalized Fourier coefficients. Then $v_\perp$ is orthogonal to all x_k, $k = 1, 2, \ldots, n$, and hence

$$\left\|v - \sum_{k=1}^{n} a_k x_k\right\|^2 = \left\|v_\perp + \sum_{k=1}^{n} (c_k - a_k) x_k\right\|^2 = \|v_\perp\|^2 + \sum_{k=1}^{n} |c_k - a_k|^2 \|x_k\|^2.$$

In particular,

$$\left\|v - \sum_{k=1}^{n} a_k x_k\right\|^2 \geq \|v_\perp\|^2,$$

and equality holds if and only if $a_k = c_k$ for $k = 1, \ldots, n$. This is the *least squares approximation*:

Proposition 13.21. *Given $v \in V$ and an orthogonal set $\{x_1, \ldots, x_n\}$, the choice of a_k that minimizes the error $\|v - \sum_{k=1}^{n} a_k x_k\|^2$ of the approximation of v by $\sum_{k=1}^{n} a_k x_k$ is given by the generalized Fourier coefficients, c_k.*

Thus, even if the generalized Fourier series $\sum_{k=1}^{\infty} c_k x_k$ does not converge to the function, its partial sums give the best possible approximations of v in the precise sense of this proposition.

We now actually turn to completeness statements. We prove in the next chapter that all of the orthogonal sets listed in Corollary 13.16 are complete. Moreover, for the Laplacian we have the following result, which we prove in Chapter 18 after developing more advanced tools:

Theorem 13.22. *Suppose Ω is a bounded domain in $\mathbb{R}^n$ with smooth boundary. For the Laplacian, defined on the subspace D of $C^0(\overline{\Omega})$ given by (i), (ii) or (iii) of Proposition 13.18, the eigenspace for each eigenvalue $\lambda \in \mathbb{R}$ is finite dimensional, the eigenvalues can be arranged in an increasing sequence, tending to infinity, $\lambda_1 \leq \lambda_2 \leq \ldots$, and choosing an orthogonal basis $x_{k,j}$, $j = 1, \ldots, N_k$, of the λ_j-eigenspace, the orthogonal set*

$$\{x_{k,j} : k \in \mathbb{Z}, \ k \geq 1, \ j = 1, \ldots, N_k\}$$

is complete.

The analogue of this theorem for variable coefficient symmetric elliptic second order differential operators with appropriate boundary conditions is also valid; indeed, the same proof works.

There is one more issue that should be observed. For the generalized Fourier series of v as in (13.6), and for $n < m$,

$$(13.7) \qquad \|v_n - v_m\|^2 = \sum_{k=n+1}^{m} |c_k|^2 \|x_k\|^2.$$

Since $\sum_{k=1}^{\infty} |c_k|^2 \|x_k\|^2$ converges, the differences of the partial sums, on the right-hand side of (13.7), go to 0 as $n, m \to \infty$. Thus, by (13.7), the same holds for the differences of the partial sums of $\sum_{k=1}^{\infty} c_k x_k$. In an ideal world, this ought to imply that the series $\sum_{k=1}^{\infty} c_k x_k$ converges. However, this need not be a case.

Definition 13.23. Suppose that V is a normed vector space.

We say that a sequence $\{v_n\}_{n=1}^{\infty}$ is a *Cauchy sequence* if $\|v_n - v_m\| \to 0$ as $n, m \to \infty$. Explicitly, this means that given any $\epsilon > 0$ there is $N > 0$ such that for $n, m \geq N$, $\|v_n - v_m\| < \epsilon$, i.e. beyond a certain point, all elements of the sequence are within ϵ of each other.

We say that a normed vector space V is *complete* if every Cauchy sequence converges.

Thus, we see that for any inner product space and any orthogonal set inside it, the generalized Fourier series for any $v \in V$ is Cauchy. If V is complete, it will thus converge to some $v' \in V$ (not necessarily to v though; recall the $\mathbb{R}^2$ example).

Now, an example of a complete normed vector space is, for Ω as above, $V = C^0(\overline{\Omega})$ with the C^0-norm:

$$\|f\|_{C^0} = \sup_{x \in \overline{\Omega}} |f(x)|;$$

see Problem 3.6. Another example of a complete normed vector space is ℓ^2, discussed in example (iv) above; see Problem 13.7. Unfortunately, an example of an incomplete normed vector space is $V = C^0(\overline{\Omega})$ with the L^2-norm, i.e. our inner product space; see Problem 13.8. (On the flipside, the C^0-norm does not arise from an inner product; see Problem 13.1.)

It turns out that every incomplete normed vector space V can be completed, i.e. there is a complete normed vector space $\hat{V}$ and an inclusion map $\iota : V \to \hat{V}$ which is linear such that for $v \in V$, $\|\iota(v)\|_{\hat{V}} = \|v\|_V$, and such that for any $v \in \hat{V}$, there is a sequence $\{v_n\}_{n=1}^{\infty}$ in V such that $\iota(v_n) \to v$ in $\hat{V}$ (i.e. the image of V under ι is dense in $\hat{V}$).

This completion is (essentially) unique. Note that if $\{v_n\}_{n=1}^{\infty}$ is Cauchy in V, then $\{\iota(v_n)\}_{n=1}^{\infty}$ is Cauchy in $\hat{V}$, so it converges. Moreover, ι is one-to-one, so one can simply think of elements of V as elements of $\hat{V}$. It is also useful to note that if V is an inner product space, then so is $\hat{V}$.

The actual construction of completion is very similar to how the real numbers are constructed from the rationals. There are Cauchy sequences of rationals which do not converge inside $\mathbb{Q}$; for instance, take truncated decimal expressions for $\sqrt{2}$. One constructs $\mathbb{R}$ by considering the set of such Cauchy sequences, more precisely equivalence classes of Cauchy sequences: two Cauchy sequences $\{v_n\}_{n=1}^\infty$ and $\{v_n'\}_{n=1}^\infty$ are equivalent if $\|v_n - v_n'\| \to 0$ as $n \to \infty$. For instance, in the case of $\mathbb{R}$, one wants the decimal and binary expansions of $\sqrt{2}$ to be equivalent; i.e. give the same element of $\mathbb{R}$, namely $\sqrt{2}$, which is exactly what is accomplished.

For readers desiring further details, the additional material in Chapter 5 provides some, including further references.

We thus make a bold definition. Suppose that Ω is a bounded domain in $\mathbb{R}^n$. We let $L^2(\Omega)$ *be the completion* $\hat{V}$ *of* $V = C^0(\overline{\Omega})$ *in the* $L^2(\Omega)$-*norm.* Thus, $L^2(\Omega)$ is a complete inner product space and $C^0(\overline{\Omega})$ can be regarded as a subset of $L^2(\Omega)$. Similarly, we let $L^1(\Omega)$ be the completion $\hat{V}$ of $V = C^0(\overline{\Omega})$ in the $L^1(\Omega)$-norm, which is thus a complete normed vector space.

We also let $L^2(\mathbb{R}^n)$ be the completion $\hat{V}$ of

$$V_N = \{f \in C^0(\mathbb{R}^n) : \ (1 + |x|)^N f \text{ is bounded}\},$$

with the inner product

$$\langle f, g \rangle = \int_{\mathbb{R}^n} f(x)\, \overline{g(x)}\, dx,$$

where we fixed some $N > n/2$. It is straightforward to show that the completion is independent of the choice of such an N, for if $n/2 < N' < N$, then any element of $V_{N'}$ can be approximated by elements of V_N in the $L^2(\mathbb{R}^n)$-norm.

While this definition of $L^2(\Omega)$ may sound radical indeed, elements of $L^2(\Omega)$ are actually distributions, i.e. very familiar objects. To see this, first note that if $\phi \in C_c^\infty(\Omega)$, $f \in L^2(\Omega)$, we can define

$$\iota_f(\phi) = \langle f, \overline{\phi} \rangle.$$

By Cauchy-Schwarz, we have

$$|\iota_f(\phi)| \le \|f\|_{L^2} \|\phi\|_{L^2} = \|f\|_{L^2} \Big(\int_\Omega |\phi|^2\, dx \Big)^{1/2} \le \|f\|_{L^2} \Big(\int_\Omega 1\, dx \Big)^{1/2} \sup_{x \in \Omega} |\phi(x)|,$$

so ι_f is indeed a distribution, i.e. it is continuous $C_c^\infty(\Omega) \to \mathbb{C}$ (the linearity being clear). If $f \in C^0(\overline{\Omega})$, this agrees with the usual definition since then

$$\langle f, \overline{\phi} \rangle = \int_\Omega f\phi\, dx.$$

Thus, $\iota : L^2(\Omega) \to \mathcal{D}'(\Omega)$. In order to be able to consider $L^2(\Omega)$ as a subset of $\mathcal{D}'(\Omega)$, we also need that ι be injective. But this is easy to see: first, for $f \in L^2(\Omega)$ there is a sequence $\phi_n \in \mathcal{C}_c^\infty(\Omega)$ such that $\phi_n \to f$ in $L^2(\Omega)$. Indeed, the existence of such a sequence with ϕ_n replaced by $f_n \in C^0(\overline{\Omega})$ follows from the definition of $L^2(\Omega)$ as a completion, so one merely needs to check that the f_n can be approximated within, say, $1/n$, in L^2-norm by some ϕ_n; but this is straightforward, and now all integrals are simply Riemann integrals. Now, for this sequence ϕ_n, by the continuity of the inner product,

$$\langle f, \overline{\phi_n} \rangle \to \langle f, \overline{f} \rangle = \|f\|_{L^2}^2.$$

Now, if $f \neq 0$, then the right-hand side is non-zero, so in particular for sufficiently large n, $\iota_f(\phi_n) = \langle f, \overline{\phi_n} \rangle \neq 0$, showing that ι_f is *not* the zero distribution, so ι is injective. Hence we can indeed simply think of $L^2(\Omega)$ as a subspace of distributions.

We also remark that if V is an inner product space and $\{x_n\}_{n=1}^\infty$ is a complete orthogonal set in V, then it is also complete in the completion $\hat{V}$ of V. This follows immediately since for $v \in V$ we can take a sequence $v_n \in V$ such that $v_n \to v$, and use the fact that the generalized Fourier series of v_n converges to v_n, as well as the fact that the generalized Fourier series of v converges to some v' to show that $v' = v$.

Problems

Problem 13.1.

(i) Show the *parallelogram law* in inner product spaces V:

$$\|v + w\|^2 + \|v - w\|^2 = 2(\|v\|^2 + \|w\|^2), \ v, w \in V.$$

Draw a picture illustrating $v \pm w$ in terms of v and w on a parallelogram with two adjacent sides given by v and w.

(ii) Show that for Ω a bounded domain in $\mathbb{R}^n$, $V = C^0(\overline{\Omega})$ with the C^0-norm, $\sup_{x \in \overline{\Omega}} |f(x)|$, is *not* an inner product space, i.e. there exists no inner product on V such that the norm with respect to this inner product is the C^0-norm. (*Hint:* Can you find a counterexample to the parallelogram law?)

Note: it can be shown that the norm on a normed vector space arises from an inner product if and only if it satisfies the parallelogram law.

Problem 13.2. Let $\phi(x) = |x|$ on $[-\pi, \pi]$. Let

$$f(x) = a_0 + a_1 \cos x + a_2 \cos(2x) + b_1 \sin x + b_2 \sin(2x).$$

With what choice of the coefficients a_j and b_j is the L^2 error $\|f - \phi\|$ minimal? (Here $\|f - \phi\|^2 = \int_{-\pi}^{\pi} |f(x) - \phi(x)|^2 \, dx$.)

Problem 13.3. Let $V = C([0, \ell])$ with the inner product $\langle f, g \rangle = \int_0^\ell f(x)\overline{g(x)}\,dx$. If $D \subset V$ is a subspace, and $A : D \to V$ is symmetric, we say that A is positive if $\langle Av, v \rangle \geq 0$ for all $v \in D$.

(i) Show that if A is positive then all eigenvalues λ of A are ≥ 0.

(ii) Show that $A = -\frac{d^2}{dx^2}$ with domain

$$D = \{f \in C^2([0, \ell]) : f(0) = f'(\ell) = 0\}$$

is symmetric and is positive.

(iii) Show that $A = \frac{d^4}{dx^4}$ with domain

$$D = \{f \in C^4([0, \ell]) : f(0) = f'(0) = f(\ell) = f'(\ell) = 0\}$$

is symmetric and is positive.

Problem 13.4.

(i) Using the general 'separated' solution you found in Problem 12.1, solve the wave equation

$$u_{tt} = c^2 u_{xx}, \ u(0, t) = 0, \ u_x(\ell, t) = 0,$$

with initial conditions

$$u(x, 0) = 0, \ u_t(x, 0) = x(x - \ell)^2.$$

(*Hint:* See Problem 13.3, part (ii).)

(ii) Using the general 'separated' solution you found in Problem 12.1, solve the heat equation

$$u_t = k u_{xx}, \ u(0, t) = 0, \ u_x(\ell, t) = 0, \ u(x, 0) = x(x - \ell)^2.$$

You may assume throughout that the generalized Fourier series you construct converges to the function it is supposed to represent.

Problem 13.5. The *2-torus* is the set $\mathbb{T}^2 = \mathbb{S}^1 \times \mathbb{S}^1$, where one may write the variables as (θ_1, θ_2), or (θ, ω) or (x, y) (which notation we'll adopt). Consider the operator

$$A = -\frac{d^2}{dx^2} - \frac{d^2}{dy^2}$$

defined on $D = C^2(\mathbb{T}^2)$, with $V = C^0(\mathbb{T}^2)$. Note that one can think of the functions on $\mathbb{T}^2$ as 2π-periodic functions on $\mathbb{R}^2$, periodic both in x and y, i.e. functions f with

$$f(x + 2\pi, y) = f(x, y) = f(x, y + 2\pi)$$

for all $x, y \in \mathbb{R}$. (The torus can be thought of as the surface of a donut, though in physical space $\mathbb{R}^3$ there would be some relative stretching/compression of the sides of the torus!)

(i) Find the eigenfunctions $A : D \to V$, and show that they can be chosen to be orthogonal with respect to the inner product

$$\langle f, g \rangle = \int_0^{2\pi} \int_0^{2\pi} f(x, y) \, \overline{g(x, y)} \, dx \, dy.$$

(*Hint:* Look for product type eigenfunctions $u : u(x, y) = X(x)Y(y)$.)

(ii) Solve the Helmholtz equation on the 2-torus:

$$u_{xx} + u_{yy} - u = f(x, y),$$

where f is a given function. You may leave your answer as a series, with coefficients calculated in terms of f, and you may also assume that the series you calculate actually converges.

Hint: Expand both u and f in terms of eigenfunctions of A.

Problem 13.6 (*Gram-Schmidt orthogonalization*). Show that if $v_1, \ldots, v_n$ are linearly independent vectors in an inner product space V, then there exist vectors $x_1, \ldots, x_n \in V$ such that for all $m \leq n$, $\text{span}\{v_1, \ldots, v_m\} = \text{span}\{x_1, \ldots, x_m\}$, and such that $x_1, \ldots, x_n$ are orthonormal, i.e. $\langle x_i, x_j \rangle = 0$ if $i \neq j$ and $\|x_j\|^2 = 1$ for all j.

Note: Notice that this procedure works equally well if one has a sequence $v_1, v_2, \ldots$ of linearly independent vectors, where linear independence means that no *finite* linear combination of these other than the trivial one is 0.

Hint: Start by letting $x_1 = \frac{v_1}{\|v_1\|}$, and then let y_2 be the orthogonal projection of v_2 to the orthocomplement of $\text{span}\{x_1\}$, i.e. $(v_2)_\perp$ with $\perp$ taken relative to x_1, and $x_2 = \frac{y_2}{\|y_2\|}$. Then proceed by letting y_3 be the orthogonal projection of v_3 to the orthocomplement of $\text{span}\{x_1, x_2\}$, etc.; see the discussion of generalized Fourier coefficients to see how to compute these (thereby showing that these exist). Make sure you check that one never has to divide by 0!

Problem 13.7. In this problem we consider ℓ^2 of example (iv), consisting of square summable sequences; parts (ii)-(iii) are more difficult, but help in the understanding of ℓ^2, and more generally of sequence spaces.

(i) Show that ℓ^2 is indeed an inner product space, i.e. check carefully properties (i)-(iii) of the definition of an inner product space.

(ii) Suppose that $\{x_n\}_{n=1}^\infty$ is a Cauchy sequence of elements of ℓ^2, so for each n, x_n is a sequence $\{x_{n,j}\}_{j=1}^\infty$ which lies in ℓ^2, and

$$\|x_n - x_m\|^2 = \sum_{j=1}^\infty |x_{n,j} - x_{m,j}|^2 \to 0$$

as $n, m \to \infty$. Show that for any $\epsilon > 0$ there exist J and N such that for $n \geq N$, $\sum_{j \geq J} |x_{n,j}|^2 < \epsilon^2$, i.e. for large n, the 'tails' of the

x_n are uniformly small in this square summable sense. *Hint:* Take N such that for $n, m \geq N$, $\|x_n - x_m\| < \epsilon/2$, and take J such that $\sum_{j=J}^{\infty} |x_{N,j}|^2 < \epsilon^2/4$.

(iii) Show that ℓ^2 is a complete inner product space (i.e. is complete with the norm induced by the inner product).

    Hint: Consider a Cauchy sequence $\{x_n\}_{n=1}^{\infty}$ of elements of ℓ^2, as above. Then $\{x_n\}_{n=1}^{\infty}$ is also bounded by Problem 3.5; say $\|x_n\| \leq M$ for all n. Show first that for each j, $\{x_{n,j}\}_{n=1}^{\infty}$ is Cauchy (i.e. the jth elements of the x_n form a Cauchy sequence as j varies) and thus converges to some $x_{\infty,j}$ by the completeness of $\mathbb{C}$. Then show, using the boundedness of $\{x_n\}_{n=1}^{\infty}$, that $\sum_{j=1}^{\infty} |x_{\infty,j}|^2 \leq M^2 < \infty$, so $\{x_{\infty,j}\}_{j=1}^{\infty} \in \ell^2$. Finally, show that $x_n \to x$ in ℓ^2, i.e.

$$\|x_n - x\|^2 = \sum_{j=1}^{\infty} |x_{n,j} - x_{\infty,j}|^2 \to 0$$

as $n \to \infty$, using (ii), namely that the tails of the x_n are uniformly small.

Problem 13.8. Show that $V = C^0([0,1])$ equipped with the L^2-norm is not complete by considering the sequence

$$f_n(x) = \begin{cases} 0, & 0 \leq x \leq 1/2, \\ n(x - 1/2), & 1/2 \leq x \leq 1/2 + 1/n, \\ 1, & 1/2 + 1/n \leq x \leq 1 \end{cases}$$

in V, showing that it is Cauchy and that it does not converge to any element of V.

Chapter 14

Convergence of the Fourier series and the Poisson formula on disks

1. Notions of convergence

We now discuss convergence of the Fourier series on compact intervals I. 'Convergence' depends on the notion of convergence we use, such as

(i) L^2: $u_j \to u$ in L^2 if $\|u_j - u\|_{L^2} \to 0$ as $j \to \infty$;

(ii) *uniform*, or C^0: $u_j \to u$ uniformly if

$$\|u_j - u\|_{C^0} = \sup_{x \in I} |u_j(x) - u(x)| \to 0;$$

(iii) *uniform with all derivatives*, or C^∞: $u_j \to u$ in C^∞ if for all non-negative integers k, $\sup_{x \in I} |\partial^k u_j(x) - \partial^k u(x)| \to 0$.

(iv) *pointwise*: $u_j \to u$ pointwise if for each $x \in I$, $u_j(x) \to u(x)$, i.e. for each $x \in I$, $|u_j(x) - u(x)| \to 0$.

Note that pointwise convergence is too weak for most purposes, so e.g. just because $u_j \to u$ pointwise, it does not follow that $\int_I u_j(x)\,dx \to \int u(x)\,dx$. This would follow, however, if one assumes uniform convergence, or indeed L^2- (or L^1-) convergence, since

$$\left| \int_I u_j(x)\,dx - \int_I u(x)\,dx \right| = \left| \int_I (u_j(x) - u(x))\,dx \right|$$
$$= |\langle u_j - u, 1 \rangle| \leq \|u_j - u\|_{L^2} \|1\|_{L^2}.$$

Note also that uniform convergence implies L^2-convergence since

$$\|u_j - u\|_{L^2} = \left(\int_I |u_j - u|^2\,dx \right)^{1/2}$$
$$\leq \left(\int_I \|u_j - u\|_{C^0}^2\,dx \right)^{1/2} = \left(\int_I 1\,dx \right)^{1/2} \|u_j - u\|_{C^0},$$

so if $u_j \to u$ uniformly, it also converges in L^2. Uniform convergence also implies pointwise convergence directly from the definition. On the other hand, convergence in C^∞ implies uniform convergence directly from the definition.

On the failure of convergence side: the uniform limit of a sequence of continuous functions is continuous, so in view of the continuity of the complex exponentials, sines and cosines, the various Fourier series cannot converge uniformly *unless the limit is continuous*. On the other hand, even if the limit is continuous, the convergence may not be uniform; understanding the conditions under which it is uniform, is one of our first tasks.

There are two issues regarding convergence: first, whether the series in question converges at all (in whatever sense we are interested in) and second, whether it converges to the desired limit, in this case the function whose Fourier series we are considering. The first part is easier to answer: we have already seen that even the generalized Fourier series converges in L^2 (but not necessarily to the function!).

Now consider uniform convergence. Recall that the typical way one shows convergence of a series is to show that each term in absolute value is $\leq M_n$, where M_n is a non-negative constant, such that $\sum_n M_n$ converges. Similarly, one shows that a series $\sum_n u_n(x)$ converges uniformly by showing that there are non-negative constants M_n such that $\sup_{x \in I} |u_n(x)| \leq M_n$ and such that $\sum_n M_n$ converges. The *Weierstrass M-test* is the statement that under these assumptions the series $\sum_n u_n(x)$ converges uniformly. In particular, if one shows that for $n \neq 0$ this boundedness holds with $M_n = C/|n|^s$ where $C > 0$ and $s > 1$, the Weierstrass M-test shows that the series converges uniformly. (Since a finite number of terms do not affect convergence, one can always ignore a finite number of terms if it is convenient.)

2. Uniform convergence of the Fourier transform

So suppose that ϕ is a function on $[-\ell, \ell]$ and that its 2ℓ-periodic extension, denoted by ϕ_{ext}, is a C^k, $k \geq 1$, integer. The *full Fourier series* is

$$(14.1) \qquad \sum_{n=-\infty}^{\infty} C_n e^{in\pi x/\ell}, \quad C_n = \frac{1}{2\ell} \int_{-\ell}^{\ell} \phi(x) \, e^{-in\pi x/\ell} \, dx.$$

This gives us the bound

$$(14.2) \quad |C_n| \leq \frac{1}{2\ell} \int_{-\ell}^{\ell} |\phi(x) \, e^{-in\pi x/\ell}| \, dx = \frac{1}{2\ell} \int_{-\ell}^{\ell} |\phi(x)| \, dx \leq \sup_{x \in [-\ell,\ell]} |\phi(x)|,$$

so the coefficients are bounded; but it is not clear if they have any decay, hence the uniform convergence of the Fourier series is unclear.

Now we integrate by parts, putting additional derivatives on ϕ. Thus, for $n \neq 0$,

$$C_n = \frac{1}{2\ell} \left(\frac{\ell}{in\pi} \right) \int_{-\ell}^{\ell} (\partial_x \phi)(x) \, e^{-in\pi x/\ell} \, dx,$$

since the boundary terms vanish in view of the fact that ϕ_{ext} is C^k. Note that the right-hand side looks like the original expression for C_n, except that ϕ has been replaced by $\partial_x \phi$ and that the factor $\frac{\ell}{in\pi}$ appears in front. Thus, repeating this argument k times, we deduce that

$$C_n = \frac{1}{2\ell} \left(\frac{\ell}{in\pi} \right)^k \int_{-\ell}^{\ell} (\partial_x^k \phi)(x) \, e^{-in\pi x/\ell} \, dx.$$

In particular, using the analogue of the estimate (14.2) to estimate the integral, we deduce that

$$|C_n| \leq \left(\frac{\ell}{|n|\pi} \right)^k \sup_{x \in [-\ell,\ell]} |(\partial_x^k \phi)(x)| = \frac{C}{|n|^k}, \quad C = \left(\frac{\ell}{\pi} \right)^k \sup_{x \in [-\ell,\ell]} |(\partial_x^k \phi)(x)|.$$

In particular, for $k \geq 2$, we deduce by the Weierstrass M-test that the Fourier series converges uniformly.

It is possible to improve this conclusion as follows. Note that

$$C_n = \left(\frac{\ell}{in\pi} \right)^k C_{k,n}, \quad n \neq 0,$$

where $C_{k,n}$ is the Fourier coefficient of $\partial_x^k \phi$. Thus, as long as $\partial_x^k \phi$ is continuous, or indeed L^2, we have from Bessel's inequality, with $X_n(x) = e^{in\pi x/\ell}$, that

$$2\ell \sum_{n \in \mathbb{Z}} |C_{k,n}|^2 = \sum_{n \in \mathbb{Z}} |C_{k,n}|^2 \|X_n\|^2 \leq \|\partial_x^k \phi\|_{L^2}^2.$$

Now, in order to apply the Weierstrass M-test we need that $\sum_{n\in\mathbb{Z}}|C_n|$ converges. We now use the Cauchy-Schwarz inequality for sequences, with inner product on the sequences given by

$$\langle\{a_n\},\{b_n\}\rangle_{\ell^2} = \sum_{n\in\mathbb{Z}} a_n\overline{b_n}.$$

Below by $\left\{\left(\frac{\ell}{|n|\pi}\right)^k\right\}$ we mean the sequence whose $n=0$ entry is 0. We thus obtain using $k > 1/2$ that

$$\sum_{n\in\mathbb{Z},\,n\neq 0}|C_n| = \sum_{n\in\mathbb{Z},\,n\neq 0}\left(\frac{\ell}{|n|\pi}\right)^k|C_{k,n}| = \left\langle\{|C_{k,n}|\},\left\{\left(\frac{\ell}{|n|\pi}\right)^k\right\}\right\rangle$$

$$\leq \|\{|C_{k,n}|\}\|_{\ell^2}\left\|\left\{\left(\frac{\ell}{|n|\pi}\right)^k\right\}\right\|_{\ell^2}$$

$$= \left(\sum_{n\in\mathbb{Z}}|C_{k,n}|^2\right)^{1/2}\left(\sum_{n\in\mathbb{Z},\,n\neq 0}\left(\frac{\ell}{|n|\pi}\right)^{2k}\right)^{1/2}$$

$$\leq \frac{1}{\sqrt{2\ell}}\|\partial_x^k\phi\|_{L^2}\left(\sum_{n\in\mathbb{Z},\,n\neq 0}\left(\frac{\ell}{|n|\pi}\right)^{2k}\right)^{1/2},$$

so it indeed converges. We thus deduce by the Weierstrass M-test that the Fourier series converges uniformly if merely $k \geq 1$.

Now, the term-by-term m-times differentiated Fourier series is

$$\sum_{n=-\infty}^{\infty} C_n\left(\frac{in\pi}{\ell}\right)^m e^{in\pi x/\ell}, \quad C_n = \frac{1}{2\ell}\int_{-\ell}^{\ell}\phi(x)\,e^{-in\pi x/\ell}\,dx,$$

so for $n \neq 0$ its coefficients satisfy the estimate

$$\left|C_n\left(\frac{in\pi}{\ell}\right)^m\right| \leq \left(\frac{\ell}{|n|\pi}\right)^{k-m}\sup_{x\in[-\ell,\ell]}|(\partial_x^k\phi)(x)| = \frac{C}{|n|^{k-m}},$$

$$C = \left(\frac{\ell}{\pi}\right)^k\sup_{x\in[-\ell,\ell]}|(\partial_x^k\phi)(x)|.$$

We thus deduce that the term-by-term m-times differentiated Fourier series converges uniformly for $m \leq k-2$, and thus (by the standard analysis theorem) the uniform limit of the Fourier series is actually $k-2$-times differentiable, with the derivative given by the term-by-term differentiated series. Modifying this argument as above, using Bessel's inequality, we in fact get that the Fourier series is actually $k-1$-times differentiable, with the derivative given by the term-by-term differentiated series.

In the extreme case, when the 2ℓ-periodic extension of ϕ is $\mathcal{C}^\infty$, this tells us that in fact the Fourier series converges in $\mathcal{C}^\infty$. Notice the similarities with the Fourier transform, both in the calculations above (how integration by parts gave us the necessary decay), and also how well behaved $\mathcal{C}^\infty$ is! In fact, one way of writing our results thus far, which is similar to what we did for the Fourier transform, is that the map

$$T : \phi \mapsto \sum_{n=-\infty}^{\infty} C_n e^{in\pi x/\ell}, \ C_n = \frac{1}{2\ell} \int_{-\ell}^{\ell} \phi(x) \, e^{-in\pi x/\ell} \, dx$$

is continuous as a map

$$T : L^2([-\ell, \ell]) \to L^2([-\ell, \ell]),$$

and also as

$$T : C_{\text{per}}^1([-\ell, \ell]) \to C_{\text{per}}^0([-\ell, \ell]),$$

as well as

$$T : C_{\text{per}}^\infty([-\ell, \ell]) \to C_{\text{per}}^\infty([-\ell, \ell]),$$

where the subscript per states that the 2ℓ-periodic extensions of these functions must have the stated regularity properties. Note the loss of derivatives as a map on C^1, or more generally C^k. We would like to say that this is the identity map in each case; by the density of $C_{\text{per}}^\infty([-\ell, \ell])$ in these spaces it would suffice to show this in the last case.

A more systematic way of achieving the same conclusion regarding convergence is the following. The functions $X_n(x) = e^{in\pi x/\ell}$ are eigenfunctions of $A = -i\frac{d}{dx}$ with periodic boundary conditions and with eigenvalue $\lambda_n = n\pi/\ell$. Now, in general, suppose that X_n are orthogonal eigenfunctions of a symmetric operator $A : D \to V$ on an inner product space V with eigenvalue λ_n (which are thus real), and suppose that $\phi, A\phi, \ldots, A^k\phi \in D$. Then for n such that $\lambda_n \neq 0$, the generalized Fourier coefficients

$$C_n = \frac{\langle \phi, X_n \rangle}{\|X_n\|^2}$$

satisfy

$$C_n = \lambda_n^{-1} \frac{\langle \phi, \lambda_n X_n \rangle}{\|X_n\|^2} = \lambda_n^{-1} \frac{\langle \phi, AX_n \rangle}{\|X_n\|^2} = \lambda_n^{-1} \frac{\langle A\phi, X_n \rangle}{\|X_n\|^2}.$$

Repeating k-times, we deduce that

$$C_n = \lambda_n^{-k} \frac{\langle A^k\phi, X_n \rangle}{\|X_n\|^2}.$$

Thus,

$$|C_n| \leq |\lambda_n|^{-k} \|A^k\phi\|_{L^2} \|X_n\|_{L^2}^{-1}.$$

In the special case $X_n(x) = e^{in\pi x/\ell}$, we get the estimate

$$|C_n| \le \left(\frac{\ell}{|n|\pi}\right)^k \|\partial^k \phi\|_{L^2} \frac{1}{2\ell} = \frac{C}{|n|^k}, \quad C = \frac{1}{2\ell}\left(\frac{\ell}{\pi}\right)^k \|\partial^k \phi\|_{L^2},$$

which is almost the same estimate we had beforehand, but with a slightly different constant, and the norm of the derivative we use is weaker, not the uniform (or sup) norm, but the L^2-norm. But now note that this argument works even for instance for the cosine and sine Fourier series, using that both the sine and cosine are bounded by 1 in absolute value, i.e. we do not have to write down these cases separately.

3. What does the Fourier series converge to?

We have thus shown that under appropriate assumptions, depending on the notion of convergence, the various kinds of Fourier series all converge. Note that we may need stronger assumptions than the kind of convergence we would like, for instance we needed to know something about derivatives of ϕ to conclude uniform convergence. However, we still need to discuss *what* the Fourier series converges to!

Note that when the Fourier sine series,

$$\sum_{n=1}^{\infty} B_n \sin(n\pi x/\ell), \quad B_n = \frac{2}{\ell} \int_0^\ell \phi(x) \sin(n\pi x/\ell)\,dx,$$

converges, as it does say in L^2 when $\phi \in L^2$, or uniformly if ϕ is C^2 and satisfies the boundary conditions $\phi(0) = 0 = \phi(\ell)$, then it converges to an odd 2ℓ-periodic function since each term $\sin(n\pi x/\ell)$ is such. Similarly, the Fourier cosine series, when it converges, converges to an even 2ℓ-periodic function. Moreover, note that if a function Φ is odd on $[-\ell, \ell]$, its full Fourier series,

$$A_0' + \sum_{n=1}^{\infty} A_n' \cos(n\pi x/\ell) + B_n' \sin(n\pi x/\ell),$$

$$A_0' = \frac{1}{2\ell} \int_{-\ell}^{\ell} \Phi(x)\,dx, \quad A_n' = \frac{1}{\ell} \int_{-\ell}^{\ell} \Phi(x) \cos(n\pi x/\ell)\,dx,$$

$$B_n' = \frac{1}{\ell} \int_{-\ell}^{\ell} \Phi(x) \sin(n\pi x/\ell)\,dx, \quad n \ge 1,$$

satisfies

$$A_0' = 0, \quad A_n' = 0, \quad B_n' = \frac{2}{\ell} \int_0^\ell \Phi(x) \sin(n\pi x/\ell)\,dx,$$

i.e. if Φ is the odd 2ℓ-periodic extension of ϕ, then $B_n' = B_n$, with B_n being the Fourier sine coefficient of ϕ on $[0, \ell]$. In view of an analogous argument for the Fourier cosine series, convergence issues for both the Fourier sine and cosine series on $[0, \ell]$ can be reduced to those for the full Fourier series on $[-\ell, \ell]$, so we only consider the latter. Moreover, the change of variables of $\theta = \frac{n\pi x}{\ell}$ preserves all the notions of convergence, so it suffices to consider the Fourier series on $[-\pi, \pi]$. (The general case works directly, by the same arguments, but we have to write less after this rescaling.)

One way of proving the convergence of the Fourier series is to proceed similarly to the Fourier transform. We write $\mathcal{C}^\infty(\mathbb{S}^1) = \mathcal{C}^\infty_{\mathrm{per}}([-\pi, \pi])$, and $s(\mathbb{Z})$ for sequences $\{c_n\}_{n \in \mathbb{Z}}$ such that $|n|^k |c_n|$ is bounded for all $k \geq 0$ (s stands for 'Schwartz sequences'), with the notion of convergence that a sequence $\{\{c_{j,n}\}_{n \in \mathbb{Z}}\}_{j=1}^\infty$ converges to $\{c_n\}_{n \in \mathbb{Z}}$ in $s(\mathbb{Z})$ as $j \to \infty$ if for all $k \geq 0$ integers,

$$\sup_{n \in \mathbb{Z}} |n|^k |c_{j,n} - c_n| \to 0$$

as $j \to \infty$. Thus, the map

$$\mathcal{F} : \mathcal{C}^\infty(\mathbb{S}^1) \to s(\mathbb{Z}), \quad (\mathcal{F}\phi)_n = \frac{1}{2\pi} \int_{-\pi}^{\pi} \phi(x) e^{-in\theta} \, d\theta$$

is continuous and satisfies

$$\mathcal{F}\frac{d\phi}{d\theta} = in\mathcal{F}\phi$$

and

$$(\mathcal{F}(e^{i\theta}\phi))_n = \frac{1}{2\pi} \int_{-\pi}^{\pi} \phi(x) e^{-i(n-1)\theta} \, d\theta = (\mathcal{F}\phi)_{n-1}.$$

Similarly, with

$$\mathcal{F}^{-1} : s(\mathbb{Z}) \to \mathcal{C}^\infty(\mathbb{S}^1), \quad \mathcal{F}^{-1}\{c_n\}(\theta) = \sum_{n=-\infty}^{\infty} c_n e^{in\theta}$$

we have

$$\mathcal{F}^{-1}\{inc_n\} = \frac{d}{d\theta}\mathcal{F}^{-1}\{c_n\},$$

while

$$\mathcal{F}^{-1}\{c_{n-1}\} = \sum_{n=-\infty}^{\infty} c_{n-1} e^{in\theta} = \sum_{n=-\infty}^{\infty} c_n e^{i(n+1)\theta} = e^{i\theta}\mathcal{F}^{-1}\{c_n\}.$$

Thus, the map $T = \mathcal{F}^{-1}\mathcal{F} : \mathcal{C}^\infty(\mathbb{S}^1) \to \mathcal{C}^\infty(\mathbb{S}^1)$ satisfies

$$T\frac{d}{d\theta} = \frac{d}{d\theta}T, \quad Te^{i\theta} = e^{i\theta}T,$$

in analogy with properties of the Fourier transform. We again have a lemma:

Lemma 14.1. *Suppose $T : \mathcal{C}^\infty(\mathbb{S}^1) \to \mathcal{C}^\infty(\mathbb{S}^1)$ is linear and commutes with $e^{i\theta}$ and $\frac{d}{d\theta}$. Then T is a scalar multiple of the identity map, i.e. there exists $c \in \mathbb{C}$ such that $Tf = cf$ for all $f \in \mathcal{C}^\infty(\mathbb{S}^1)$.*

Proof. Let $\omega \in \mathbb{R}$. We show first that if $\phi(\omega) = 0$ and $\phi \in \mathcal{C}^\infty(\mathbb{S}^1)$, then $(T\phi)(\omega) = 0$. Indeed, if we let

$$\phi_1(\theta) = \phi(\theta)/(e^{i\theta} - e^{i\omega}),$$

then by Taylor's theorem (or L'Hopital's rule) ϕ_1 is $\mathcal{C}^\infty$, and it is 2π-periodic as both the denominator and the numerator are. Thus, $\phi = (e^{i\theta} - e^{i\omega})\phi_1$, and hence

$$T\phi = (e^{i\theta} - e^{i\omega})(T\phi_1),$$

where we used the fact that T is linear and commutes with multiplication by $e^{i\theta}$. Substituting in $\theta = \omega$ yields $(T\phi)(\omega) = 0$ indeed.

Thus, fix $\omega \in \mathbb{R}$, and let $g \in \mathcal{C}^\infty(\mathbb{S}^1)$ be the function $g \equiv 1$. Let $c(\omega) = (Tg)(\omega)$; thus $c \in \mathcal{C}^\infty(\mathbb{S}^1)$. We claim that for $f \in \mathcal{C}^\infty(\mathbb{S}^1)$,

$$(Tf)(\omega) = c(\omega)f(\omega).$$

Indeed, let $\phi(\theta) = f(\theta) - f(\omega)g(\theta)$, so $\phi(\omega) = f(\omega) - f(\omega)g(\omega) = 0$. Thus, $0 = (T\phi)(\omega) = (Tf)(\omega) - f(\omega)(Tg)(\omega) = (Tf)(\omega) - c(\omega)f(\omega)$, proving our claim.

We have not used the fact that T commutes with ∂_θ yet. But

$$c(\omega)(\partial_\theta f)(\omega) = T(\partial_\theta f)(\omega) = \partial_\theta(Tf)|_{\theta=\omega} = \partial_\theta(c(\theta)f(\theta))|_{\theta=\omega}$$
$$= (\partial_\theta c)(\omega)f(\omega) + c(\omega)(\partial_\theta f)(\omega).$$

Comparing the two sides and taking f such that f never vanishes yields

$$(\partial_\theta c)(\omega) = 0$$

for all ω. Thus, c is a constant, proving the lemma. $\square$

The constant c can be evaluated by applying T to a single function, e.g. to $f \equiv 1$, which yields $(\mathcal{F}f)_n = 0$ if $n \neq 0$, $(\mathcal{F}f)_0 = 1$ and hence $Tf = f$, so T is indeed the identity map.

Now, if $\phi \in L^2([-\pi, \pi])$, we take ϕ_n whose 2π-periodic extensions are $\mathcal{C}^\infty$, which converge to ϕ in L^2. Since $T\phi_n = \phi_n$ and $T : L^2([-\pi, \pi]) \to L^2([-\pi, \pi])$ is continuous, we deduce that

$$T\phi = \lim_{n\to\infty} T\phi_n = \lim_{n\to\infty} \phi_n = \phi,$$

i.e. the Fourier series converges in L^2 to ϕ.

If $\phi \in C^1(\mathbb{S}^1)$, then we have seen that the Fourier series of ϕ converges uniformly to some function ψ, and as $C^1(\mathbb{S}^1) \subset L^2(\mathbb{S}^1)$, it converges in L^2 to ϕ. But uniform convergence implies L^2 convergence, so in fact the Fourier series converges in L^2 to ψ, so $\psi = \phi$, and thus we conclude that the Fourier series of ϕ converges uniformly to ϕ.

As we could reduce all the Fourier series that we have considered to the full Fourier series on $[-\pi, \pi]$, we deduce the following result:

Theorem 14.2. *The following orthogonal sets of functions are complete in the respective vector spaces V, and thus the corresponding Fourier series of any $\phi \in V$ converges in L^2 to ϕ.*

(i) *In $V = L^2([0, \ell])$, with the standard inner product,*

$$f_n(x) = \sin\left(\frac{n\pi x}{\ell}\right), \ n \geq 1, \ n \in \mathbb{Z}.$$

(ii) *In $V = L^2([0, \ell])$, with the standard inner product,*

$$f_0(x) = 1, \ f_n(x) = \cos\left(\frac{n\pi x}{\ell}\right), \ n \geq 1, \ n \in \mathbb{Z}.$$

(iii) *In $V = L^2([-\ell, \ell])$, with the standard inner product,*

$$f_n(x) = e^{in\pi x/\ell}, \ n \in \mathbb{Z}.$$

(iv) *In $V = L^2([-\ell, \ell])$, with the standard inner product,*

$$f_0(x) = 1, \ f_n(x) = \cos\left(\frac{n\pi x}{\ell}\right), \ n \geq 1, \ n \in \mathbb{Z},$$

$$g_n(x) = \sin\left(\frac{n\pi x}{\ell}\right), \ n \geq 1, \ n \in \mathbb{Z}.$$

Moreover, if the appropriate extension, namely

(i) *odd, 2ℓ-periodic,*

(ii) *even, 2ℓ-periodic,*

(iii) *2ℓ-periodic,*

(iv) *2ℓ-periodic*

of ϕ is C^1, then the corresponding Fourier series converges to ϕ uniformly, and if the appropriate extension is C^∞, then the convergence is in C^∞.

We discuss a more traditional proof, using the Dirichlet kernel, in the Appendix.

4. The Dirichlet problem on the disk

We also connect these results to solving the Dirichlet problem for the Laplacian on the disk of radius r_0,

$$D = \mathbb{B}^2_{r_0} = \{x \in \mathbb{R}^2 : |x| < r_0\};$$

namely,

$$\Delta u = 0, \ x \in D,$$

$$u|_{\partial D} = h,$$

with h a given function on ∂D. The general separated solution was

$$(14.3) \qquad u(r, \theta) = A_0 + \sum_{n=1}^{\infty} r^n (A_n \cos(n\theta) + B_n \sin(n\theta)),$$

and using orthogonality, we have deduced that the coefficients must be

$$A_0 = \frac{1}{2\pi} \int_0^{2\pi} h(\theta) \, d\theta,$$

$$A_n = \frac{1}{\pi r_0^n} \int_0^{2\pi} h(\theta) \cos(n\theta) \, d\theta,$$

$$B_n = \frac{1}{\pi r_0^n} \int_0^{2\pi} h(\theta) \sin(n\theta) \, d\theta.$$

Now, if h is C^1, then we have seen that its Fourier series converges uniformly to h, i.e. $u(r_0, \theta) = h(\theta)$. Also, if h is C^2, then $|A_n|, |B_n| \leq C/r_0^n n^2$ for $n \geq 1$. This suffices to conclude that for $r < r_0$,

$$|r^n (A_n \cos(n\theta) + B_n \sin(n\theta))| \leq \frac{2C}{n^2} \left(\frac{r}{r_0}\right)^n.$$

By the Weierstrass M-test, the series (14.3) converges uniformly on the closed disk, i.e. for $r \in [0, r_0]$, and in addition the term-by-term m-times differentiated series still converges uniformly in $[0, \rho]_r$ for any $\rho < r_0$ since the factors of n^m this differentiation gives are counterbalanced by $\left(\frac{\rho}{r_0}\right)^n$, and $\sum_{n=1}^{\infty} n^m \alpha^n$ converges whenever $m \in \mathbb{R}$ and $\alpha \in (0, 1)$, so u is a C^∞ function of r and θ for $r \in [0, r_0)$. A simple modification of our argument for uniform convergence of the Fourier series of C^1 functions would in fact extend this conclusion to h merely C^1. As $v = r^n \cos(n\theta)$ and $v = r^n \sin(n\theta)$ solve $\Delta v = 0$ for $r \in (0, r_0)$, u itself satisfies $\Delta u = 0$ for $r \in (0, r_0)$—the origin may a priori be a problem, since polar coordinates are singular there. That this is not the case follows from $r^n e^{in\theta} = (re^{i\theta})^n = (x + iy)^n$, so taking real and imaginary parts shows that $r^n \cos(n\theta)$ and $r^n \sin(n\theta)$ are simply polynomials in D. In particular, the term-by-term differentiated series, differentiated with respect to x or y, actually still converges in $|x|^2 + |y|^2 < r_0$, uniformly in $|x|^2 + |y|^2 \leq \rho$, $\rho < r_0$, so in fact u is C^∞ at the origin as well and solves $\Delta u = 0$ for each summand.

Thus, for at least $h \in C^2$ (and as a simple argument shows, $h \in C^1$ in fact), we have solved the PDE: we found $u \in C^2(D) \cap C^0(\overline{D})$, solving our problem. It turns out that one can write the solution in a more explicit form, which enables one to obtain a better result. Namely, for $r \leq \rho$, where $\rho < r_0$, in view of the absolute convergence of the series (with plenty of room left, so one could even differentiate arbitrarily many times and have absolute and uniform convergence),

$$u(r,\theta) = \frac{1}{2\pi} \int_{-\pi}^{\pi} h(\omega)\, d\omega$$

$$+ \sum_{n=1}^{\infty} \left(\frac{r}{r_0}\right)^n \frac{1}{\pi} \int_{-\pi}^{\pi} h(\omega)\left(\cos(n\omega)\cos(n\theta) + \sin(n\omega)\sin(n\theta)\right) d\omega$$

$$= \frac{1}{2\pi} \int_{-\pi}^{\pi} h(\omega)\left(1 + 2\sum_{n=1}^{\infty} \left(\frac{r}{r_0}\right)^n (\cos(n\omega)\cos(n\theta) + \sin(n\omega)\sin(n\theta))\right) d\omega.$$

Now, denoting $\frac{1}{2\pi}$ times the term in parentheses by K,

$$2\pi K(r,\theta,\omega) = 1 + 2\sum_{n=1}^{\infty} \left(\frac{r}{r_0}\right)^n (\cos(n\omega)\cos(n\theta) + \sin(n\omega)\sin(n\theta))$$

$$= 1 + 2\sum_{n=1}^{\infty} \left(\frac{r}{r_0}\right)^n \cos(n(\theta - \omega))$$

(14.4)

$$= 1 + \sum_{n=1}^{\infty} \left(\frac{r}{r_0}\right)^n \left(e^{in(\theta-\omega)} + e^{-in(\theta-\omega)}\right)$$

$$= \sum_{n=0}^{\infty} \left(\frac{r}{r_0}\right)^n e^{in(\theta-\omega)} + \sum_{n=1}^{\infty} \left(\frac{r}{r_0}\right)^n e^{-in(\theta-\omega)}.$$

Both sums are those of a geometric series, so the summation yields

$$2\pi K(r,\theta,\omega) = \frac{1}{1 - \left(\frac{r}{r_0}\right)e^{i(\theta-\omega)}} + \left(\frac{r}{r_0}\right)e^{-i(\theta-\omega)} \frac{1}{1 - \left(\frac{r}{r_0}\right)e^{-i(\theta-\omega)}}$$

$$= \frac{1 - \left(\frac{r}{r_0}\right)e^{-i(\theta-\omega)}}{1 - 2\left(\frac{r}{r_0}\right)\cos(\theta-\omega) + \left(\frac{r}{r_0}\right)^2} + \frac{\left(\frac{r}{r_0}\right)e^{-i(\theta-\omega)} - \left(\frac{r}{r_0}\right)^2}{1 - 2\left(\frac{r}{r_0}\right)\cos(\theta-\omega) + \left(\frac{r}{r_0}\right)^2}$$

$$= \frac{1 - \left(\frac{r}{r_0}\right)^2}{1 - 2\left(\frac{r}{r_0}\right)\cos(\theta-\omega) + \left(\frac{r}{r_0}\right)^2},$$

where the second equality was bringing the two terms to a common denominator. We thus deduce that for $r < r_0$,

$$u(r,\theta) = \frac{1}{2\pi} \int_{-\pi}^{\pi} h(\omega) \frac{r_0^2 - r^2}{r_0^2 - 2r_0 r\cos(\theta-\omega) + r^2}\, d\omega.$$

Rewriting in terms of Euclidean coordinates, if $y = r_0\omega$, $x = r\theta$, then the numerator is $r_0^2 - |x|^2$ while the denominator is $|x - y|^2$, so altogether we have

(14.5) $$u(x) = \frac{1}{2\pi r_0} \int_{\partial D} h(y) \frac{r_0^2 - |x|^2}{|x - y|^2} \, dS(y).$$

This is called the *Poisson formula*. In particular,

$$u(0) = \frac{1}{2\pi r_0} \int_{\partial D} h(y) \, dS(y),$$

i.e. since the circumference of D is $2\pi r_0$, the value of u in the center is the *average value of h*, which is called the *mean value property* of solutions of $\Delta u = 0$.

Now, the Poisson formula would give an independent (i.e. new, not relying on Lemma 14.1) way of proving that the Fourier series of h sums to h. Namely, one can show directly, using (14.5), that for $h \in C^1$ the Fourier series, which we already know converges uniformly to some function, converges to h. Indeed, in view of the uniform convergence of the series (14.3) on the closed disk, at $r = r_0$ it converges to

$$\lim_{r \to r_0-} u(r, \theta) = \lim_{r \to r_0-} \int_{-\pi}^{\pi} h(\omega) K_P(r, \theta - \omega) \, d\omega,$$

$$K_P(r, \psi) = \frac{1}{2\pi} \frac{r_0^2 - r^2}{r_0^2 - 2r_0 r \cos\psi + r^2};$$

see Figure 14.1 for the graph of K_P for fixed values of r. We claim that this limit is $h(\theta)$. Using the series definition of K, (14.4), and noting that

$$K(r, \theta, \omega) = K_P(r, \theta - \omega),$$

so

$$K_P(r, \psi) = \frac{1}{2\pi} \left(1 + 2 \sum_{n=1}^{\infty} \left(\frac{r}{r_0} \right)^n \cos(n\psi) \right),$$

one sees that by the orthogonality of cosines to the constant function 1, for each $r < r_0$,

$$\int_{-\pi}^{\pi} K_P(\psi) \, d\psi = 1,$$

and K_P is 2π-periodic in ψ, so

$$u(r, \theta) - h(\theta) = \int_{-\pi}^{\pi} K_P(r, \theta - \omega) \, (h(\omega) - h(\theta)) \, d\omega$$

$$= \int_{-\pi}^{\pi} K_P(r, \omega) \, (h(\theta - \omega) - h(\theta)) \, d\omega,$$

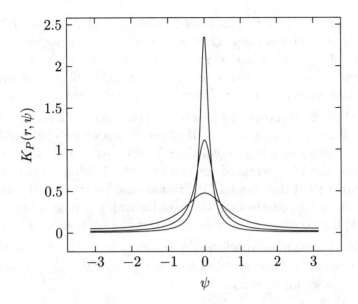

Figure 14.1. The Poisson kernel $K_P(r, \psi)$ for $r = \frac{1}{2}, \frac{3}{4}, \frac{7}{8}$, with $r_0 = 1$. As $r \to r_0$, the peak at the origin becomes higher and sharper, and away from the origin K_P tends to 0.

where the last equality involved a change of variables; however, in view of the 2π-periodicity, the domain of integration did not need to change (i.e. the integral over $[-\pi, \pi]$ is the same as that over $[\theta - \pi, \theta + \pi]$). Now, the advantage of the *Poisson kernel*, K_P, over the Dirichlet kernel discussed below is that for $r < r_0$ one has $K_P \geq 0$, directly from its definition. Moreover, $K_P(r, \psi) \to 0$ as $r \to r_0$ uniformly outside $[-\delta, \delta]$ for any $\delta > 0$. Thus,

$$\int_{-\pi}^{\pi} K_P(r, \omega) \left(h(\theta - \omega) - h(\theta) \right) d\omega$$

$$= \int_{-\delta}^{\delta} K_P(r, \omega) \left(h(\theta - \omega) - h(\theta) \right) d\omega$$

$$+ \int_{[-\pi, \pi] \setminus [-\delta, \delta]} K_P(r, \omega) \left(h(\theta - \omega) - h(\theta) \right) d\omega.$$

Now even if h is merely continuous, given any $\epsilon > 0$, choosing δ sufficiently small, $|h(\theta - \omega) - h(\theta)| < \epsilon$ for all ω with $|\omega| \leq \delta$, and thus *for any* $r \in (0, r_0)$, the first term satisfies

$$\left| \int_{-\delta}^{\delta} K_P(r, \omega) \left(h(\theta - \omega) - h(\theta) \right) d\omega \right| \leq \epsilon \int_{-\delta}^{\delta} K_P(r, \omega) \, d\omega$$

$$\leq \epsilon \int_{-\pi}^{\pi} K_P(r, \omega) \, d\omega = \epsilon.$$

On the other hand, by the uniform convergence of $K_P(r,\omega)$ to 0 outside $(-\delta, \delta)$, the second term also goes to 0 as $r \to r_0$, in particular is $< \epsilon$ if r is sufficiently large. Thus, even if just h is continuous, $u(r,\theta) \to h(\theta)$ uniformly as $r \to r_0-$. In particular, if h is just C^1, the partial sums of the Fourier series converge to h uniformly, giving the promised proof.

Note that the standard definition of convergence of a series, via partial sums, is a kind of a regularization. If all terms of a series $\sum u_n$ are bounded, a different regularization is to consider $\sum r^n u_n$ for $r < 1$, when the series converges by the Weierstrass M-test, and let $r \to 1$. This is a better behaved regularization than the standard definition, and is called *Abel summability*. Thus, for h merely continuous, the Fourier series is Abel summable to h, but it need not converge uniformly to h (there are actual counterexamples).

Another possible approach to the convergence of a series is to replace the partial sums by their average, so e.g. replace $S_N = \sum_{n=1}^{N} u_n$ by $s_M = \frac{1}{M} \sum_{N=1}^{M} S_N$. So, for instance,

$$s_3 = \frac{1}{3}(S_1 + S_2 + S_3) = u_1 + \frac{2}{3}u_2 + \frac{1}{3}u_3,$$

and more generally

$$s_M = u_1 + \frac{M-1}{M}u_2 + \ldots + \frac{1}{M}u_M.$$

Thus, the subsequent terms u_n are taken with lesser and lesser weight, but for any fixed n, the weight $\frac{M-n+1}{M}$ tends to 1 as $M \to \infty$. Correspondingly, this is yet another regularization of the sum of the series. This notion is called *Cesàro summability* and is discussed for the Fourier series in Problem 14.7.

Additional material: The Dirichlet kernel

We now consider a more classical proof of the convergence of the Fourier series. Suppose now that ϕ is a function on $[-\pi, \pi]$ whose 2π-periodic extension, Φ, is C^1. Then the Fourier series of ϕ is given by (14.1), with $\ell = \pi$. We work out partial sums of the series,

$$(14.6) \qquad S_N(\theta) = \sum_{n=-N}^{N} C_n e^{in\theta}, \quad C_n = \frac{1}{2\pi} \int_{-\pi}^{\pi} \phi(\omega) e^{-in\omega} d\omega.$$

Thus,

$$S_N(\theta) = \sum_{n=-N}^{N} \frac{1}{2\pi} \int_{-\pi}^{\pi} \phi(\omega) \, e^{in(\theta-\omega)} \, d\omega$$

$$= \int_{-\pi}^{\pi} \frac{1}{2\pi} \sum_{n=-N}^{N} e^{in(\theta-\omega)} \phi(\omega) \, d\omega$$

$$= \int_{-\pi}^{\pi} K_N(\theta - \omega) \phi(\omega) \, d\omega, \quad K_N(\psi) = \frac{1}{2\pi} \sum_{n=-N}^{N} e^{in\psi}.$$

Now, the sum for K_N is a geometric series, so using the summation formula for a geometric series,

(14.7)
$$K_N(\psi) = \frac{1}{2\pi} \sum_{n=-N}^{N} e^{in\psi} = \frac{1}{2\pi} e^{-iN\psi} \frac{1 - e^{i(2N+1)\psi}}{1 - e^{i\psi}}$$

$$= \frac{1}{2\pi} \frac{e^{i(N+1/2)\psi} - e^{-i(N+1/2)\psi}}{e^{i\psi/2} - e^{-i\psi/2}} = \frac{1}{2\pi} \frac{\sin\left((N+1/2)\psi\right)}{\sin\left(\psi/2\right)}.$$

Note that the right-hand side has the form $0/0$ at $\psi = 2k\pi$, $k \in \mathbb{Z}$, but by L'Hopital's rule it is continuous, with value $K_N(0) = \frac{1}{2\pi}(2N+1)$, and it is indeed C^∞ at these points. Moreover, K_N has the additional property, which is clear from the series formula since $e^{in\psi}$ is orthogonal to 1 in $L^2([-\pi, \pi])$ for $n \neq 0$,

$$\int_{-\pi}^{\pi} K_N(\psi) \, d\psi = \langle K_N, 1 \rangle = \frac{1}{2\pi} \langle 1, 1 \rangle = 1$$

for all N, and, as all the summands in the sum are 2π-periodic,

$$K_N(\psi + 2\pi) = K_N(\psi).$$

One calls K_N the *Dirichlet kernel*; see Figure 14.2.

Now, we need to analyze the difference

$$S_N(\theta) - \phi(\theta) = \int_{-\pi}^{\pi} K_N(\theta - \omega) \phi(\omega) \, d\omega - \phi(\theta) \int_{-\pi}^{\pi} K_N(\theta - \omega) \, d\omega$$

$$= \int_{-\pi}^{\pi} K_N(\theta - \omega)(\phi(\omega) - \phi(\theta)) \, d\omega.$$

It is convenient to rewrite the integral by a change of variables. In view of the 2π-periodicity of the integrand, we do not need to change the limits of the integral:

$$S_N(\theta) - \phi(\theta) = \int_{-\pi}^{\pi} K_N(\psi)(\Phi(\theta - \psi) - \Phi(\theta)) \, d\psi.$$

Note that this integral, much like the one before, is very similar to a convolution on $\mathbb{R}$; the difference is that we merely integrate over an interval of

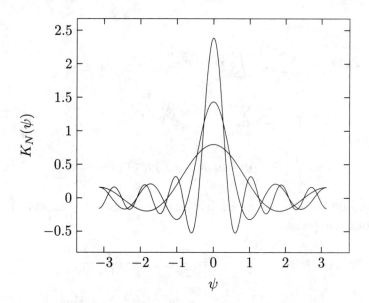

Figure 14.2. The Dirichlet kernel K_N for $N = 2, 4, 7$. As N gets larger, the peak at the origin becomes higher and sharper, but there are still relatively large (closely spaced) oscillations away from 0.

length 2π. Since the integrand is 2π-periodic, this is best thought of as the convolution of K_N and the function $u_\theta(\psi) = \Phi(\psi) - \Phi(\theta)$ on the circle, $\mathbb{S}^1$.

Now, roughly, the K_N are approximations to δ_0. However, this is in a weaker sense than e.g. $\chi_N(x) = N\chi(x/N)$, where $\chi \geq 0$ vanishes outside $[-1, 1]$, or indeed the Poisson kernel K_P above is such, since $\int_{-\pi}^{\pi} |K_N(\psi)| \, d\psi$ is *not* bounded by a fixed constant C (independent of N). It is the oscillatory nature of K_N that makes the analysis more involved.

Writing out $K_N(\psi)$ as in (14.7), an orthogonality argument can be used to deduce the decay of $S_N(\theta) - \phi(\theta)$ as $N \to \infty$. Namely, suppose that Φ is C^1. Then by Taylor's theorem, the function

$$v_\theta(\psi) = \frac{\Phi(\theta - \psi) - \Phi(\theta)}{\sin(\psi/2)}$$

is continuous. Thus,

$$S_N(\theta) - \phi(\theta) = \frac{1}{2\pi} \int_{-\pi}^{\pi} \sin\left((N + 1/2)\psi\right) v_\theta(\psi) \, d\psi.$$

Now, the functions

$$Z_n(\psi) = \sin\left((n + 1/2)\psi\right), \ n \geq 0, \ n \in \mathbb{Z},$$

are orthogonal to each other on $[-\pi, \pi]$ (and indeed on $[0, \pi]$), since they are eigenfunctions of the symmetric operator $-\frac{d^2}{d\theta^2}$ with Neumann boundary condition at $-\ell$ and ℓ (as $\cos((n+1/2)\pi) = 0$) with distinct eigenvalues $(n+1/2)^2$. Moreover,

$$\int_{-\pi}^{\pi} Z_n(\psi)^2 \, d\psi = \pi.$$

Thus,

$$S_N(\theta) - \phi(\theta) = \frac{1}{2\sqrt{\pi}} \frac{\langle v_\theta, Z_N \rangle}{\|Z_N\|_{L^2}^2} \|Z_N\|_{L^2}.$$

But we have shown that for *any* orthogonal set $\{x_n\}_{n=0}^{\infty}$ in an inner product space V, and any $v \in V$ with generalized Fourier coefficients c_n, $\sum |c_n|^2 \|x_n\|^2$ converges, and $c_n = \frac{\langle v, x_n \rangle}{\|x_n\|^2}$. Thus,

$$|S_N(\theta) - \phi(\theta)|^2 = \frac{1}{4\pi} |c_N|^2 \|Z_N\|_{L^2}^2, \quad c_N = \frac{\langle v_\theta, Z_N \rangle}{\|Z_N\|_{L^2}^2}.$$

But as $\sum_N |c_N|^2 \|Z_N\|_{L^2}^2$ converges, its terms go to 0, i.e. $|c_N|^2 \|Z_N\|_{L^2}^2 \to 0$ as $N \to \infty$, so we deduce that $|S_N(\theta) - \phi(\theta)| \to 0$ as $N \to \infty$, proving the desired convergence.

Since we already know that for functions ϕ whose 2π-periodic extensions are C^1 the Fourier series converges uniformly (and thus pointwise) to some function, and we just showed pointwise convergence of the series to ϕ, we deduce that for such ϕ the Fourier series converges uniformly to ϕ.

Problems

Problem 14.1. Let γ_n be a sequence of constants tending to ∞. Consider the following continuous functions f_n, $n \geq 2$, on $[0, 1]$:

$$f_n(x) = \begin{cases} 2n\gamma_n x, & 0 \leq x \leq \frac{1}{2n}, \\ 2n\gamma_n(\frac{1}{n} - x), & \frac{1}{2n} \leq x \leq \frac{1}{n}, \\ 0, & \frac{1}{n} \leq x \leq 1. \end{cases}$$

   (i) Sketch the graph of a few f_n.

   (ii) Show that $f_n \to 0$ pointwise.

   (iii) Show that the convergence is not uniform.

   (iv) Show that $f_n \to 0$ in L^2 if $\gamma_n = n^{1/4}$.

   (v) Show that f_n does not converge in L^2 if $\gamma_n = n$.

Problem 14.2. Let $\phi(x) = x(\ell - x)$ on $[0, \ell]$.

   (i) Find the Fourier sine series of ϕ and state what it converges to for $x \in \mathbb{R}$.

   (ii) Find the Fourier cosine series of ϕ and state what it converges to for $x \in \mathbb{R}$.

(iii) Compare the decay rates of the coefficients of the two series as $n \to \infty$. Why do the coeffients decay faster in one of the cases?

Problem 14.3. For both of the following functions discuss whether the Fourier sine series converges uniformly or in L^2:

(i) $\phi(x) = x$ on $[0, \ell]$,

(ii) $\phi(x) = x(\ell - x)^2$ on $[0, \ell]$.

Justify your answer by quoting the relevant convergence theorems. You do not need to compute the respective Fourier series.

Problem 14.4. In this problem we consider the inhomogeneous wave equation on $[0, \ell]$ with inhomogeneous Dirichlet boundary conditions:

$$u_{tt} - c^2 u_{xx} = f, \ u(0, t) = h(t), \ u(\ell, t) = j(t), \ u(x, 0) = \phi(x), \ u_t(x, 0) = \psi(x).$$

(i) Suppose $h(t) = 0 = j(t)$ for all t. For each t, we can expand $u(x, t)$ (which is a function of x only then) in a Fourier sine series in x, with coefficients depending on t:

$$u(x, t) = \sum_{n=1}^{\infty} u_n(t) \sin(n\pi x/\ell),$$

and we can obtain a similar expansion for f.

(ii) Assuming that you can differentiate term by term, i.e. that the odd 2ℓ-periodic extension of u is well behaved, find the Fourier sine series of u_{xx} and u_{tt}.

(iii) Derive an ODE for $u_n(t)$ by substituting the Fourier series into the PDE. Find the initial conditions satisfied by $u_n(t)$ and solve the ODE.

(iv) Now do not assume that $h(t) = 0 = j(t)$. We could solve the problem by subtracting from u a function F that satisfies the boundary conditions to place ourselves into the previous scenario. Instead, we proceed as follows.

As u does *not* satisfy the homogeneous boundary conditions that sine does, the Fourier coefficients will decay slowly, and term by term differentiation is not allowed to calculate u_{xx}. Instead, let $w(x, t) = u_{xx}(x, t)$, and expand w in the Fourier sine series:

$$w(x, t) = \sum_{n=1}^{\infty} w_n(t) \sin(n\pi x/\ell).$$

Calculate the coefficients $w_n(t)$ in terms of w, and then integrate by parts twice to obtain them in terms of $u_n(t)$, $h(t)$ and $j(t)$.

(v) Obtain an ODE for $u_n(t)$ and solve it.

Problem 14.5. Consider the wave equation on a ring of length 2ℓ. We let x be the arclength variable along the ring, $x \in [-\ell, \ell]$. We would like to understand wave propagation along the ring, so consider the wave equation with *periodic boundary conditions*:

$$u_{tt} = c^2 u_{xx}, \ u(-\ell, t) = u(\ell, t), \ u_x(-\ell, t) = u_x(\ell, t).$$

(i) Find the general 'separated' solution.

(ii) Find the solution with initial condition

$$u(x, 0) = 0, \ u_t(x, 0) = \cos(2\pi x/\ell) - \sin(\pi x/\ell), \ x \in [-\ell, \ell].$$

(iii) Give an alternative method of solution by extending u to be a 2ℓ-periodic function in x on all of $\mathbb{R}$ and using d'Alembert's formula.

(iv) How do singularities of u propagate? That is, if the only singularity of the initial data is at some x_0 (i.e. they are C^∞ elsewhere), where can u be singular? Interpret this physically.

Problem 14.6. Consider the 2-torus, $\mathbb{T}^2 = \mathbb{S}^1_x \times \mathbb{S}^1_y$, as in Problem 13.5. Recall that functions on $\mathbb{T}^2$ can be thought of as functions which are 2π-periodic in both x and y.

(i) Show that the (full) Fourier series of a function ϕ on $\mathbb{T}^2$,

$$\sum_{n,m=-\infty}^{\infty} c_{nm} e^{inx+imy}, \ c_{nm} = \frac{1}{(2\pi)^2} \int_{-\pi}^{\pi} \int_{-\pi}^{\pi} e^{-inx-imy} \phi(x, y) \, dx \, dy,$$

converges to the function ϕ in C^∞ if ϕ is C^∞, and in L^2 if ϕ is L^2. (*Hint:* Show first that if ϕ is C^∞, then for any k there is a constant A_k such that $|c_{nm}| \le \frac{A_k}{(1+|n|)^k(1+|m|)^k}$ for all n, m, i.e. the Fourier coefficients decay rapidly in both n and m.)

(ii) Show that the (full) Fourier series of a C^4 function ϕ on $\mathbb{T}^2$ converges uniformly to ϕ. Can you weaken the C^4 regularity hypothesis?

Problem 14.7. In this problem we consider another way of summing the Fourier series by *averaging the partial sums*, i.e. considering

$$s_{M+1}(\theta) = \frac{1}{M+1} \sum_{N=0}^{M} S_N(\theta),$$

where $S_N(\theta)$ is as in (14.6), namely the Nth partial sum of the Fourier series of a function ϕ. It suffices for us to assume ϕ is a function on $[-\pi, \pi]$ whose 2π-periodic extension is C^0.

(i) Write

$$\sigma_{M+1}(\theta) = \int_{-\pi}^{\pi} \tilde{K}_M(\theta - \omega) \, \phi(\omega) \, d\omega,$$

calculating $\tilde{K}_M$ explicitly, namely show that

$$\tilde{K}_M(\psi) = \frac{1}{2\pi(M+1)} \frac{\sin^2((M+\frac{1}{2})\psi)}{\sin^2(\psi/2)}.$$

The function $\tilde{K}_M$ is called the *Fejér kernel*. (*Hint:* It is useful to start from the Dirichlet kernel expression for S_N, written in terms of complex exponentials.)

(ii) Show that $\int_{-\pi}^{\pi} \tilde{K}_M(\psi)\,d\psi = 1$ and that for any $\delta > 0$, $\tilde{K}_M \to 0$ uniformly on $[-\pi,\pi] \setminus (-\delta,\delta)$.

(iii) Show that for ϕ with continuous 2π-periodic extension,

$$s_M(\theta) = \int_{-\pi}^{\pi} \tilde{K}_M(\theta - \omega)\phi(\omega)\,d\omega$$

converges uniformly to ϕ as $M \to \infty$. (*Hint:* Argue as done in the chapter for the Poisson kernel.)

What this shows is that although the partial sums of the Fourier series do not necessarily converge uniformly for a merely continuous ϕ, the *averaged* partial sums do. As for Abel summation, this is a better regularization of the sum of an infinite series than the standard definition; this is called *Cesàro summability*.

Chapter 15

Bessel functions

1. The definition of Bessel functions

Bessel functions are one of the best known special functions and arise for instance in the study of vibrating circular membranes. As we have seen in Chapter 12, if D is the disk $D = \{x \in \mathbb{R}^2 : |x| < r_0\}$, then separation of variables for the wave equation

$$u_{tt} - c^2 \Delta u = 0$$

with Dirichlet or Neumann boundary conditions gives rise to the eigenvalue equation

$$-\Delta X = \lambda X,$$

with the same boundary condition. One can then proceed (as at the end of Chapter 12 where Laplace's equation on the disk was solved) using polar coordinates, so the disk is identified with $[0, r_0)_r \times \mathbb{S}^1_\theta$. Write $X(r, \theta) = R(r)\Theta(\theta)$ and recall that the Laplacian in polar coordinates is

$$\Delta = \partial_r^2 + r^{-1}\partial_r + r^{-2}\partial_\theta^2.$$

Thus, substitution into the PDE yields

$$-R''\Theta - r^{-1}R'\Theta - r^{-2}R\Theta'' = \lambda R\Theta,$$

so dividing by $R\Theta$ and multiplying by r^2 yields

$$-\frac{\Theta''}{\Theta} = \frac{r^2 R'' + rR' + \lambda r^2 R}{R}.$$

Now the left-hand side is independent of r and the right-hand side of θ, so they must be both equal to a constant, μ, and thus we have two ODE's:

$$r^2 R'' + r R' + \lambda r^2 R = \mu R,$$

(15.1)

$$- \Theta'' = \mu \Theta.$$

We now argue just as in the case of Laplace's equation in Chapter 12 to conclude that $\mu = n^2$, $n \geq 0$ an integer,

$$\Theta_n(\theta) = A_n \cos(n\theta) + B_n \sin(n\theta), \ n > 0,$$

$$\Theta_0(\theta) = A_0.$$

So it remains to deal with the radial equation, $r^2 R'' + r R' + \lambda r^2 R = \mu R$, where the interesting case is $\mu = n^2$ with n an integer. The ODE can be rewritten as

(15.2) $$r(r R')' + \lambda r^2 R - n^2 R = 0,$$

where we keep the n^2 notation, but n being an integer is not necessary for the arguments below. This is a regular singular ODE, with its singular point at 0. The reason for the 'singularity' is that the coefficient of R'' vanishes there, so the Picard iteration is not applicable to solving the ODE near $r = 0$, though of course it works near any $\tilde{r} > 0$ (but the solution might in principle only exist on a small interval around $\tilde{r}$, excluding 0). The reason for 'regular' is that each differentiation comes with multiplication by r. As explained in Chapter 12, if $\lambda = 0$ (the case of interest there), a logarithmic change of variables reduces this ODE to a constant coefficient, one which is easily solved explicitly by exponentials. The same approach does not work here since the $\lambda r^2 R$-term becomes $\lambda e^{2t} F(t)$, so the resulting ODE is not a constant coefficient. So instead we work directly by attempting to find the solution in a series around $r = 0$.

Before proceeding we discuss a change of variables to reduce the number of parameters (λ, n currently). Namely, let $\rho = \sqrt{\lambda} r$. Then let $Y(\rho) = R(\rho/\sqrt{\lambda})$, which is possible if $\lambda > 0$, or more generally if $\lambda \neq 0$, but in that case one needs to be careful about *where* the equation is being solved. By the chain rule, $\frac{d}{dr} = \sqrt{\lambda} \frac{d}{d\rho}$, so $r \frac{d}{dr} = \rho \frac{d}{d\rho}$, and the ODE for $R(r) = Y(\sqrt{\lambda} r)$ becomes

$$\rho(\rho Y')' + \rho^2 Y - n^2 Y = 0,$$

i.e. the $\lambda \neq 0$ case can be reduced to $\lambda = 1$ by a change of variables. Correspondingly, one could simply consider the $\lambda = 1$ equation, which is called Bessel's equation; its solutions are called *Bessel functions*.

We do not change our notation for now, i.e. we keep λ unchanged, but we will keep in mind that eventually we can reduce to the $\lambda = 1$ case. Before going further, we note that the term $\lambda r^2 R$ has an additional r^2 beyond what is necessary in the 'regular singular' scheme, namely that each derivative has a matching power of r at least, which would allow even a term like λR. Of course, this still makes the ODE regular singular (greater powers of r are fine, as are a matching power of r times a general C^∞ function of r), but any term with additional powers of r beyond the requirement can be expected to be less important at $r = 0$ (since they have additional vanishing there). Thus, one expects that the leading order solutions of the ODE behave as those of the ODE in which this term is dropped:

$$(15.3) \qquad r(rR_0')' - n^2 R_0 = 0.$$

But for this ODE the change of variables $t = \log r$, $F(t) = R_0(e^t)$, as in Chapter 12, yields the constant coefficient equation

$$F'' - n^2 F = 0,$$

so

$$F(t) = Ae^{nt} + Be^{-nt}, \ n \neq 0,$$
$$F(t) = A + Bt, \ n = 0.$$

Returning to R_0,

$$R_0(r) = Ar^n + Br^{-n}, \ n \neq 0,$$
$$R_0(r) = A + B\log r, \ n = 0.$$

Recall that for now we want to study solutions of the ODE in general, without imposing that n is an integer, or even the lack of blow-up at the origin, so we allow all of these asymptotic behaviors as $r \to 0$.

Now, R_0 does not solve the ODE (15.2), though it does solve (15.3). Since R_0 is an arbitrary linear combination of r^n and r^{-n} (when $n \neq 0$), we consider these separately. We then expect that by adding correction terms to R_0 we may obtain an actual solution of (15.2). Now, a power r^s (where now $s = \pm n$) produces an error term λr^{s+2} in (15.2), i.e. the right-hand side of (15.2) is replaced by this (it is no longer 0), while a power r^σ when subsituted into (15.3) produces $(\sigma^2 - n^2)r^\sigma$. Thus, the former can be solved away, modulo an error that vanishes to an even higher order at $r = 0$, by subtracting $\frac{\lambda}{(s+2)^2 - n^2} r^{s+2}$ from the first approximate solution (R_0 in our case), provided $s + 2 \neq \pm n$.

To make this systematic, let us try to write

$$(15.4) \qquad R(r) = r^s \left(1 + \sum_{k=1}^\infty a_k r^{2k}\right) = r^s + \sum_{k=1}^\infty a_k r^{s+2k}, \ s = \pm n.$$

In general, for regular singular ODEs one expects to use all non-negative integer powers in the second, correction, term; here we only use the even ones since these are the only ones generated by our 'error term' $\lambda r^2 R$. Substituting this into the ODE (15.2) yields

$$\lambda r^{s+2} + \sum_{k=1}^{\infty} a_k((s+2k)^2 - s^2)r^{s+2k} + \sum_{k=1}^{\infty} a_k \lambda r^{s+2k+2} = 0,$$

i.e. relabelling the second series, replacing any occurrence of k by $k-1$,

$$\lambda r^{s+2} + \sum_{k=1}^{\infty} a_k((s+2k)^2 - s^2)r^{s+2k} + \sum_{k=2}^{\infty} a_{k-1} \lambda r^{s+2k} = 0.$$

Equating the coefficient of r^{s+2k} with 0 yields

$$\lambda + a_1((s+2)^2 - s^2) = 0, \ k = 1,$$

and

$$a_k((s+2k)^2 - s^2) + a_{k-1}\lambda = 0, \ k \geq 2.$$

These can be solved recursively, starting with $k = 1$, giving

$$a_1 = -\frac{\lambda}{(s+2)^2 - s^2}$$

and

(15.5) $$a_k = -\frac{\lambda}{(s+2k)^2 - s^2} a_{k-1}, \ k \geq 2.$$

This gives us the coefficients of the series (15.4), assuming that $(s+2k)^2 \neq s^2$ for any $k \geq 1$, i.e. that $s + 2k \neq -s$ for any $k \geq 1$, i.e. that $k \neq -s$ for $k \geq 1$, i.e. that s is not a negative integer. Here recall that our original solutions were $r^{\pm n}$, where $n \geq 0$ is an integer so, in particular, we did allow negative exponents, in which case our solutions would have to be modified by logarithmic terms.

So now we need to study whether the series $\sum_{k=1}^{\infty} a_k r^{2k}$ converges at least near $r = 0$. (Of course, if $\lambda = 0$, the series has only the $k = 0$-term, r^s, so there are no convergence issues; thus one may take $\lambda \neq 0$.) But this is straightforward using the ratio test: recall that a series $\sum_{k=1}^{\infty} c_k$ converges if there is $0 < \alpha < 1$ such that $|\frac{c_{k+1}}{c_k}| < \alpha$ for sufficiently large k (this is proved by comparison with the geometric series); this gives in our case that we have convergence provided $|\frac{a_{k+1}}{a_k}|r^2 < \alpha$ for large k. But in our case,

$$\frac{a_{k+1}}{a_k} = \frac{-\lambda}{(s+2k+2)^2 - s^2} \to 0 \text{ as } k \to \infty,$$

so in fact the series converges for *all* $r \in \mathbb{R}$, and indeed for all $r \in \mathbb{C}$. Further, by the standard power series result (indeed, this is a consequence of the Weierstrass M-test), if the series converges at some $r'_0 > 0$, then the convergence is uniform on $[-r'_1, r'_1]$ for any $r'_1 < r'_0$. In our case we can take r'_0 arbitrary, so the convergence is uniform on compact intervals, and indeed on compact subsets of $\mathbb{C}$. In addition, the term-by-term differentiated series has the same properties, even when done arbitrarily many times, so in fact the series gives a C^∞ function whose derivatives can be calculated by term-by-term differentiation. Since the ODE was solved exactly when the derivatives are evaluated by term-by-term differentiation, the same calculation now gives that we have *solved* the ODE. We give a name to the solutions when $\lambda = 1$: the *Bessel function* $J_n(r)$, when n is not a negative integer, is the solution of Bessel's equation (15.2) with $\lambda = 1$ such that $2^n n! J_n(r)$ is of the form of the form (15.4) with $s = n$; see Figure 15.1. Correspondingly, the solution for general $\lambda \neq 0$ of the form (15.4) with $s = n$ is $2^n n! J_n(\sqrt{\lambda} r)$. Here $2^n n!$ is the standard normalizing factor; it could be dropped in the definition of a Bessel function, but it is present in the standard normalization. Also, when n is not an integer, $n!$ is replaced by the generalization of the factorial, namely the Γ function, $\Gamma(n+1)$.

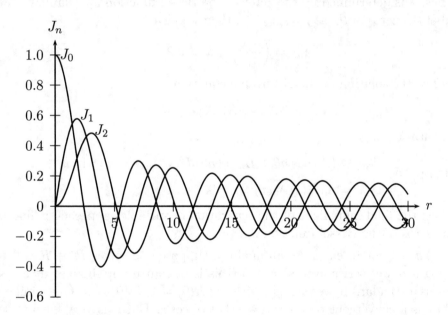

Figure 15.1. The Bessel functions J_0, J_1 and J_2. Note that the order of vanishing at the origin increases with the order n of the Bessel function J_n.

Notice that for n not an integer, we have two linearly independent solutions of the ODE (15.2) on $(0, \infty)$, namely $J_n(\sqrt{\lambda} r)$ and $J_{-n}(\sqrt{\lambda} r)$, so by general ODE theory, this being a second order ODE, every solution is a linear combination of these. Thus, $2^n \Gamma(n+1) J_n(\sqrt{\lambda} r)$ is indeed the *unique* solution of (15.2) of the form (15.4) with $s = n$. If $n \geq 0$ is an integer, one can make a similar construction, adding logarithmic terms, to construct a solution similar to (15.4) with $s = -n$ (or a logarithmic term if $s = 0$), and the conclusion would again be the uniqueness of $2^n n! J_n(\sqrt{\lambda} r)$ of the form (15.4); see Problem 15.1.

2. The zeros of Bessel functions

At this point we can argue similarly to the case of Laplace's equation in Chapter 12, so in particular exclude solutions of Bessel's equation that blow up at $r = 0$. However, we still need to deal with the boundary condition. Thus, for Dirichlet boundary conditions at $r = r_0$ we have

$$r(rR')' + \lambda r^2 R - n^2 R = 0, \ R(r_0) = 0,$$

which means that

$$R(r) = J_n(\sqrt{\lambda} r), \ J_n(\sqrt{\lambda} r_0) = 0.$$

Thus, λ is determined by the zeros of the Bessel function J_n. Namely, if we label the zeros of J_n as $\rho_{n1}, \rho_{n2}, \ldots$, then we have

$$\lambda_{nk} = \left(\frac{\rho_{nk}}{r_0}\right)^2, \ k = 1, 2, \ldots,$$

and so the solution of the eigenvalue equation

$$-\Delta X = \lambda X, \ X|_{r=r_0} = 0$$

is, with $\lambda = \lambda_{nk}$,

(15.6)
$$X_{nk} = (A_n \cos n\theta + B_n \sin n\theta) J_n(\sqrt{\lambda_{nk}} r), \ n \geq 1, \ n \in \mathbb{Z},$$
$$X_{0k} = A_0 J_0(\sqrt{\lambda_{0k}} r), \ n = 0.$$

Thus, the final ingredient of the separation of variables approach is finding the zeros of the Bessel function J_n.

This is not so easy. As an analogy, suppose one solves $R'' + R = 0$ by using a power series around $r = 0$ (this is of course not singular at all, so this is a standard power series), with say $R(0) = 1$, $R'(0) = 0$. One similarly obtains a convergent power series—that of $\cos r$. But it is very hard to see *directly from the power series* that this has any zeros! A standard way of doing this is by considering $f(r) = \cos^2 r + \sin^2 r$, so $f'(r) = 0$ from the Taylor series (as $\cos' = -\sin$, $\sin' = \cos$ from the series, or indeed from the ODE), i.e. f is constant, but one checks that $\cos 0 = 1 > 0$, $\cos 2 < 0$ directly from the series, so it must have a zero on $(0, 2)$ by the intermediate value theorem. One then also shows that if the smallest zero is r_1, then $\cos$ and $\sin$ are periodic with period $4r_1$; then π is defined as $2r_1$.

Now, in order to analyze Bessel's equation, let us consider the ODE again from scratch:

(15.7)
$$r^2 R'' + r R' + \lambda r^2 R - n^2 R = 0.$$

We are now interested in what happens for large r. For $\lambda \neq 0$ (since we already understand $\lambda = 0$, where $r^{\pm n}$ have no zeros!), λr^2 is *much* bigger than n^2 for r large, so we expect we can drop $n^2 R$ to approximate the solution. This leaves

$$R_0'' + r^{-1} R_0' + \lambda R_0 = 0$$

as an approximate equation to solve. Now if we drop the $r^{-1} R_0'$-term as well on the grounds that it still has a decaying factor, namely r^{-1}, while R_0'' and R_0 have no such factor, we get the ODE

$$\tilde{R}_0'' + \lambda \tilde{R}_0 = 0,$$

whose solutions are linear combinations of $\cos(\sqrt{\lambda} r)$ and $\sin(\sqrt{\lambda} r)$, or indeed to simplify matters below $e^{i\sqrt{\lambda} r}$ and $e^{-i\sqrt{\lambda} r}$. So in analogy with the series considered above, consider solutions R of the form $R_\pm = e^{\pm i\sqrt{\lambda} r} f_\pm(r)$. As

$$R_+' = e^{i\sqrt{\lambda} r}(f_+'(r) + i\sqrt{\lambda} f_+(r)),$$

so

$$R_+'' = e^{i\sqrt{\lambda} r}(f_+''(r) + 2i\sqrt{\lambda} f_+'(r) - \lambda f_+(r)),$$

substituting into the ODE (15.7) gives, after division by $e^{i\sqrt{\lambda} r}$,

$$f_+''(r) + 2i\sqrt{\lambda} f_+'(r) - \lambda f_+(r) + r^{-1}(f_+'(r) + i\sqrt{\lambda} f_+(r)) + (\lambda - n^2 r^{-2}) f_+(r) = 0,$$

i.e.

$$f_+''(r) + (2i\sqrt{\lambda} + r^{-1}) f_+'(r) + (i\sqrt{\lambda} r^{-1} - n^2 r^{-2}) f_+(r) = 0.$$

While this is not a regular singular ODE, we can try to see what happens if we substitute in a power series for f_+; in this case the series is around $r = \infty$, so it should be in increasing powers of r^{-1}. Now, substituting in r^s gives powers of r^{s-1} and r^{s-2}, with $2i\sqrt{\lambda} s + i\sqrt{\lambda} r^{-1}$ being the coefficient of r^{s-1} (which is larger as $r \to \infty$), so to make it vanish we need $s = -1/2$. Since the other term has one order of additional decay, a reasonable attempt is to find f_+ of the form

$$(15.8) \qquad f_+(r) = r^s \left(1 + \sum_{k=1}^{\infty} b_k r^{-k}\right) = r^s + \sum_{k=1}^{\infty} b_k r^{s-k}, \quad s = -1/2.$$

Substituting this gives

$$(s(s-1) + s - n^2) r^{s-2} + \sum_{k=1}^{\infty} b_k i\sqrt{\lambda} (2(s-k) + 1) r^{s-k-1}$$

$$+ \sum_{k=1}^{\infty} b_k ((s-k)(s-k-1) + (s-k) - n^2) r^{s-k-2} = 0,$$

i.e.

$$(s^2-n^2)r^{s-2}+\sum_{k=1}^{\infty} b_k i\sqrt{\lambda}(2(s-k)+1)r^{s-k-1}+\sum_{k=1}^{\infty} b_k((s-k)^2-n^2)r^{s-k-2} = 0,$$

which after a change of the index in the last term (replacing k by $k-1$) gives, with $s=1/2$ still,

$$(s^2 - n^2)r^{s-2} + \sum_{k=1}^{\infty} b_k i\sqrt{\lambda}(2(s-k)+1)r^{s-k-1}$$

$$+ \sum_{k=2}^{\infty} b_{k-1}((s-k+1)^2 - n^2)r^{s-k-1} = 0.$$

This gives

$$i\sqrt{\lambda}(2(s-k)+1)b_1 + (s^2 - n^2) = 0, \;\; k=1,$$

and

$$i\sqrt{\lambda}(2(s-k)+1)b_k + ((s-k+1)^2 - n^2)b_{k-1} = 0, \;\; k \geq 2.$$

Notice that

$$\frac{b_k}{b_{k-1}} = \frac{(s-k+1)^2 - n^2}{i\sqrt{\lambda}(2(s-k)+1)} = i\frac{(k-1/2)^2 - n^2}{2\sqrt{\lambda}k},$$

so $|b_k/b_{k-1}| \to \infty$ as $k \to \infty$ unless b_k is 0 from some point on; i.e. unless n is of the form $l+1/2$ where l is an integer, that is, n is a half-integer, in which case the terms from $k=l+1$ vanish if $l \geq 0$. If $l < 0$, the vanishing is from $k=-l$ on. This means that the series $\sum_{k=1}^{\infty} b_k r^{-k}$ does *not* converge for *any* $r > 0$ unless n is a half-integer.

Notice that if n is a half-integer the series is finite, so there are no convergence issues, and indeed one concludes that the finite series (15.8) gives a solution of Bessel's ODE, $R_+(r) = e^{i\sqrt{\lambda}r}f_+(r)$. One similarly obtains $R_-(r)$, so $R_{\pm}(r)$ are two linearly independent solutions of Bessel's equation; therefore $J_{\pm}(\sqrt{\lambda}r)$ are both linear combinations of these. In the simplest case, $n = 1/2$, we have $R_{\pm}(r) = e^{\pm i\sqrt{\lambda}r}r^{-1/2}$. Taking real parts and imaginary parts, or equivalently $\frac{R_+(r)+R_-(r)}{2}$ and $\frac{R_+(r)-R_-(r)}{2i}$, we get

$$r^{-1/2}\cos(\sqrt{\lambda}r), r^{-1/2}\sin(\sqrt{\lambda}r).$$

Note that the former is of the form $r^{-1/2}\sum_{k=0}^{\infty} a_k r^{2k}$ and the latter is $r^{1/2}\sum_{k=0}^{\infty} a_k r^{2k}$, in view of the vanishing of sine at the origin, with $a_0 = 1$, resp. $a_0 = \sqrt{\lambda}$, in the two cases. This shows

$$J_{-1/2}(r) = \frac{2^{1/2}}{\Gamma(1/2)}r^{-1/2}\cos r, \;\; J_{1/2}(r) = \frac{1}{2^{1/2}\Gamma(3/2)}r^{-1/2}\sin r.$$

As $\Gamma(1/2) = \sqrt{\pi}$, $\Gamma(3/2) = \frac{1}{2}\sqrt{\pi}$ (see Problem 15.2), we have

(15.9) $$J_{-1/2}(r) = \sqrt{\frac{2}{\pi r}} \cos r, \quad J_{1/2}(r) = \sqrt{\frac{2}{\pi r}} \sin r.$$

In spite of the lack of convergence when n is not a half-integer, we nonetheless have:

Proposition 15.1. *There exist solutions of* (15.2) *of the form*

$$g_\pm(r) = e^{\pm i\sqrt{\lambda}r} f_\pm(r)$$

and

$$f_\pm(r) \sim r^{-1/2}\left(1 + \sum_{k=1}^{\infty} b_{k,+} r^{-k}\right), \quad r \to \infty,$$

where $\sim$ *means that if we stop at the* $k = K - 1$-*term the difference of the LHS and the RHS decays like* $r^{-1/2-K}$, *and the* j*th derivative of the difference decays like* $r^{-1/2-K-j}$.

Note that $g_\pm$ are certainly linearly independent over $(0, \infty)$, and thus every solution of Bessel's ODE is a linear combination of them. In particular, we conclude

Theorem 15.2. *For suitable constants* α_n, β_n, $J_n(\rho)$ *has the form*

$$J_n(\rho) = \alpha_n e^{i\rho} f_+(\rho) + \beta_n e^{-i\rho} f_-(\rho)$$

and

$$f_\pm(\rho) \sim \rho^{-1/2}\left(1 + \sum_{k=1}^{\infty} b_{k,+} \rho^{-k}\right), \quad \rho \to \infty,$$

where $\sim$ *means that if we stop at the* $k = K - 1$-*term the difference decays like* $\rho^{-1/2-K}$, *and the* j*th derivative of the difference decays like* $\rho^{-1/2-K-j}$.

An expansion as in the theorem is called an *asymptotic expansion*. We cannot prove Proposition 15.1, and thus the theorem, directly here, and thus *for now* we will *assume* Proposition 15.1 and its consequence. However, in Chapter 16 we show these results for n a non-negative integer by using a new method, that of a *stationary phase*, and the Fourier transform. For now we note that the theorem implies that for some A_n, ω_n,

(15.10) $$J_n(\rho) = A_n \sin(\rho - \omega_n) \rho^{-1/2}\left(1 + \sum_{k=1}^{\infty} b_k \rho^{-k}\right).$$

In fact it can be shown, as we shall see in Chapter 16, that

(15.11) $$\omega_n = n\frac{\pi}{2} - \frac{\pi}{4}, \quad A_n = \sqrt{\frac{2}{\pi}}.$$

(Notice that for $n = 1/2$, this agrees with (15.9).) In particular for large r, $J_n(\rho)$ has zeros near those of sine, i.e. at $\rho - \omega_n$ approximately $j\pi$, j a (sufficiently large!) positive integer; that is, for large enough ρ, there is essentially one zero in each interval of length π.

In summary, we conclude that the separated solutions of the Dirichlet eigenvalue equation have the form (15.6), and those of the wave equation with Dirichlet boundary condition correspondingly have the form (15.12)

$$u(r,\theta,t) = \sum_{k=1}^{\infty} (A_{0k}\cos(\sqrt{\lambda_{0k}}ct) + C_{0k}\sin(\sqrt{\lambda_{0k}}ct))J_0(\sqrt{\lambda_{0k}}r)$$

$$+ \sum_{n=1}^{\infty}\sum_{k=1}^{\infty} \Big((A_{nk}\cos n\theta + B_{nk}\sin n\theta)\cos(\sqrt{\lambda_{nk}}ct)$$

$$+ (C_{nk}\cos n\theta + D_{nk}\sin n\theta)\sin(\sqrt{\lambda_{nk}}ct)\Big)J_n(\sqrt{\lambda_{nk}}r),$$

$$\lambda_{nk} = \left(\frac{\rho_{nk}}{r_0}\right)^2,$$

where ρ_{nk} is the kth 0 of J_n, and $\rho_{nk} \approx k\pi$ for large k. What remains is to express the constants $A_{nk}, B_{nk}, C_{nk}, D_{nk}$ in terms of initial conditions

$$u(r,\theta,0) = \phi(r,\theta), \quad u_t(r,\theta) = \psi(r,\theta).$$

As usual, this amounts to showing an eigenfunction expansion, i.e. that ϕ (and similarly ψ) can be expressed as a series

$$\phi(r,\theta) = \sum_{n=0}^{\infty}\sum_{k=1}^{\infty} X_{nk}$$

$$= \sum_{k=1}^{\infty} A_{0k}J_0(\sqrt{\lambda_{0k}}r) + \sum_{n=1}^{\infty}\sum_{k=1}^{\infty}(A_{nk}\cos n\theta + B_{nk}\sin n\theta)J_n(\sqrt{\lambda_{nk}}r).$$

Since Δ is a symmetric operator with Dirichlet boundary conditions, we know a priori that for different eigenvalues, the corresponding eigenfunctions are orthogonal to each other. While for fixed n the λ_{nk} are distinct (they are the rescaled zeros of J_n), for different n's this need not be the case. To deal with this we observe that the inner product is

$$\langle f,g\rangle = \int_0^{r_0}\int_0^{2\pi} f(r,\theta)\overline{g(r,\theta)}\, r\, d\theta\, dr,$$

so in fact for different n the X_{nk} are automatically orthogonal to each other as the trigonometric functions sine and cosine are such. (So this follows after doing the θ integral, before the r integral is done.) Thus, $J_n(\sqrt{\lambda_{nk}}r)\cos n\theta$ and $J_n(\sqrt{\lambda_{nk}}r)\sin n\theta$ are all orthogonal to each other for distinct values of

(n, k) (with only the cosine allowed if $n = 0$), and as a result, provided these eigenfunctions are actually complete, the solution is given by

$$A_{nk} = \frac{\langle \phi, J_n(\sqrt{\lambda_{nk}}r) \cos n\theta \rangle}{\| J_n(\sqrt{\lambda_{nk}}r) \cos n\theta \|^2},$$

$$B_{nk} = \frac{\langle \phi, J_n(\sqrt{\lambda_{nk}}r) \sin n\theta \rangle}{\| J_n(\sqrt{\lambda_{nk}}r) \sin n\theta \|^2}.$$

Similarly,

$$C_{nk} = \frac{1}{\sqrt{\lambda_{nk}}c} \frac{\langle \psi, J_n(\sqrt{\lambda_{nk}}r) \cos n\theta \rangle}{\| J_n(\sqrt{\lambda_{nk}}r) \cos n\theta \|^2},$$

$$D_{nk} = \frac{1}{\sqrt{\lambda_{nk}}c} \frac{\langle \psi, J_n(\sqrt{\lambda_{nk}}r) \sin n\theta \rangle}{\| J_n(\sqrt{\lambda_{nk}}r) \sin n\theta \|^2}.$$

Typically, the inner products are not easy to evaluate explicitly, though numerically the integrals can be done easily. But, for instance, if one is given

$$\phi(r, \theta) = J_0(\sqrt{\lambda_{01}}r), \quad \psi(r, \theta) = 0,$$

i.e. only the base frequency λ_{01} is excited by the initial condition, then the solution is simply

$$u(r, \theta, t) = J_0(\sqrt{\lambda_{01}}r) \cos(\sqrt{\lambda_{01}}ct).$$

Now the remaining question is the *completeness* of our eigenfunctions, i.e. whether every element of $L^2(D)$ can be expanded in terms of these. As usual, it suffices to check this for a dense subset (i.e. one whose elements can approximate any element of $L^2(D)$), e.g. C^∞ functions on D vanishing near the origin $r = 0$ and near $r = r_0$. For such a function ϕ, we can expand ϕ in a Fourier series in the angular variable:

$$\phi(r, \theta) = A_0(r) + \sum_{n=1}^{\infty} A_n(r) \cos(n\theta) + B_n(r) \sin(n\theta),$$

with, say,

$$A_n(r) = \frac{1}{\pi} \int_0^{2\pi} \phi(r, \theta) \cos(n\theta) \, d\theta, \quad n \geq 1,$$

so differentiation under the integral sign shows $A_n(r), B_n(r)$ are C^∞ functions on $[0, r_0]$ which vanish near $r = 0$ and $r = r_0$. Thus, the completeness question reduces to whether for each fixed n, the Bessel functions $J_n(\sqrt{\lambda_{nk}}r)$ are complete in $L^2([0, r_0], r\, dr)$, where the $r\, dr$ denotes that the inner product we are considering is $\int f(r)\overline{g(r)}\, r\, dr$. This in fact is true, though harder to show than that of the Fourier series.

3. Higher dimensions

Bessel functions arise also in higher dimensional problems. For example, in three dimensions, using ω as the variable on $\mathbb{S}^2$, one separates variables by writing

$$X(r, \omega) = R(r)\Omega(\omega).$$

The spherical coordinate version of the Laplacian is

$$\Delta = \partial_r^2 + 2r^{-1}\partial_r + \Delta_{\mathbb{S}^2},$$

where $\Delta_{\mathbb{S}^2}$ is the Laplacian on the sphere, i.e. a differential operator in the ω variables, which one can express explicitly using spherical coordinates (θ, ϕ). Thus,

$$-\Delta X = \lambda X$$

becomes, similarly to (15.1),

$$r^2 R'' + 2r R' + \lambda r^2 R = \mu R,$$
$$-\Delta_{\mathbb{S}^2}\Omega = \mu\Omega.$$

Thus, one has to first solve the eigenvalue equation for the Laplacian on the sphere, and then an ODE in R which is very similar to Bessel's ODE, except that the coefficient of rR' is 2. It turns out that the eigenvalues of the negative of the spherical Laplacian are

$$\mu = l(l+1), \ l \geq 0, \ l \in \mathbb{Z},$$

and the corresponding eigenfunctions are called spherical harmonics, $\Omega(\theta, \phi) = Y_l^m(\theta, \phi)$, $|m| \leq l$, m an integer. (Thus, the $l(l+1)$-eigenspace is $2l+1$-dimensional.)

Now, in fact the radial ODE is not only similar to Bessel's ODE, but can be reduced to it. In fact, we can do this more generally for the ODE

(15.13) $$r^2 R'' + (d-1)r R' + \lambda r^2 R = \mu R,$$

which is the radial part of the d-dimensional Laplacian. To do so, simply write

$$R(r) = r^{-(d-2)/2}\tilde{R}(r).$$

Substituting this and multiplying by $r^{(d-2)/2}$ yields

(15.14) $$r(r\tilde{R}')' + \left(\lambda r^2 - \mu - \frac{(d-2)^2}{4}\right)\tilde{R} = 0,$$

which is Bessel's ODE with the previous role of $\mu = n^2$ being played by $\mu + \frac{(d-2)^2}{4}$.

Concretely, when $d = 3$, writing $\mu = l(l+1)$ as above, as

$$\mu + \frac{(d-2)^2}{4} = l(l+1) + \frac{1}{4} = \left(l + \frac{1}{2}\right)^2,$$

this means that the relevant Bessel functions are

$$J_{l+1/2}(\sqrt{\lambda}r),$$

and the actual eigenfunctions for the Laplacian on $B_{r_0} = \{x \in \mathbb{R}^3 : |x| < r_0\}$ with Dirichlet boundary condition are

$$r^{-1/2}J_{l+1/2}(\sqrt{\lambda_{l+1/2,k}}r)Y_l^m(\theta,\phi), \ \lambda_{l+1/2,k} = \left(\frac{\rho_{l+1/2,k}}{r_0}\right)^2,$$

where, as before, $\rho_{l+1/2,k}$ is the kth zero of $J_{l+1/2}$. Recall from the discussion leading to (15.9) that these Bessel functions have explicit formulae in terms of trigonometric function, with the $l = 0$ case given by (15.9), so in fact in this case the formulae are rather simple, and for $l = 0$ the zeros of the Bessel functions can be determined explicitly.

Problems

Problem 15.1. We want to find a second linearly independent solution of Bessel's ODE (15.2) when n is an integer.

(i) First assume $n > 0$ is an integer, and take $s = -n$. Then the iterative procedure (15.5) works for $k < n$. In order to deal with $k = n$, add a logarithmic term, i.e. consider a modified series

$$R(r) = r^s(1 + \sum_{k=1}^{\infty} a_k r^{2k} + \sum_{k=n}^{\infty} b_k r^{2k} \log r), \ s = -n.$$

Find a recursive expression for both the a_k and the b_k, and show that the resulting series converge for all values of r.

(ii) Now consider $n = 0$. In this case one expects a logarithmic leading term since the ODE $r(rR')' = 0$, which describes the leading behavior at 0, has $\log r$ as one of its solutions as seen in Chapter 12. So consider

$$R(r) = \log r + \sum_{k=1}^{\infty} a_k r^{2k} + \sum_{k=1}^{\infty} b_k r^{2k} \log r,$$

and find a recursive expression for both the a_k and the b_k, and show that the resulting series converge for all values of r.

Problem 15.2. The gamma function Γ is defined by

$$\Gamma(s) = \int_0^{\infty} t^{s-1}e^{-t} \, dt$$

for $s > 0$.

(i) Show that $\Gamma(s + 1) = s\Gamma(s)$.

(ii) Show that if n is an integer ≥ 0, then $n! = \Gamma(n + 1)$.

(iii) Show that $\Gamma(1/2) = \sqrt{\pi}$. (*Hint:* change variables, letting $x = t^{1/2}$.)

Problem 15.3. Solve the Neumann problem for the wave equation on the disk $B_{r_0} = \{x \in \mathbb{R}^2 : |x| < r_0\}$:

$$u_{tt} - c^2 \Delta u = 0$$

with

$$u_r|_{r=r_0} = 0$$

and with initial conditions

$$u(r, \theta, 0) = \phi(r, \theta), \ u_t(r, \theta, 0) = \psi(r, \theta), \ r < r_0.$$

Note: This problem involves the zeros of the derivatives of the Bessel functions. Show that there are infinitely many, and find their asymptotic location for large r.

Problem 15.4. Show that the substitution $R(r) = r^{-(d-2)/2}\tilde{R}(r)$ into (15.13) yields the ODE (15.14).

Problem 15.5. Solve the Dirichlet problem for the wave equation on the ball $B_{r_0} = \{x \in \mathbb{R}^3 : |x| < r_0\}$:

$$u_{tt} - c^2 \Delta u = 0$$

with

$$u|_{r=r_0} = 0$$

and with initial conditions

$$u(r, \theta, \phi, 0) = r^{-1}\sin(\pi r/r_0), \ u_t(r, \theta, \phi, 0) = 0, \ r < r_0.$$

Find the solution explicitly in terms of trigonometric functions.

Hint: The constant function 1 is an eigenfunction of the spherical Laplacian with eigenvalue 0.

Chapter 16

The method of
stationary phase

In this chapter we discuss a method that is very useful in determining the asymptotic behavior of functions of interest. A typical example is the inverse Fourier transform of a delta distribution on the circle $\mathbb{S}^1$ in $\mathbb{R}^2$,

$$u = \mathcal{F}^{-1}f, \; f = \delta_{|\xi|=1};$$

cf. Problem 9.5 where the $\mathbb{R}^3$ calculation was done. Note that

$$f(\phi) = \int_{\mathbb{S}^1} \phi(\omega)\,d\omega,$$

where we write the variable on $\mathbb{S}^1$ as ω. Here we can think of the circle as the interval $[-\pi, \pi]$ (rather than as a subset of $\mathbb{R}^2$), and in terms of this

$$f(\phi) = \int_{-\pi}^{\pi} \phi(\omega)\,d\omega.$$

As discussed at the end of Chapter 9, the inverse Fourier transform is the distribution f applied to $(2\pi)^{-2}e^{ix\cdot\xi}$, and since the point in $\mathbb{S}^1 \subset \mathbb{R}^2$ corresponding to $\omega \in [-\pi, \pi]$ is $(\cos\omega, \sin\omega)$, writing $x = r(\cos\theta, \sin\theta)$,

$$x \cdot \xi = r(\cos\theta\cos\omega + \sin\theta\sin\omega) = r\cos(\omega - \theta).$$

Thus, this inverse Fourier transform is simply the integral

$$(2\pi)^{-2} \int_{-\pi}^{\pi} e^{ir\cos(\omega-\theta)}\,d\omega.$$

Directly from the inverse Fourier transform u is $\mathcal{C}^\infty$ (as the inverse Fourier transform of a compactly supported distribution is such); of course this can be seen from the integral as well. Then the question is how this behaves as $r \to \infty$. Note that the integral is highly oscillatory for large r; a small change in ω may change the sign of the exponential: the phase $r\cos(\omega - \theta)$ changing by π accomplishes this sign change, which for large r means a small change in $\cos(\omega - \theta)$, thus in ω. Thus, there may be a lot of cancellation, so the asymptotics of this integral is not very clear; it could be decaying rather fast as $r \to \infty$.

Note that u solves the Helmholtz equation $(\Delta + 1)u = 0$ (since $(|\xi|^2 - 1)f = 0$ in the sense of tempered distributions). Moreover, as f is rotationally symmetric, so is u. This follows from the integral formula as well: replacing θ by some θ' does not change the integral as seen by a change of variables. Indeed,

$$(2\pi)^{-2} \int_{-\pi}^{\pi} e^{ir\cos(\omega - \theta)} \, d\omega = (2\pi)^{-2} \int_{-\pi}^{\pi} e^{ir\cos(\omega')} \, d\omega',$$

as follows by letting $\omega' = \omega - \theta$ and using the 2π-periodicity of the integrand (so that the interval of integration does not have to change). Thus, $u = u(r)$ is a function of r only, so the Helmholtz equation it satisfies becomes the Bessel equation we considered in Chapter 15 with $n = 0$ and $\lambda = 1$. Since u remains bounded as $r \to 0$, due to being actually C^∞ even at the origin, we conclude that $u(r)$ is a multiple of $J_0(r)$. To find which multiple, we evaluate the inverse Fourier transform at $r = 0$, which gives $(2\pi)^{-1}$, so we conclude that

$$u(r) = (2\pi)^{-1} J_0(r).$$

Correspondingly, finding the asymptotics of $u(r)$ as $r \to \infty$ gives us the asymptotic behavior of J_0:

$$J_0(r) = (2\pi)^{-1} \int_{-\pi}^{\pi} e^{ir\cos(\omega - \theta)} \, d\omega.$$

Now, to get a feeling for this problem it is useful to consider more general integrals of the form

$$I(r) = \int_{\mathbb{R}} e^{irf(\omega)} \phi(\omega) \, d\omega,$$

where $\phi \in C_c^\infty(\mathbb{R})$ (complex valued), and $f \in C^\infty(\mathbb{R})$ is real valued, and we want to understand the limit $r \to \infty$. Here f is called the *phase function*, while ϕ is the *amplitude*. As explained, this integrand should be highly oscillatory in general, but of course not if f is constant! In fact, the key question is whether the derivative of f vanishes. Indeed, if not, we can integrate by parts as follows:

$$\frac{d}{d\omega} e^{irf(\omega)} = irf'(\omega) e^{irf(\omega)},$$

so if f' does not vanish on supp ϕ (the integrand is 0 away from supp ϕ, so we need not be concerned with the behavior of f' there), then

$$e^{irf(\omega)} = \frac{1}{irf'(\omega)} \frac{d}{d\omega} e^{irf(\omega)}.$$

Substituting this into $I(r)$ gives

$$I(r) = \int_{\mathbb{R}} \frac{1}{irf'(\omega)} \frac{d}{d\omega} (e^{irf(\omega)}) \phi(\omega) \, d\omega,$$

so integration by parts gives, using the compact support of ϕ (so there are no boundary terms),

$$I(r) = \int_{\mathbb{R}} e^{irf(\omega)} \frac{d}{d\omega} \left(\frac{i}{rf'(\omega)} \phi(\omega) \right) d\omega.$$

Now, $\frac{i}{f'(\omega)} \phi(\omega)$ is just another C^∞ function of compact support as f' does not vanish on supp ϕ; hence so is its derivative,

$$\phi_1(\omega) = \frac{d}{d\omega} \left(\frac{i}{f'(\omega)} \phi(\omega) \right).$$

Hence

$$I(r) = r^{-1} \int_{\mathbb{R}} e^{irf(\omega)} \phi_1(\omega) \, d\omega,$$

and one can repeat this argument. Doing so k times gives

$$I(r) = r^{-k} \int_{\mathbb{R}} e^{irf(\omega)} \phi_k(\omega) \, d\omega,$$

with ϕ_k being C^∞ of compact support. Thus,

$$|I(r)| \le r^{-k} \int_{\mathbb{R}} |e^{irf(\omega)} \phi_k(\omega)| \, d\omega = r^{-k} \int_{\mathbb{R}} |\phi_k(\omega)| \, d\omega \le C_k r^{-k},$$

where C_k is a constant that depends on k but not on r (recall that ϕ_k has compact support!). Thus, $I(r)$ is rapidly decreasing, in the sense that it is decaying faster than any power of r^{-1} as $r \to \infty$.

This means that the only points ω which may contribute to the asymptotic behavior of $I(r)$ as $r \to \infty$ (up to a rapidly decaying, i.e. decaying faster than any power of r^{-1} as $r \to \infty$, error) are the critical points of f, i.e. the points where f' vanishes. These are also called *stationary points* of f, and correspondingly the calculation of the asymptotics of $I(r)$ is called the *method of stationary phase*.

Now let us suppose that ω_0 is a stationary point of f, so $f'(\omega_0) = 0$, and let us suppose that it is non-degenerate, so $f''(\omega_0) \ne 0$. (In this one dimensional setting, ω_0 is a maximum or minimum of f depending on the sign of the second derivative; in higher dimensions it could be a saddle point.) Taylor's theorem with second order remainder gives

$$f(\omega) = f(\omega_0) + (\omega - \omega_0)^2 \int_0^1 (1-t) f''(\omega_0 + t(\omega - \omega_0))\, dt$$

$$= f(\omega_0) + \frac{1}{2}(\omega - \omega_0)^2 F(\omega, \omega_0),$$

where

$$F(\omega, \omega_0) = 2 \int_0^1 (1-t) f''(\omega_0 + t(\omega - \omega_0))\, dt.$$

Note that $F(\omega_0, \omega_0) = f''(\omega_0) \neq 0$ and F is a C^∞ function of ω, so if $f''(\omega_0) > 0$, then $\sqrt{F}$ is also a C^∞ function for ω near ω_0; otherwise if $f''(\omega_0) < 0$, $\sqrt{-F}$ is such. Assuming $f''(\omega_0) > 0$ (with the other case being similar), we let

$$s = S(\omega) = (\omega - \omega_0)\sqrt{F(\omega, \omega_0)},$$

which is a C^∞ change of variables, with $\omega = \Omega(s)$ being the inverse change of variables and $s = 0$ corresponding to $\omega = \omega_0$. Changing the variables in the integral, *assuming that ϕ is supported sufficiently close to ω_0 so that the change of variables is valid,*

$$I(r) = \int_{\mathbb{R}} e^{irf(\omega_0)} e^{irs^2/2} \phi(\Omega(s)) |\Omega'(s)|\, ds = e^{irf(\omega_0)} \int_{\mathbb{R}} e^{irs^2/2} \phi(\Omega(s)) |\Omega'(s)|\, ds,$$

note that $|\Omega'(0)| = |S'(\omega_0)|^{-1} = |f''(\omega_0)|^{-1/2}$. Now, $\phi(\Omega(s))|\Omega'(s)|$ is just another compactly supported C^∞ function of s, so we may simply write it as $\tilde{\phi}(s)$ and consider integrals of the form

$$\tilde{I}_+(r) = \int_{\mathbb{R}} e^{irs^2/2} \tilde{\phi}(s)\, ds.$$

Now, since the function s^2 is a very special phase function (well behaved even as $|s| \to \infty$), we may even consider Schwartz functions $\tilde{\phi}$ in the integral (and not just compactly supported C^∞ ones). Further, if $\tilde{\phi}(s)$ is an *odd* function of s, then the integral vanishes by symmetry (as the phase function is an *even* function of s). We may write any function as a sum of an odd and an even one:

$$\tilde{\phi}(s) = \tilde{\phi}_{\text{odd}}(s) + \tilde{\phi}_{\text{even}}(s), \quad \tilde{\phi}_{\text{odd}}(s) = \frac{\tilde{\phi}(s) - \tilde{\phi}(-s)}{2}, \quad \tilde{\phi}_{\text{even}}(s) = \frac{\tilde{\phi}(s) + \tilde{\phi}(-s)}{2},$$

so in $\tilde{I}_+(r)$ we may simply replace $\tilde{\phi}$ by its even part, $\tilde{\phi}_{\text{even}}$. Note that for an even function the Taylor series at 0 consists of only even powers of s, i.e. all derivatives of odd order vanish at 0.

The same argument as above, namely integrating by parts, using the fact that

$$e^{irs^2/2} = \frac{1}{irs}\frac{d}{ds} e^{irs^2/2},$$

shows that if $\tilde{\phi}_{\text{even}} \in \mathcal{S}(\mathbb{R})$ and vanishes to order 2 at 0 (which just means $\tilde{\phi}_{\text{even}}$ vanishes at 0, since $\tilde{\phi}'_{\text{even}}$ automatically does), so $\frac{d}{ds}(s^{-1}\tilde{\phi}(s))$ is still C^∞, then

$$|\tilde{I}_+(r)| \leq C_1 r^{-1}.$$

Iterating this shows that if $\tilde{\phi}$ vanishes to order $2k$ at 0 (i.e. derivatives of order $\leq 2k - 1$ vanish at 0), then

$$|\tilde{I}_+(r)| \leq C_k r^{-k}.$$

Thus, if we want the integral for a specific function $\tilde{\phi}_{\text{even}}$, we may replace it by another function $\tilde{\psi}$ provided $\tilde{\phi}_{\text{even}} - \tilde{\psi}$ vanishes to order $2k$ at the origin, without affecting the asymptotic behavior of $\tilde{I}_+(r)$ up to errors that decay as r^{-k}. Here of course we may take k as large as we want. In particular, we may choose suitable linear combinations of Gaussians with polynomial weights:

$$s^j e^{-s^2/2}, 0 \leq j \leq 2k - 1;$$

note that for odd values of j the integral

$$\int_{\mathbb{R}} e^{irs^2/2} s^j e^{-s^2/2}\, ds$$

vanishes since the integrand is odd. So it remains to consider integrals of the form

$$\int_{\mathbb{R}} e^{irs^2/2} s^{2j} e^{-s^2/2}\, ds = \int s^{2j} e^{-(1-ir)s^2/2},\, ds,\ 0 \leq j \leq k - 1.$$

In particular, to obtain the asymptotics up to order r^{-1} decay, only the $j = 0$ term enters in our calculations, and we have calculated this as an integral of the Gaussian when we discussed the Fourier transform in Chapter 8; cf. Problem 8.5. Thus, for $j = 0$ we get

(16.1) $$\frac{\sqrt{2\pi}}{\sqrt{1-ir}}$$

for this integral, and thus

$$\left| \tilde{I}_+(r) - \frac{\sqrt{2\pi}}{\sqrt{1-ir}} \right| \leq \frac{C}{r},\ r \geq 1,$$

for some constant $C > 0$. Now in $1 - ir$, for r large, 1 is relatively small, so

$$(1 - ir)^{-1/2} = (-ir)^{-1/2}(1 + ir^{-1})^{-1/2} = e^{i\pi/4} r^{-1/2}(1 + ir^{-1})^{-1/2}$$

and

$$(1 + ir^{-1})^{-1/2} = 1 - \frac{1}{2}(ir^{-1}) + \cdots,$$

where ... denote higher powers of r^{-1}. Thus, the term 1 in the denominator affects $\frac{\sqrt{2\pi}}{\sqrt{1-ir}}$ by a term that can be absorbed into the right-hand side, at the cost of changing the constant C to a new $C' > 0$, so

(16.2) $$\left| \tilde{I}_+(r) - \frac{\sqrt{2\pi}e^{i\pi/4}}{\sqrt{r}} \right| \leq \frac{C'}{r}, \quad r \geq 1.$$

Note that if we consider

$$\tilde{I}_-(r) = \int_{\mathbb{R}} e^{-irs^2/2} \psi(s)\, ds$$

instead, the only change is that (16.1) is replaced by

$$\frac{\sqrt{2\pi}}{\sqrt{1+ir}},$$

so (16.2) becomes

$$\left| \tilde{I}_-(r) - \frac{\sqrt{2\pi}e^{-i\pi/4}}{\sqrt{r}} \right| \leq \frac{C}{r}, \quad r \geq 1.$$

Returning to I for $f(\omega) = \cos(\omega - \theta)$, the critical points of the phase $\cos(\omega - \theta)$ for $\theta = 0$, say, are $\omega = 0, \pi$, with $\omega = 0$ being a maximum (so the second derivative is negative) and $\omega = \pi$ being a minimum. One can decompose a general C^∞ function ϕ on $\mathbb{S}^1$, which we can think of as a 2π-periodic function on $\mathbb{R}$, as a function ϕ_- supported near 0 (and its integer multiples of 2π-translates in the periodic picture) but away from π, plus a function ϕ_+ supported near π (and its integer multiples of 2π-translates in the periodic picture) but away from 0, plus a function ϕ_0 supported away from both: $\phi = \phi_+ + \phi_- + \phi_0$; see Problem 5.2 for the partition of unity construction. Then ϕ_0 gives rise to rapidly decaying asymptotics as $r \to \infty$ (since the phase has no stationary points on supp ϕ_0), while the stationary phase calculation above can be applied to $\phi_\pm$. Correspondingly, for the constant function $\phi = 1$ we obtain that

$$\left| I(r) - \frac{\sqrt{2\pi}e^{-i\pi/4}}{\sqrt{r}} - \frac{\sqrt{2\pi}e^{i\pi/4}}{\sqrt{r}} \right| \leq \frac{C}{r}.$$

This says that

$$\left| J_0(r) - (2\pi)^{-1/2}r^{-1/2}(e^{ir}e^{-i\pi/4} + e^{-ir}e^{i\pi/4}) \right| \leq \frac{C}{r}.$$

Now

$$(2\pi)^{-1/2}r^{-1/2}(e^{ir}e^{-i\pi/4} + e^{-ir}e^{i\pi/4}) = \sqrt{\frac{2}{\pi r}} \frac{e^{i(r-\pi/4)} + e^{-i(r-\pi/4)}}{2}$$

$$= \sqrt{\frac{2}{\pi r}} \cos(r - \pi/4) = \sqrt{\frac{2}{\pi r}} \sin(r + \pi/4),$$

which is exactly the leading term of the asymptotics in (15.10)-(15.11)!

In general, one can work out further terms of the expansion by taking larger k (so larger j also enter). Below is one way to see what the result is for $j \geq 1$,

$$\int s^{2j} e^{-(1-ir)s^2/2} \, ds = \frac{-1}{1-ir} \int s^{2j-1} \frac{\partial}{\partial s} e^{-(1-ir)s^2/2} \, ds$$
$$= \frac{1}{1-ir} \int \frac{\partial}{\partial s} (s^{2j-1}) e^{-(1-ir)s^2/2} \, ds$$
$$= \frac{2j-1}{1-ir} \int s^{2j-2} e^{-(1-ir)s^2/2} \, ds,$$

so inductively one can reduce to the Gaussian integral we already computed (with $j = 0$). Note that each increase of j by 1 (in the exponent of s^{2j}) effectively adds a factor of r^{-1} to the decay of the asymptotics. The conclusion is that

$$J_0(r) = (2\pi)^{-1/2} r^{-1/2} (e^{ir} e^{-i\pi/4} g_+ + e^{-ir} e^{i\pi/4} g_-),$$

where

$$\left| g_\pm - \left(1 + \sum_{j=1}^{k-1} a_j r^{-j}\right) \right| \leq C r^{-k},$$

and the a_j are suitable complex numbers (constants).

In an asymptotic expansion one cannot just differentiate the expansion term-by-term in general, while in the case of stationary phase one can, so one can compute the asymptotic behavior of $J_0'(r)$, $J_0''(r)$, etc., this way. Indeed, the derivative $I'(r)$ is

$$I'(r) = \int_{\mathbb{R}} e^{if(\omega)} if(\omega)\phi(\omega) \, d\omega,$$

so ϕ is replaced by $if(\omega)\phi(\omega)$, but otherwise one has a similar expansion from the stationary phase for $I'(r)$ (as well as higher derivatives of I). To see that the terms are the same as for the term-by-term derivative, perform the change of variables, substituting in f, to get (localized near critical points with $f''(\omega_0) > 0$, for instance)

$$I'(r) = ie^{irf(\omega_0)} \int_{\mathbb{R}} e^{irs^2/2} \left(f(\omega_0) + \frac{s^2}{2}\right) \phi(\Omega(s)) |\Omega'(s)| \, ds$$
$$= if(\omega_0) \int_{\mathbb{R}} e^{irs^2/2} \phi(\Omega(s)) |\Omega'(s)| \, ds$$
$$\quad + e^{irf(\omega_0)} \int_{\mathbb{R}} e^{irs^2/2} \frac{is^2}{2} \phi(\Omega(s)) |\Omega'(s)| \, ds$$
$$= \frac{\partial}{\partial r} (e^{irf(\omega_0)}) \tilde{I}_+(r) + e^{irf(\omega_0)} \int_{\mathbb{R}} e^{irs^2/2} \frac{is^2}{2} \phi(\Omega(s)) |\Omega'(s)| \, ds.$$

Now the first term on the right is just the derivative of the $e^{irf(\omega_0)}$ factor in the asymptotic expansion of $I(r)$, times $\tilde{I}_+(r)$, while the second term is the derivative of $\tilde{I}_+(r)$, times $e^{irf(\omega_0)}$, so one just needs to check that the asymptotic expansion of $\tilde{I}_+(r)$ can be term-by-term differentiated. But in this one can replace $\tilde{\phi}(s) = \phi(\Omega(s))|\Omega'(s)|$ by any equivalent function $\tilde{\psi}(s)$ in the sense that the difference vanishes to high order, for instance by a sum of Gaussians times s^{2j}, and then the conclusion follows from

$$\int_{\mathbb{R}} e^{irs^2/2}\frac{is^2}{2}e^{-s^2/2}s^{2j}\,ds = \frac{\partial}{\partial r}\left(\int_{\mathbb{R}} e^{irs^2/2}e^{-s^2/2}s^{2j}\,ds\right),$$

and the term-by-term differentiability of the *convergent* series of $(1-ir)^{j-1/2}$ in powers of r^{-1}.

The Bessel functions J_n can be handled similarly by considering the inverse Fourier transform

$$u = \mathcal{F}^{-1}f, \; f = \phi(\omega)\delta_{\xi=1}, \; \phi(\omega) = e^{in\omega}, \; n \geq 0, \; n \in \mathbb{Z},$$

where we used polar coordinates (ρ, ω) in ξ so that $\xi = \rho(\cos\omega, \sin\omega)$. The case $n = 0$ was the case considered above that gave us the J_0-asymptotics. For all integers n, u still solves the Helmholtz equation for the same reason as in the $n = 0$ case, and it is still $\mathcal{C}^\infty$, including at the origin. Further,

$$u = (2\pi)^{-2}\int_\pi^\pi e^{ir\cos(\omega-\theta)}e^{in\omega}\,d\omega = (2\pi)^{-2}e^{in\theta}\int_\pi^\pi e^{ir\cos(\omega-\theta)}e^{in(\omega-\theta)}\,d\omega$$

$$= (2\pi)^{-2}e^{in\theta}\int_\pi^\pi e^{ir\cos(\omega')}e^{in\omega'}\,d\omega'$$

shows that it is of the form

$$u = e^{in\theta}R(r),$$

so from Chapter 15, $R(r)$ solves the Bessel equation with $\lambda = 1$ and parameter n. Since it is bounded at $r = 0$ (as u is actually smooth), we conclude that R is a multiple of $J_n(r)$—this is where $n \geq 0$ is used (for $n < 0$, we would get $J_{-n}(r)$ at this point). The value of this can be computed by computing the nth derivative of u in r at the origin, using the integral formula; recall that for J_n this derivative is 2^{-n} due to the normalization of J_n. On the other hand, the method of stationary phase can be applied to the integral

$$\int_\pi^\pi e^{ir\cos\omega}e^{in\omega}\,d\omega.$$

Comparison of these two gives the Bessel function asymptotics; see Problem 16.1 for details.

Problems

Problem 16.1. Use the stationary phase method to compute the asymptotic behavior of $J_n(r)$ as $r \to \infty$ when n is a positive integer, and use this to show (15.10)-(15.11) for such n.

Problem 16.2. Use the stationary phase method to compute the asymptotic behavior at spacetime infinity of solutions of the Schrödinger equation from Problem 8.9:

With $\hbar > 0$ fixed, solve the (mass 1) Schrödinger equation

$$i\hbar u_t = -\frac{\hbar^2}{2}\Delta_x u, \ u(x,0) = \psi(x), \ \psi \in \mathcal{S}(\mathbb{R}^n),$$

for a free particle in $\mathbb{R}_x^n$ using the Fourier transform, and find the asymptotic behavior of u if one keeps $v = \frac{x}{t}$ fixed and lets $t \to \pm\infty$. You may leave your answer in terms of the Fourier transform of ψ evaluated at a point expressed in terms of v. Draw a picture of the spacetime showing the lines along which the limit is taken.

Chapter 17

Solvability via duality

1. The general method

We now return to solving PDE using duality arguments and energy estimates. Before getting into details, we note that the ideal type of well-posedness result we would like is the following. We are given a PDE (including various additional conditions), and we would like to have spaces of functions X and Y such that there is a unique solution of the PDE in X which depends continuously on the data in Y. To be specific, let's consider the linear PDE $Pu = f$, with any other condition we may want to have made homogeneous (e.g. homogeneous Dirichlet BC or homogeneous initial conditions)—which we have seen can be done (cf. Problem 11.3 for initial conditions)—and we want to find function spaces X and Y, which should have a notion of convergence, e.g. be normed spaces, such that

(i) (Existence) For any $f \in Y$ there exists $u \in X$ such that $Pu = f$.

(ii) (Uniqueness) For any $f \in Y$ there is at most one $u \in X$ such that $Pu = f$.

(iii) (Stability) If $f_j \to f$ in Y, then the unique solutions, u_j, resp. u, in X, of $Pu_j = f_j$, $Pu = f$, satisfy $u_j \to u$ in X.

Note that we do not say that P itself maps all of X into Y, i.e. for $u \in X$, Pu need not lie in Y at this point. Of course, Pu needs to have some meaning, so a minimal requirement is that X and Y are both subspaces of $\mathcal{D}'(\Omega)$ for some Ω (e.g. $\Omega = \mathbb{R}^n$), or Ω a bounded domain in $\mathbb{R}^n$, or $\Omega = \mathbb{T}^n = (\mathbb{S}^1)^n$, $\mathcal{C}^\infty$ functions on which, recall, can be identified with 2π-periodic $\mathcal{C}^\infty$ functions in every coordinate on $\mathbb{R}^n$. Also, to make this problem meaningful, Y should be relatively large—typically it should include $\mathcal{C}_c^\infty(\Omega)$ at least. (If P is linear and $Y = \{0\}$, it is trivial to find an X: $X = \{0\}$ will do.)

Now, to get a feeling for what this means, note that if we can solve a PDE in a space X (let's keep Y fixed for now), then we can also solve it in a bigger space $\tilde{X} \supset X$, since the solution u in X also lies in $\tilde{X}$. On the other hand, if there is at most one solution of the PDE in a space X, then there is at most one solution in a smaller space $\tilde{X} \subset X$ since if u_1 and u_2 are solutions in $\tilde{X}$, then they are solutions in X, thus by uniqueness in X they are equal. Thus, there is some tension between existence and uniqueness: for the former, one could try to increase the space to make the problem easier, but then may lose uniqueness for the latter one may try to make the space smaller, but then may lose existence. To be extreme, it may be easy to show uniqueness if $X = C_c^\infty(\Omega)$ (e.g. using the maximum principle or energy estimates), but existence may not hold, while it may be relatively easy to show existence when $X = \mathcal{D}'(\Omega)$, but uniqueness may not hold.

As a concrete example, if $Y = C_c^\infty(\mathbb{R})$, $P = \frac{d}{dx}$, then with $X = C_c^\infty(\mathbb{R})$ one has at most one solution of $Pu = f$, $f \in Y$ (since by the fundamental theorem of calculus one can write u as the indefinite integral of f from $-\infty$), but may not have any solutions (this indefinite integral may not vanish for large x), while if $X = \mathcal{D}'(\mathbb{R})$, or indeed if $X = C^\infty(\mathbb{R})$, there is always a solution (the aforementioned indefinite integral), but it is not unique (one can always add a constant). If however one lets X be the subspace of $C^\infty(\mathbb{R})$ consisting of functions u which vanish near $-\infty$ (i.e. there exists $R > 0$ such that if $x < -R$, then $u(x) = 0$) one still has uniqueness, and one also has existence: the solution is the previously mentioned indefinite integral.

If (i) and (ii) hold, then there is a solution operator for the PDE, $S : Y \to X$, given by the requirement that Sf is the unique $u \in X$ such that $Pu = f$. Thus, $PSf = f$, so S is a *right inverse* for P. If in addition (iii) holds, then $f_j \to f$ in Y implies that $Sf_j \to Sf$ in X, i.e. S is continuous. Note that in view of $PSf = f$, S is automatically one-to-one, but it need *not* be onto. However, if $P : X \to Y$ actually, then S is onto, since given $u \in X$, u solves $Pu = f$ where $f = Pu \in Y$, so $u = Sf = SPu$. In fact, if S is onto, then for any $u \in X$ there is an $f \in Y$ such that $Sf = u$, and thus $Pu = PSf = f$, so $P : X \to Y$. Hence $SPu = Sf = u$, so S is also a *left inverse* for P, and thus $P : X \to Y$ is invertible.

To see what existence entails, by duality arguments, note that by the definition of distributional derivatives, for $\Omega \subset \mathbb{R}^n$ open, $u \in \mathcal{D}'(\Omega)$,

$$Pu(\phi) = u(P^t \phi), \ \phi \in C_c^\infty(\Omega),$$

where

$$P^t = \sum_{|\alpha| \le m} (-1)^{|\alpha|} \partial^\alpha a_\alpha, \ P = \sum_{|\alpha| \le m} a_\alpha \partial^\alpha.$$

Thus, $Pu = f$ states exactly that

$$u(P^t\phi) = f(\phi) \text{ for all } \phi \in C_c^\infty(\Omega).$$

Thus, if we want a distributional solution u, i.e. a continuous linear map $u : C_c^\infty(\Omega) \to \mathbb{C}$, such that $u(P^t\phi) = f(\phi)$ for all $\phi \in C_c^\infty(\Omega)$, we do not have *any choice* as to what u does on the range of $P^t : C_c^\infty(\Omega) \to C_c^\infty(\Omega)$, which we denote by $\mathrm{Ran}_{C_c^\infty(\Omega)} P^t$. Moreover, if $P^t\phi_1 = P^t\phi_2$ then we must have $f(\phi_1) = f(\phi_2)$ in order to have any chance of finding such a u. Since we do not want to restrict the possible f's (we want to allow at least $C_c^\infty(\Omega)$ for Y), this means that P^t needs to be one-to-one to have a chance for a solution. Then, we define at first a linear functional

$$(17.1) \qquad\qquad \ell : \mathrm{Ran}_{C_c^\infty(\Omega)} P^t \to \mathbb{C}$$

by

$$(17.2) \qquad\qquad \ell(\psi) = f(\phi), \text{ where } \phi \text{ is s.t. } P^t\phi = \psi$$

(note that ϕ is unique by the assumed injectivity of P^t), and the question is whether this is continuous and whether we can extend it to all of $C_c^\infty(\Omega)$. Continuity follows if we know that $\psi_j \to \psi$ in $C_c^\infty(\Omega)$ implies $\phi_j \to \phi$ in $C_c^\infty(\Omega)$ (where $\psi_j = P^t\phi_j$), which is a kind of an energy estimate for the operator P^t: one estimates ϕ_j by $P^t\phi_j$. Once we have such a continuity, we can extend ℓ to a distribution (i.e. a continuous linear map defined *for all* $\phi \in C_c^\infty(\Omega)$) by abstract functional analysis; this is easier to explain in the special context of inner product spaces, which we do below. Note that if $\mathrm{Ran}_{C_c^\infty(\Omega)} P^t$ is dense in $C_c^\infty(\Omega)$, i.e. for every $\psi \in C_c^\infty(\Omega)$ there exist $\phi_j \in C_c^\infty(\Omega)$ such that $P^t\phi_j \to \psi$, then we do not have any choice in defining the extension u applied to ψ, since $u(\psi) = \lim_{j\to\infty} u(P^t\phi_j)$, and the latter is already fixed! In this case well-definedness and continuity can be checked as well; see Lemma 17.10 for a similar argument in the simpler context of normed vector spaces. Thus, we have the following conclusion:

Proposition 17.1. *Suppose that*

(i) *P^t is injective on $C_c^\infty(\Omega)$,*

(ii) *if $\psi_j \to \psi$, then $\phi_j \to \phi$ in $C_c^\infty(\Omega)$, where $\psi_j = P^t\phi_j$, $\psi = P^t\phi$, $\phi_j, \phi \in C_c^\infty(\Omega)$,*

(iii) *$\mathrm{Ran}_{C_c^\infty(\Omega)} P^t$ is dense in $C_c^\infty(\Omega)$.*

Then for any $f \in \mathcal{D}'(\Omega)$ there exists a unique $u \in \mathcal{D}'(\Omega)$ such that $Pu = f$.

If we drop (iii), we cannot expect uniqueness since we have some freedom in choosing the extension u of ℓ. However, as already mentioned, existence still holds, thanks to a functional analytic result.

Proposition 17.2 (Hahn-Banach). *Suppose that V is a normed vector space, or $V = \mathcal{C}_c^\infty(\Omega)$, or $V = \mathcal{S}(\mathbb{R}^n)$. Suppose that D is a subspace of V. A continuous linear functional $\ell : D \to \mathbb{C}$ can be extended to a continuous linear functional $u : V \to \mathbb{C}$.*

Remark 17.3. For the readers interested in what the natural assumption on V is here: it is that V is a *locally convex topological vector space*. This includes all of the examples in the statement of the proposition. However, it is important to keep in mind that for such a space, in general, continuity is *not* the same as sequential continuity (i.e. $\phi_j \to \phi$ implies $\ell(\phi_j) \to \ell(\phi)$), although in all the enumerated cases they are the same.

Note that u is determined on $\overline{D}$ by taking $\psi_j \in D$ converging to $\psi \in \overline{D}$, but it is usually not determined on all of V.

Thus, we conclude:

Proposition 17.4. *Suppose that*

(i) *P^t is injective on $\mathcal{C}_c^\infty(\Omega)$,*

(ii) *if $\psi_j \to \psi$, then $\phi_j \to \phi$ in $\mathcal{C}_c^\infty(\Omega)$, where $\psi_j = P^t\phi_j$, $\psi = P^t\phi$, $\phi_j, \phi \in \mathcal{C}_c^\infty(\Omega)$.*

Then for any $f \in \mathcal{D}'(\Omega)$ there exists $u \in \mathcal{D}'(\Omega)$ such that $Pu = f$.

Although Proposition 17.2 has *no assumptions* of completeness, the easiest setting to prove this result is that of complete inner product spaces, i.e. *Hilbert spaces*, for there one can first extend ℓ to a linear functional L on $\overline{D}$ (this always works) and then use the *orthogonal projection* from V to $\overline{D}$ to extend L further to u; see Lemma 17.16. Namely, if the orthogonal projection is $\pi : V \to \overline{D}$, then one defines $u = L \circ \pi$, which satisfies all requirements. Of course, one needs to show that the orthogonal projection actually makes sense, i.e. one can write any $v \in V$ uniquely as $v = v_\parallel + v_\perp$, with $v_\parallel \in \overline{D}$ and $v_\perp$ orthogonal to $\overline{D}$, which is where completeness comes in. (In the special case when $\overline{D}$ is finite dimensional, this already follows from our explicit formulae, without need for completeness: (13.6) is the orthogonal projection of v to the span of $\{x_1, \ldots, x_n\}$. However, this is much too weak for us here!)

The context of inner product spaces also has the virtue in that we worked in the biggest possible space, $\mathcal{D}'(\Omega)$, to get existence; as already remarked, this may be too large to get uniqueness. In addition, if we work with a boundary value problem, e.g. $\Omega \subset \mathbb{R}^n$ is a domain, if u is merely a distribution, the boundary condition does not make any sense. Thus, we want to discuss this procedure in the context of inner product spaces.

But first, let's relax the convergence assumptions in (ii) of Proposition 17.1. So suppose that there is a subspace Z of $\mathcal{D}'(\Omega)$, which is equipped with a norm $\|.\|_Z$, such that

(i) $\mathcal{C}_c^\infty(\Omega) \subset Z$, with convergence in $\mathcal{C}_c^\infty(\Omega)$ implying convergence in Z (i.e. the inclusion of $\mathcal{C}_c^\infty(\Omega)$ into Z is a continuous map), and

(ii) $\mathcal{C}_c^\infty(\Omega)$ dense in Z.

Then any continuous linear map $f : Z \to \mathbb{C}$ in particular restricts to a continuous linear map $f : \mathcal{C}_c^\infty(\Omega) \to \mathbb{C}$ (for if $\phi_j \to \phi$ in $\mathcal{C}_c^\infty(\Omega)$, then they also converge in Z by our assumption, so $f(\phi_j) \to f(\phi)$ in $\mathbb{C}$ by the Z-continuity of f), so continuous linear maps $f : Z \to \mathbb{C}$ actually define distributions. One writes Z^* for the set of these maps f, as already mentioned in the introduction:

Definition 17.5. For a normed vector space Z, its dual Z^* is the set of continuous linear maps $f : Z \to \mathbb{C}$.

Lemma 17.6. *If $f \in Z^*$ vanishes as a distribution (i.e. $f(\phi) = 0$ for all $\phi \in \mathcal{C}_c^\infty(\Omega)$), then it vanishes (i.e. $f(\phi) = 0$ for all $\phi \in Z$).*

Proof. This follows immediately from the assumed density of $\mathcal{C}_c^\infty(\Omega)$ in Z: if $f(\psi) = 0$ for all $\psi \in \mathcal{C}_c^\infty(\Omega)$, and $\phi \in Z$, take $\phi_j \in \mathcal{C}_c^\infty(\Omega)$, $\phi_j \to \phi$ in Z; then $f(\phi) = \lim_{j\to\infty} f(\phi_j) = 0$, where the first equality is the consequence of the Z-continuity of f. $\qquad\square$

Thus, one can think of Z^* as a subspace of $\mathcal{D}'(\Omega)$; the map $j : Z^* \to \mathcal{D}'(\Omega)$ is injective. (This is similar to how we can think of continuous functions as distributions.)

As a first step, suppose that for Z as above, and some P, we know that

(i) P^t is injective on $\mathcal{C}_c^\infty(\Omega)$ and

(ii) if $\psi_j \to \psi$ in $\mathcal{C}_c^\infty(\Omega)$, then $\phi_j \to \phi$ in Z, where $\psi_j = P^t\phi_j$, $\psi = P^t\phi$.

(Note that $\phi_j \to \phi$ in Z is a weaker statement than, i.e. follows immediately from, $\phi_j \to \phi$ in $\mathcal{C}_c^\infty(\Omega)$.) If $f \in Z^*$, let ℓ be as in (17.1)-(17.2). Then $\ell(\psi) = f(\phi)$, $\psi \in \mathrm{Ran}_{\mathcal{C}_c^\infty(\Omega)} P^t$, shows that $\ell(\psi_j) = f(\phi_j) \to f(\phi) = \ell(\psi)$, so ℓ is continuous, and hence using the Hahn-Banach theorem we can solve $Pu = f$ for $f \in Z^*$, getting $u \in \mathcal{D}'(\Omega)$. Thus, we made a weaker assumption than in Proposition 17.4 (because we are only assuming $\phi_j \to \phi$ in Z), and we got a weaker conclusion (solvability for only $f \in Z^*$, not for all $f \in \mathcal{D}'(\Omega)$), but at least we still have a solvability statement!

Now suppose that we also have a subspace W of $\mathcal{D}'(\Omega)$, again equipped with a norm $\|.\|_W$, such that

(i) $\mathcal{C}_c^\infty(\Omega) \subset W$, with convergence in $\mathcal{C}_c^\infty(\Omega)$ implying convergence in W, and

(ii) $\mathcal{C}_c^\infty(\Omega)$ dense in W.

Thus, W^* is a subspace of $\mathcal{D}'(\Omega)$, much as in the case of Z discussed above. Suppose also that for some P we know that

(i) P^t is injective on $\mathcal{C}_c^\infty(\Omega)$ and

(ii) if $\psi_j \to \psi$ in W, then $\phi_j \to \phi$ in Z, where $\psi_j = P^t\phi_j$, $\psi = P^t\phi$.

Notice that the hypotheses in (ii) are stronger than in the previous paragraph, since $\psi_j \to \psi$ in $\mathcal{C}_c^\infty(\Omega)$ would imply $\psi_j \to \psi$ in W. Now if $f \in Z^*$, again let ℓ be as in (17.1)-(17.2). Then $\ell(\psi) = f(\phi)$ shows that if $\psi_j \to \psi$ in W, then $\ell(\psi_j) = f(\phi_j) \to f(\phi) = \ell(\psi)$, so ℓ is continuous when one considers W-convergence in its domain. Note that now we have normed spaces, so our continuity statement for P^t is equivalent to the estimate that for some $C > 0$,

$$\|\phi\|_Z \le C\|P^t\phi\|_W;$$

such an estimate also gives that P^t is injective. Using the Hahn-Banach theorem we can solve $Pu = f$ for $f \in Z^*$ with $u \in W^*$. Note the prevalant appearances of the dual spaces! The gain over the original result, Proposition 17.4, is precision: if we know more about f (not just that it is a distribution), then we know more about the solution u we get (namely, not just that it is a distribution). We state this as a result.

Proposition 17.7. *Let W, Z be as above. Suppose that*

$$\|\phi\|_Z \le C\|P^t\phi\|_W, \quad \phi \in \mathcal{C}_c^\infty(\Omega).$$

Then for any $f \in Z^*$ there exists $u \in W^*$ such that $Pu = f$.

2. An example: Laplace's equation

The simplest kind of energy estimate we had was for the variable coefficient 'Laplacian'

$$\Delta_c = \nabla \cdot c(x)^2 \nabla,$$

where Ω is a *bounded* domain, $c \in C^1(\overline{\Omega})$, $c > 0$, namely

$$\Delta_c u = f, \ u \in C^2(\overline{\Omega}), \ u|_{\partial\Omega} = 0 \Rightarrow \|u\|_{H^1(\Omega)}^2 \le C\|f\|_{L^2(\Omega)}^2,$$

where

$$(17.3) \ \|u\|_{H^1(\Omega)}^2 = \|u\|_{L^2(\Omega)}^2 + \sum_{j=1}^n \|cD_{x_j}u\|_{L^2(\Omega)}^2 = \int_\Omega (|u|^2 + c(x)^2|\nabla u|^2)\, dx.$$

You may note that different choices of c give different norms, but (see Problem 17.1) these are equivalent to each other, i.e. if we *temporarily* denote by H_c^1 the norm corresponding to a particular c, then given $c, c' \in C^1(\overline{\Omega})$ both of which are positive, there are constants $C_1, C_2 > 0$ such that for $u \in C^2(\overline{\Omega})$, $u|_{\partial\Omega} = 0$, one has

$$C_1 \|u\|_{H_c^1} \leq \|u\|_{H_{c'}^1} \leq C_2 \|u\|_{H_c^1}.$$

In particular, they are equivalent to the norm with $c \equiv 1$, which is the *standard* H^1-norm. The reason we put the c into the norm is because then we can express the operator in terms of the inner product inducing the norm, which is particularly convenient.

As discussed in Chapter 7, this gives uniqueness and stability with

(17.4) $$X = \{u \in C^2(\overline{\Omega}) : u|_{\partial\Omega} = 0\}, \ Y = C^0(\overline{\Omega})$$

equipped with the norms

$$\|u\|_X = \|u\|_{H^1} = \left(\|u\|_{L^2(\Omega)}^2 + \sum_{j=1}^n \|cD_{x_j}u\|_{L^2(\Omega)}^2 \right)^{1/2}, \ \|f\|_Y = \|f\|_{L^2(\Omega)},$$

(for purposes of stability: norms do not matter for existence and uniqueness). However, existence is *false*, even if $c \equiv 1$: there exists $f \in Y$ such that there is no $u \in X$ with $\Delta u = f$ (though it's not so easy to write such an f and show that it works!). Thus, we would like to enlarge the space X, but not so much as to lose uniqueness or stability.

The simplest way of getting existence is dualizing as above. Note that $\Delta_c^t = \Delta_c$, so the energy estimate gives that

$$\|\phi\|_X \leq C' \|\Delta_c^t \phi\|_Y,$$

so Δ_c^t is injective on $C_c^\infty(\Omega)$, and if $\psi_j \to \psi$ in the H^1-norm, then $\phi_j \to \phi$ in the L^2-norm. Correspondingly we deduce:

Proposition 17.8. *For any $f \in Y^*$ there exists $u \in X^*$ such that $\Delta_c u = f$. Here the dual spaces are with respect to the H^1-, resp. the L^2-norms, i.e. X^* consists of continuous linear maps from X to $\mathbb{C}$ which are continuous in the H^1-norm, and similarly for Y.*

While this solves the PDE, as already mentioned it is not clear whether we have solved the problem since the boundary conditions may not make any sense! However, in a slightly different context we would be set. For instance, let

$$P = \Delta_c - q = \nabla \cdot c(x)^2 \nabla - q,$$

$c \in C^1(\mathbb{R}^n)$, $0 < c_1 \leq c \leq c_2$, $0 < q_1 \leq q \leq q_2$ with c_1, c_2, q_1, q_2 constants; here q is useful as we do not have a Poincaré inequality! Now let $N > n/2$, and let

$$X = \{u \in C^2(\mathbb{R}^n) : (1 + |x|)^N \partial^\alpha u \text{ is bounded for } |\alpha| \leq 2\},$$

$$Y = \{u \in C^0(\mathbb{R}^n) : (1 + |x|)^N u \text{ is bounded}\},$$

equipped with the norms

$$\|u\|_X = \|u\|_{H^1} = \left(\|\sqrt{q}u\|^2_{L^2(\mathbb{R}^n)} + \sum_{j=1}^{n} \|cD_{x_j}u\|^2_{L^2(\mathbb{R}^n)} \right)^{1/2}, \quad \|f\|_Y = \|f\|_{L^2(\mathbb{R}^n)}.$$

Then for $f \in Y$,

$$(17.5) \qquad Pu = f, \ u \in X \Rightarrow \|u\|^2_{H^1(\mathbb{R}^n)} \leq C\|f\|^2_{L^2(\mathbb{R}^n)},$$

since

$$\int_{\mathbb{R}^n} Pu\bar{v} \, dx = - \int_{\mathbb{R}^n} (c(x)^2 \nabla u \cdot \overline{\nabla v} + qu\bar{v}) \, dx$$

so, in particular, with $v = u$,

$$\|u\|^2_{H^1} = - \int_{\mathbb{R}^n} Pu\bar{u} \, dx \leq \|f\|_{L^2}\|u\|_{L^2} \leq \epsilon\|u\|^2_{L^2} + \epsilon^{-1}\|f\|^2_{L^2}.$$

Hence rearrangement and

$$\|u\|^2_{L^2} \leq q_1^{-1}\|\sqrt{q}u\|^2_{L^2} \leq q_1^{-1}\|u\|^2_{H^1}$$

gives

$$(1 - \epsilon q_1^{-1})\|u\|^2_{H^1} \leq \epsilon^{-1}\|f\|^2_{L^2},$$

and hence (17.5), if we take $\epsilon > 0$ sufficiently small, e.g. $\epsilon = q_1/2$. We thus deduce

Proposition 17.9. *For any $f \in Y^*$ there exists $u \in X^*$ such that $Pu = f$.*

3. Inner product spaces and solvability

While here there are no boundary conditions to be concerned with, it would still be nice to have a more concrete description of the spaces Y^* and X^*, and in particular to relate them to spaces on which we have stability and uniqueness. This is most easily done via completions and a stronger use of the inner product structure. One reason completions work very nicely is the following result on extending linear maps defined on subspaces. Recall also that if $\tilde{X}$ is the completion of X, then X is dense in $\tilde{X}$.

Lemma 17.10 (Bounded linear transformation, or BLT, theorem). *Suppose that $\tilde{X}, Y$ are normed vector spaces, Y is complete, and X is a subspace of $\tilde{X}$ and X is dense in $\tilde{X}$. Suppose that $T : X \rightarrow Y$ is a continuous linear (respectively conjugate linear) map. Then there is a unique continuous extension $\tilde{T} : \tilde{X} \rightarrow Y$ of T (extension means that for $x \in X$, $\tilde{T}x = Tx$), and it is linear (resp. conjugate linear).*

Proof. Uniqueness is immediate, for if $x \in \tilde{X}$, then by the density of X in $\tilde{X}$ we have $x_j \in X$ such that $x_j \to x$, and then by the continuity of $\tilde{T}$, $\tilde{T}x_j \to \tilde{T}x$. But $\tilde{T}x_j = Tx_j$, so $\tilde{T}x = \lim_{j\to\infty} Tx_j$ is determined by the sequence $\{Tx_j\}_{j=1}^{\infty}$.

For existence, we note that if $x_j \to x$, then $\{x_j\}$ is Cauchy. We claim that $\{Tx_j\}$ is Cauchy. Indeed, T is continuous, so there is some $C > 0$ such that $\|Tz\|_Y \le C\|z\|_X$ for all $z \in X$ (see Lemma 1.3). Thus,

$$\|Tx_j - Tx_k\|_Y = \|T(x_j - x_k)\|_Y \le C\|x_j - x_k\|_X,$$

and the right-hand side goes to 0 as $j, k \to \infty$ as $\{x_j\}$ is Cauchy, and so $\{Tx_j\}$ is Cauchy, and thus converges as Y is complete. Now we would like to define $\tilde{T}x = \lim_{j\to\infty} Tx_j$. We already saw that the limit exists, but it may a priori depend on the sequence $\{x_j\}$. To see that this is not the case, suppose $\{x_j'\}$ is another sequence in X with limit x; then $\|x_j - x_j'\|_X \to 0$ as $j \to \infty$, hence

$$\|Tx_j - Tx_j'\|_Y = \|T(x_j - x_j')\|_Y \le C\|x_j - x_j'\|_X \to 0$$

as $j \to \infty$, and thus $\lim Tx_j = \lim Tx_j'$. Correspondingly $\tilde{T} : \tilde{X} \to Y$ is well defined.

One also checks (conjugate) linearity easily using the linearity of limits. To see continuity, it suffices to show that $\|\tilde{T}x\|_Y \le C\|x\|_{\tilde{X}}$, which already holds if $x \in X$. But if $x_j \to x$ in $\tilde{X}$ with $x_j \in X$, then $Tx_j \to \tilde{T}x$ in Y by the definition of $\tilde{T}$. Since the norm is continuous on Y, $\|Tx_j\|_Y \to \|\tilde{T}x\|_Y$. Also, as the norm is continuous on $\tilde{X}$, $\|x_j\|_{\tilde{X}} \to \|x\|_{\tilde{X}}$. Now letting $j \to \infty$ on both sides in $\|Tx_j\|_Y \le C\|x_j\|_{\tilde{X}}$ yields $\|\tilde{T}x\|_Y \le C\|x\|_{\tilde{X}}$, giving the desired continuity. $\qquad\square$

Returning to the boundary problem, i.e. to Ω, in view of the norms on X and Y, given in (17.4), it is natural to enlarge X and Y by completing them. Recall that the completion of $C^0(\overline{\Omega})$ is $L^2(\Omega)$. Similarly, for $u \in C^2(\overline{\Omega})$ let

$$\|u\|_{H^1(\Omega)}^2 = \int_{\Omega}(|u|^2 + |c\nabla u|^2)\,dx.$$

This is a norm arising from an inner product, namely

$$\langle u, v \rangle = \int_{\Omega}(u\overline{v} + c^2\nabla u \cdot \overline{\nabla v})\,dx = \langle u, v \rangle_{L^2} + \langle c\nabla u, c\nabla v \rangle_{L^2}.$$

Then the energy estimate states that

$$\|u\|_{H^1(\Omega)} \le C\|\Delta_c u\|_{L^2(\Omega)}.$$

Definition 17.11. With X as in (17.4) let $H_0^1(\Omega)$ be the completion of X with respect to the H^1-norm and let $H^1(\Omega)$ be the completion of $C^2(\overline{\Omega})$ with respect to the H^1-norm. These spaces are called *Sobolev spaces*.

Note that elements of X vanish at $\partial\Omega$, which is what is referred to by the subscript 0. Thus, $H_0^1(\Omega)$ is a subspace of $H^1(\Omega)$: any Cauchy sequence in X is a Cauchy sequence in $C^2(\overline{\Omega})$; the difference between X and $C^2(\overline{\Omega})$ is that elements of the former are required to vanish at $\partial\Omega$.

It is useful to note that by the Poincaré inequality,

$$\int_\Omega |u|^2\, dx \le C \int_\Omega |c\nabla u|^2\, dx,$$

on X the H^1-norm is equivalent to the norm

$$\|u\|_{H_0^1} = \left(\int_\Omega |c\nabla u|^2\, dx \right)^{1/2},$$

i.e. we have estimates $\|u\|_{H_0^1} \le \|u\|_{H^1} \le (1+C)^{1/2}\|u\|_{H_0^1}$. That is, the notion of convergence, etc., is the same with respect to these two norms. Note that the H_0^1-norm is also given by an inner product,

$$\langle u, v\rangle_{H_0^1} = \langle c\nabla u, c\nabla v\rangle_{L^2}.$$

These hold at first for $u, v \in X$, but extend immediately to the completion $H_0^1(\Omega)$ due to Lemma 17.10, which shows that $\partial_j : X \to C^0(\overline{\Omega})$. Hence $\partial_j : X \to L^2(\Omega)$ (as L^2 is just the completion of C^0, so the latter space sits inside the former) extends to a continuous linear map $\partial_j : H_0^1(\Omega) \to L^2(\Omega)$ since for $u \in X$,

$$\|\partial_j u\|_{L^2} \le C\|u\|_{H_0^1}.$$

The use of the space $H_0^1(\Omega)$ allows us to incorporate the boundary condition into the function space itself: elements of $H_0^1(\Omega)$ vanish at the boundary, as is suggested by being limits of Cauchy sequences in X. One way of making this precise is that even elements of $H^1(\Omega)$ can be restricted to $\partial\Omega$, more precisely the restriction map $r : C^2(\overline{\Omega}) \to C^0(\partial\Omega)$ extends to a *continuous* map $r : H^1(\Omega) \to L^2(\partial\Omega)$, and for $u \in H_0^1(\Omega)$, $r(u) = 0$. This is proved using the lemma: one merely has to show that

$$u \in C^2(\overline{\Omega}) \Rightarrow \|r(u)\|_{L^2(\partial\Omega)} \le C\|u\|_{H^1(\Omega)},$$

which in turn follows by an argument similar to the proof of the Poincaré lemma; see Problem 17.5.

Now we can make a well-posedness statement:

Theorem 17.12. *The PDE $\Delta_c u = f$ is well-posed for $u \in X$, $f \in Y$ with $X = H_0^1(\Omega)$, $Y = L^2(\Omega)$.*

Remark 17.13. As so defined, the solution operator S, given by $Sf = u$, is *not surjective* onto X. This could be remedied by changing Y, but we do not follow up on this. In any case, surjectivity was not one of our criteria for well-posedness!

As a first step towards this theorem, let's rewrite the energy argument. Recall that this arose from the divergence theorem: for $u \in C^2(\overline{\Omega})$,

(17.6)
$$\langle \Delta_c u, \overline{\phi} \rangle_{L^2} = \int_\Omega \Delta_c u \, \phi \, dx = \int_{\partial\Omega} c^2 \hat{n} \cdot \nabla u \, \phi \, dS(x) - \int_\Omega c^2 \nabla u \cdot \nabla \phi \, dx$$

$$= - \int_\Omega c^2 \nabla u \cdot \nabla \phi \, dx = -\langle c\nabla u, c\overline{\nabla \phi} \rangle_{L^2} = -\langle u, \overline{\phi} \rangle_{H^1_0},$$

provided $\phi \in C^1(\overline{\Omega})$, $\phi|_{\partial\Omega} = 0$. We first observe that if $\phi \in C_c^\infty(\Omega)$ (hence vanishes near $\partial\Omega$) and $u \in C^1(\overline{\Omega})$, then the definition of the distributional derivative $\Delta_c \iota_u$ is $\Delta_c \iota_u(\phi) = \iota_u(\Delta_c \phi)$, so an integration by parts (divergence theorem) using the expression on the right-hand side shows that

$$\Delta_c \iota_u(\phi) = - \int_\Omega c^2 \nabla u \cdot \nabla \phi \, dx,$$

so the right-hand side of (17.6) is the distributional derivative of u applied to ϕ.

Now suppose that $f \in L^2(\Omega)$ and we want to solve $\Delta_c u = f$ with $u \in H^1_0(\Omega)$. Thus, by (17.6) with $\overline{\phi}$ replaced by ϕ, we need $u \in H^1_0(\Omega)$ such that

(17.7) $\langle u, \phi \rangle_{H^1_0} = -\langle f, \phi \rangle_{L^2}$

for all $\phi \in C_c^\infty(\Omega)$. Note that the map

$$\ell : \phi \mapsto \langle f, \phi \rangle_{L^2}$$

is (up to a complex conjugation) just the distribution ι_f corresponding to f, and by the L^2-Cauchy-Schwarz equality,

$$|\ell(\phi)| \le \|f\|_{L^2} \|\phi\|_{L^2} \le C\|f\|_{L^2}\|\phi\|_{H^1_0},$$

with the second inequality being the Poincaré inequality. Thus,

$$\ell : H^1_0(\Omega) \to \mathbb{C}$$

is a continuous conjugate linear map. (This continuity and Lemma 17.10 show that if (17.7) holds for $\phi \in C_c^\infty(\Omega)$, then by the density of $C_c^\infty(\Omega)$ in $H^1_0(\Omega)$, this equality in fact holds for all $\phi \in H^1_0(\Omega)$.) The virtue of this observation (namely the continuity of ℓ) is that existence of $u \in H^1_0(\Omega)$ such that (17.7) holds, hence Theorem 17.12 itself, follows immediately from the following spectacular Proposition.

Proposition 17.14 (Riesz). *Suppose V is a complete inner product space and that $\ell : V \to \mathbb{C}$ is a continuous linear map. Then there exists a unique $u \in V$ such that $\ell(v) = \langle v, u \rangle$ for all $v \in V$. The analogous statement also holds if we replace linear by conjugate linear and replace $\langle v, u \rangle$ by $\langle u, v \rangle$.*

Proof. We only consider the conjugate linear case; the other case is similar.

Uniqueness is easy: if $\langle u_1, v \rangle - \langle u_2, v \rangle = 0$ for all $v \in V$, then taking $v = u_1 - u_2$ and rearranging gives $\|u_1 - u_2\|^2 = \langle u_1 - u_2, v \rangle = 0$; thus $u_1 = u_2$.

For existence, we start by remarking that if $\ell = 0$, then we can take $u = 0$, so we may assume that ℓ does not vanish identically, say $\ell(v_0) \neq 0$. Now note that if $\ell(v) = 0$, then we need $\langle u, v \rangle = 0$, i.e. u needs to be orthogonal to the nullspace of v. Now suppose that we find a vector $x \in V$ such that $\langle x, v \rangle = 0$ for all v with $\ell(v) = 0$, but $x \neq 0$. Then $\langle x, v_0 \rangle \neq 0$. Indeed, one can write any $v \in V$ as

$$(17.8) \qquad v = \frac{\overline{\ell(v)}}{\ell(v_0)} v_0 + w, \quad w = v - \frac{\overline{\ell(v)}}{\ell(v_0)} v_0,$$

so $\ell(w) = \ell(v) - \frac{\ell(v)}{\ell(v_0)} \ell(v_0) = 0$. Thus, $\langle x, w \rangle = 0$, so if $\langle x, v_0 \rangle = 0$, then $\langle x, v \rangle = 0$ follows for all $v \in V$ (because v is a linear combination of v_0 and w); in particular $v = x$, so $x = 0$, giving a contradiction. Let c be such that $c \langle x, v_0 \rangle = \ell(v_0)$, i.e.

$$c = \frac{\ell(v_0)}{\langle x, v_0 \rangle}, \quad u = cx.$$

Then for $v \in V$, using (17.8),

$$\langle u, v \rangle = c \langle x, \frac{\overline{\ell(v)}}{\ell(v_0)} v_0 \rangle = c \frac{\ell(v)}{\ell(v_0)} \langle x, v_0 \rangle = \ell(v),$$

completing the proof.

Thus, we are down to finding $x \in V$ such that $x \neq 0$ and $\langle x, v \rangle = 0$ for all v with $\ell(v) = 0$. We state this as a separate lemma; once we complete the proof of the lemma, the proof of the proposition is also complete. $\qquad\square$

Lemma 17.15. *Suppose V is complete, $\ell : V \to \mathbb{C}$ is conjugate linear and $\ell \neq 0$. Then there exists $x \in V$ such that $x \neq 0$ and $\langle x, v \rangle = 0$ for all v with $\ell(v) = 0$.*

Proof. Let $Z = \{v \in V : \ell(v) = 0\}$; this is a subspace of V since $\ell(\alpha v) = \overline{\alpha} \ell(v) = 0$ if $\ell(v) = 0$, $\alpha \in \mathbb{C}$ (i.e. conjugate linearity is just as good for this conclusion as linearity). Let

$$S = \{v \in V : \ell(v) = 1\},$$

which is non-empty as $\ell \neq 0$ (i.e. there is $v_0 \in V$ such that $\ell(v_0) \neq 0$; conjugate linearity gives a $v \in V$ such that $\ell(v) = 1$).

The idea is that the closest point of S to the origin is one which is orthogonal to vectors tangent to S, i.e. orthogonal to (vectors in) Z; see Figure 17.1. Let $\rho : S \to [0, \infty)$ be the function $\rho(v) = \|v\|^2$. Then ρ is

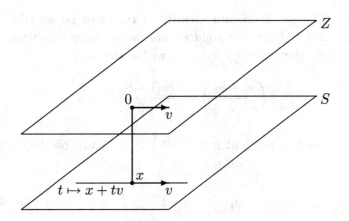

Figure 17.1. The closest point x in S to the origin 0.

bounded below, so there is a greatest lower bound A for $\rho(S) = \{\rho(v) : v \in S\}$, and thus a sequence $v_j \in S$ such that $\rho(v_j) \to A$. Note that

$$v \in S \Rightarrow \rho(v) \geq A$$

by the very definition of A. The key point is that $\{v_j\}_{j=1}^{\infty}$ is a Cauchy sequence. Once we show this, it converges to some $x \in V$ by the completeness of V. Moreover, then the continuity of the norm shows that $\rho(v_j) \to \rho(x)$, so $\rho(x) = A$, and the continuity of ℓ shows that $\ell(v_j) \to \ell(x)$. Since $\ell(v_j) = 1$, we have $\ell(x) = 1$, and thus $x \neq 0$.

We now show that for $v \in Z$, $\langle x, v \rangle = 0$. Indeed, for $v \in Z$, $x + tv \in S$, so

$$\rho(x) = A \leq \rho(x + tv) = \|x\|^2 + \bar{t}\langle x, v \rangle + t\langle v, x \rangle + |t|^2 \|v\|^2.$$

Taking t real shows that the function

$$\mathbb{R} \ni t \mapsto \|x\|^2 + 2t(\langle x, v \rangle + \langle v, x \rangle) + |t|^2 \|v\|^2$$

has a global minimum at 0 (with value given by $A = \|x\|^2$), so in particular its derivative there vanishes. Thus $\langle x, v \rangle + \langle v, x \rangle = 0$, while taking $t = is$ pure imaginary shows that the function

$$\mathbb{R} \ni s \mapsto \|x\|^2 + 2s(-\langle x, v \rangle + \langle v, x \rangle) + |s|^2 \|v\|^2$$

has a global minimum at 0, so $-\langle x, v \rangle + \langle v, x \rangle = 0$. Combining these two proves that $\langle x, v \rangle = 0$ as claimed, and thus the lemma (given the Cauchy sequence claim we still need to prove).

We finally need to prove the claimed Cauchy sequence statement. This follows from the parallelogram identity

$$\|y + z\|^2 + \|y - z\|^2 = 2\|y\|^2 + 2\|z\|^2$$

as follows. (The parallelogram identity in turn can be seen by expanding both sides after writing the squared norms as inner products; see Problem 13.1.) Note that as $\ell(v_j) = 1 = \ell(v_k)$ for any j, k,

$$\ell\left(\frac{v_j + v_k}{2}\right) = \frac{\ell(v_j) + \ell(v_k)}{2} = 1,$$

so $\frac{v_j+v_k}{2} \in S$, and in particular $\rho\left(\frac{v_j+v_k}{2}\right) \geq A$. Thus, by the parallelogram identity,

$$\|v_j - v_k\|^2 = 2\|v_j\|^2 + 2\|v_k\|^2 - 4\left\|\frac{v_j + v_k}{2}\right\|^2 \leq 2\|v_j\|^2 + 2\|v_k\|^2 - 4A.$$

But the very definition of v_j states that $v_j, v_k \to A$ as $j, k \to \infty$, so we deduce that $2\|v_j\|^2 + 2\|v_k\|^2 - 4A \to 0$ as $j, k \to \infty$. Hence $\{v_j\}_{j=1}^\infty$ is indeed a Cauchy sequence as claimed, completing the proof. $\qquad\square$

The proof of this lemma can be jazzed up easily to prove the existence of orthogonal projections in Hilbert spaces.

Lemma 17.16. *Suppose that V is a Hilbert space and $Z \subset V$ is a closed subspace; i.e. if $z_j \to v$, $z_j \in Z$, $v \in V$, then $v \in Z$. Then for any $v \in V$ there exist unique $v_\| \in Z$ and $v_\perp$ orthogonal to Z (i.e. $\langle v_\perp, z \rangle = 0$ for all $z \in Z$) such that $v = v_\| + v_\perp$. One calls $v_\|$ the orthogonal projection of v to Z.*

Remark 17.17. Note that $\|v\|^2 = \|v_\|\|^2 + \|v_\perp\|^2$ by Pythagoras' theorem, so $\|v_\|\| \leq \|v\|$. Thus the orthogonal projection $\Pi : V \to Z$, $\Pi v = v_\|$, is a continuous linear map.

Proof. Uniqueness is easy to check and does not rely on completeness, for if $v_\| + v_\perp = v = v_\|' + v_\perp'$, then $v_\| - v_\|' = v_\perp' - v_\perp$, and the left-hand side is in Z while the right-hand side is orthogonal to every element of Z, in particular to $v_\| - v_\|'$. So $\|v_\perp' - v_\perp\|^2 = \langle v_\perp' - v_\perp, v_\| - v_\|' \rangle = 0$, and thus $v_\| - v_\|' = 0$ as well.

For existence, suppose $v \in V$. The idea is again that $v_\|$ is the closest point in Z to v. So let $\rho : Z \to [0, \infty)$ be given by $\rho(z) = \|z - v\|^2$. Again, $\rho(Z)$ is bounded below and let $A = \inf \rho(Z)$. Let $v_j \in Z$ be such that $\rho(v_j) \to A$. Again, $\rho(z) \geq A$ for all $z \in Z$ by the definition of A. One again shows that $\{v_j\}_{j=1}^\infty$ is a Cauchy sequence, this time using the parallelogram identity with $y = v_j - v$, $z = v_k - v$, so $y - z = v_j - v_k$ and $\frac{y+z}{2} = \frac{v_j+v_k}{2} - v$, and $\frac{v_j+v_k}{2} \in Z$ as Z is a subspace of V. Namely, by the parallelogram identity,

$$\|v_j - v_k\|^2 = 2\|v_j - v\|^2 + 2\|v_k - v\|^2 - 4\left\|\frac{v_j + v_k}{2} - v\right\|^2$$
$$\leq 2\|v_j - v\|^2 + 2\|v_k - v\|^2 - 4A,$$

and the right-hand side again goes to 0 as $j, k \to \infty$. As $\{v_j\}$ is a Cauchy sequence in the complete space V, it converges to some $v_{\|}$. As Z is closed and $v_j \in Z$ for all j, $v_{\|} \in Z$. Let $v_{\perp} = v - v_{\|}$, so $v = v_{\perp} + v_{\|}$. It remains to check that $\langle v_{\perp}, z \rangle = 0$ for all $z \in Z$, which can be checked as above, by considering

$$\rho(v_{\|}) = A \leq \rho(v_{\|} + tz) = \|v_{\|} + tz - v\|^2 = \|v_{\perp} - z\|^2$$
$$= \|v_{\perp}\|^2 + \bar{t}\langle v_{\perp}, z \rangle + t\langle z, v_{\perp} \rangle + |t|^2 \|z\|^2,$$

which gives $\langle v_{\perp}, z \rangle = 0$ as above. $\qquad\square$

We make a definition to make this systematic:

Definition 17.18. The *orthocomplement* $Z^{\perp}$ of a subspace Z of an inner product space V is the set of vectors $w \in V$ such that $\langle w, v \rangle = 0$ for all $v \in Z$.

With this definition what we have shown is that if V is a Hilbert space and Z a closed subspace, then any $v \in V$ can be written uniquely as $v = v_{\|} + v_{\perp}$, $v_{\|} \in Z$, $v_{\perp} \in Z^{\perp}$.

Returning to our PDE, $\Delta_c u - qu = f$, if $q \geq 0$, then a similar argument to the proof of Theorem 17.12, using the norm

$$\|u\|_{H^1}^2 = \int_{\Omega} (q|u|^2 + c^2|\nabla u|^2)\, dx,$$

which is yet again equivalent to H^1-norm defined with $q = 1$, $c = 1$ on X, shows:

Theorem 17.19. *For $q \geq 0$, the PDE $\Delta_c u - qu = f$ is well-posed for $u \in \tilde{X}$, $f \in \tilde{Y}$ with $\tilde{X} = H_0^1(\Omega)$, $\tilde{Y} = L^2(\Omega)$.*

In the special case of q and c constant, we have already shown elliptic regularity, so if $f \in C^{\infty}(\Omega)$, then $u \in C^{\infty}(\Omega)$. With some additional work one can show that if $f \in C^{\infty}(\overline{\Omega})$, then $u \in C^{\infty}(\overline{\Omega})$, so this is in fact a *classical* solution of the PDE. Thus, even to obtain a classical solution we had to go through the machinery of weak solutions. In fact, a little more work would give the same conclusion even for variable (non-constant) q and c.

We finish with a statement that for Neumann boundary conditions, and also for the problem on $\mathbb{R}^n$, a similar result holds, except that we need $q > 0$ since we do not have the Poincaré inequality at our disposal:

Theorem 17.20 (See Problem 17.6). *For $q > 0$, the PDE $\Delta_c u - qu = f$ is well-posed for $u \in H^1(\Omega)$, $f \in L^2(\Omega)$, where the PDE must hold in the sense that*

$$-\langle c\nabla u, c\nabla \phi\rangle_{L^2} - \langle qu, u\rangle_{L^2} = \langle f, \phi\rangle$$

for all $\phi \in C^\infty(\overline{\Omega})$.

Note that the boundary condition simply disappeared here, for H^1 functions it does not make sense to talk about their normal derivative at the boundary. However, we allow all $\phi \in C^\infty(\overline{\Omega})$, so *if* $u \in C^2(\overline{\Omega})$, which as mentioned above holds if for instance $f \in C^\infty(\overline{\Omega})$, the integration parts of (17.6) done in reverse shows that $\hat{n} \cdot \nabla u$ vanishes at $\partial\Omega$. Thus, this should be considered a weak formulation of the Neumann boundary condition.

Theorem 17.21. *Let*

$$P = \Delta_c - q = \nabla \cdot c(x)^2 \nabla - q,$$

where $c \in C^1(\mathbb{R}^n)$, $0 < c_1 \leq c \leq c_2$, and $0 < q_1 \leq q \leq q_2$. The PDE $Pu = f$ is well-posed for $u \in H^1(\mathbb{R}^n)$, $f \in L^2(\mathbb{R}^n)$.

While we did not get to a discussion of other problems, such as the wave equation, similar methods apply. In fact, if one does not want to impose initial conditions, the Hahn-Banach argument plus the energy estimates immediately give solvability. One has to be somewhat more careful to solve the wave equation in the presence of initial conditions, but this can still be done.

Problems

Problem 17.1. We defined the H^1-norm in (17.3) in terms of the function c on $\overline{\Omega}$. Show that different (positive) choices of c give equivalent norms; i.e. if $c, c' \in C^1(\overline{\Omega})$ are positive and we denote by H_c^1 the norm corresponding to c, and similarly for $H_{c'}^1$, then there are constants $C_1, C_2 > 0$ such that for $u \in C^2(\overline{\Omega})$ with $u|_{\partial\Omega} = 0$, one has

$$C_1\|u\|_{H_c^1} \leq \|u\|_{H_{c'}^1} \leq C_2\|u\|_{H_c^1}.$$

Problem 17.2. Suppose (X, d) is a metric space with completion $\tilde{X}$ and Z is a subset of X. Show that the completion of Z in the metric d is the closure of Z in $\tilde{X}$, i.e. that this closure is complete and Z is dense in it.

Problem 17.3. Recall that with $X = \{u \in C^2(\overline{\Omega}) : u|_{\partial\Omega} = 0\}$, $H_0^1(\Omega)$ is the completion of X in the H^1-norm. Show that the completion of the subspace $Z = C_c^2(\Omega)$, consisting of C^2 functions of *compact support* in the open set Ω (i.e. elements of Z vanish near $\partial\Omega$), in the H^1-norm is also $H_0^1(\Omega)$.

Hint: By Problem 17.2, it suffices to show that Z is dense in X in the H^1-norm. Let $\rho \in C^2(\overline{\Omega})$ vanish simply at $\partial\Omega$, $\rho > 0$ in Ω. If $u \in X$ and $\chi \in C^\infty(\mathbb{R})$ with $\chi(t) = 0$ for $t < 1$, $\chi(t) = 1$ for $t > 2$, show that $\chi(\rho/\epsilon)u \to u$ in the H^1-norm as $\epsilon \to 0$ using the fact that $|u(x)| \leq C\rho(x)$, $x \in \overline{\Omega}$, by Taylor's theorem and the vanishing of u at $\partial\Omega$.

Problem 17.4. Show that the space $H^1(\Omega)$ is diffeomorphism invariant, i.e. if $\Phi : \overline{\Omega} \to \overline{\Omega'}$ is a C^∞ (in fact, C^2 suffices) diffeomorphism (invertible with C^∞ inverse), then the map $\Phi^* : C^2(\overline{\Omega'}) \to C^2(\overline{\Omega})$ given by

$$(\Phi^*u)(x) = u(\Phi(x)), \ x \in \overline{\Omega},$$

extends to a continuous linear map $\Phi^* : H^1(\Omega') \to H^1(\Omega)$, and the extension is unique.

Hint: The BLT theorem, and change of variables formula for integrals!

Problem 17.5. Show that for a bounded domain Ω with smooth boundary, the restriction map $r : C^2(\overline{\Omega}) \to C^0(\partial\Omega)$ extends to a continuous map $r : H^1(\Omega) \to L^2(\partial\Omega)$ by showing

$$u \in C^2(\overline{\Omega}) \Rightarrow \|r(u)\|_{L^2(\partial\Omega)} \leq C\|u\|_{H^1(\Omega)}.$$

Hint: Near a point $p \in \partial\Omega$, one can multiply u by a cutoff function ϕ which is identically 1 on an open set O containing p, and then the restriction $u|_{\partial\Omega \cap O}$ is the same as $\phi u|_{\partial\Omega \cap O}$. Thus, it suffices to consider the map $u \mapsto r(\phi u)$, and thus u with support near p, say in an open set U. By the use of a diffeomorphism as in Problem 17.4 one can then straighten out $\partial\Omega \cap U$ to the coordinate plane $x_n = 0$; one may assume that the support of u is in $(-1,1)^{n-1}_{x'} \times [0,1)_{x_n}$, where $x' = (x_1, \ldots, x_{n-1})$. Then one has $r(u)(x') = -\int_0^1 \partial_n u(x', x_n)\, dx_n$ by the fundamental theorem of calculus. Now use Cauchy-Schwarz.

Problem 17.6. Prove Theorem 17.20 in detail.

Problem 17.7. Prove Theorem 17.21 in detail.

Problem 17.8. Consider the PDE

$$\nabla \cdot (A(x)\nabla u) - qu = f,$$

on a bounded region Ω where $q \geq 0$ and $A(x) = (A_{ij}(x))_{i,j=1}^n$ is symmetric, positive definite; see Problem 7.8. Assume Dirichlet boundary conditions. Show that for $f \in L^2(\Omega)$ the PDE is well-posed for $u \in H_0^1(\Omega)$.

Problem 17.9. Consider the following PDE on the 2-torus $\mathbb{T}^2 = \mathbb{S}_x^1 \times \mathbb{S}_y^1$ (see Problem 13.5):

$$Pu = f, \ P = \nabla \cdot c(x,y)^2 \nabla - q(x,y),$$

$c, q > 0$. Recall that functions on $\mathbb{T}^2$ can be considered 2π-periodic on $\mathbb{R}^2$ in both the x and the y variables. Define $L^2(\mathbb{T}^2)$ to be the completion of $C^0(\mathbb{T}^2)$ in the inner product given in Problem 13.5 and $H^1(\mathbb{T}^2)$ to be the completion of $C^2(\mathbb{T}^2)$ with the norm

$$\|u\|^2_{H^1(\mathbb{T}^2)} = \|u\|^2_{L^2(\mathbb{T}^2)} + \|cD_x u\|^2_{L^2(\mathbb{T}^2)} + \|cD_y u\|^2_{L^2(\mathbb{T}^2)}.$$

(i) Show that the PDE is well-posed with $f \in L^2(\mathbb{T}^2)$, $u \in H^1(\mathbb{T}^2)$.

(ii) Is it well-posed if $q \equiv 0$?

Chapter 18

Variational problems

The variational problems we discuss here concern the minimization, possibly up to constraints, of certain expressions which correspond to finding eigenvalues of partial differential operators. As a warm-up let us consider the spectral theorem of symmetric operators A on a finite dimensional inner product space V, namely that for any such operator there is an orthonormal basis of V consisting of eigenvectors of A.

1. The finite dimensional problem

To start with, recall from Chapter 13 that for a symmetric operator all eigenvalues are real and all eigenvectors corresponding to distinct eigenvalues are orthogonal to each other. This gives a way of constructing an orthogonal set of vectors in V: suppose $\lambda_1, \ldots, \lambda_m$ are the eigenvalues of A. (Note that there can be at most n eigenvalues: for each there is a non-zero vector in the eigenspace, which are automatically orthogonal to each other; thus they are linearly independent, and there can be at most n linearly independent vectors in V.) Let $v_{j1}, \ldots, v_{jk_j}$ be an orthonormal basis of the nullspace $N(A - \lambda_j I)$ of $A - \lambda_j I$ (notice that this is exactly the λ_j-eigenspace of A); one can obtain such a basis by applying the Gram-Schmidt procedure (see Problem 13.6) to any basis. Then putting these together, we get an orthonormal collection of vectors

$$v_{11}, \ldots, v_{1k_1}, v_{21}, \ldots, v_{2k_2}, \ldots, v_{m1}, \ldots, v_{mk_m};$$

these are orthonormal because eigenvectors in different eigenspaces are automatically orthogonal to each other. In particular, this is a linearly independent collection of vectors. On the other hand, a priori it is not clear whether they span V. However, this is guaranteed by the following theorem.

Theorem 18.1 (Spectral theorem). *Suppose $A : V \to V$ is symmetric linear and V is a finite dimensional inner product space. Then V has an orthonormal basis consisting of eigenvectors of A.*

The key tool in proving this theorem is the following observation.

Definition 18.2. An *invariant subspace* of $A : V \to V$ is a subspace W of V such that $A : W \to W$, i.e. such that for all $w \in W$, $Aw \in W$.

Lemma 18.3. *Suppose that $A : V \to V$ is symmetric and W is an invariant subspace of A. Then $W^{\perp}$ is also an invariant subspace of A.*

Proof. Suppose $x \in W^{\perp}$. Then for $w \in W$, $\langle Ax, w \rangle = \langle x, Aw \rangle = 0$, since $x \in W^{\perp}$ and $Aw \in W$. This shows $Ax \in W^{\perp}$, and thus the lemma. $\square$

Notice that any eigenspace of A is an invariant subspace of A: if, say, $x \in N(A - \lambda I)$, then $Ax = \lambda x \in N(A - \lambda I)$ as well just by virtue of $N(A - \lambda I)$ being a subspace of V. Moreover, if v is any eigenvector of A, then a similar argument gives that span$\{v\}$ is an invariant subspace of A. Thus, in both cases, the orthocomplement is also invariant for A.

The proof of the theorem now reduces to showing that any symmetric A on a space of dimension ≥ 1 has a single eigenvalue. If we find an eigenvalue λ, then by taking v to be an eigenvector of A with eigenvalue λ, span$\{v\}$ is invariant for A, thus so is span$\{v\}^{\perp}$, and the latter has dimension $n-1$, so by an inductive argument we may assume that span$\{v\}^{\perp}$ has an orthonormal basis consisting of eigenvectors for A. Adding v to this set provides a desired basis of V consisting of eigenvectors of A.

In fact, one could avoid induction altogether: if $\lambda_1, \ldots, \lambda_m$ are the eigenvalues of V as above, let W be the sum of the corresponding eigenspaces:

$$W = \bigoplus_{j=1}^{m} N(A - \lambda_j I).$$

(Notice that the sum is direct since the summands are linearly independent, indeed orthogonal!) This is an invariant subspace of V such that, if $v_j \in N(A - \lambda_j I)$, then

$$A \sum_{j=1}^{m} v_j = \sum_{j=1}^{m} Av_j \in \bigoplus_{j=1}^{m} N(A - \lambda_j I).$$

Thus, $W^{\perp}$ is also an invariant subspace of A, and A is a symmetric operator on it. If $W^{\perp} \neq \{0\}$, and if we have shown that any symmetric operator on any inner product space of dimension ≥ 1 has an eigenvalue, then $A|_{W^{\perp}}$ has an eigenvalue, λ, and a corresponding eigenvector $v \neq 0$, so $Av = \lambda v$. Thus

λ is an eigenvalue of A and v is in the corresponding eigenspace, so $v \in W$, which is a contradiction with $v \in W^{\perp}$.

So the main point is to show the following lemma; once this is shown, the spectral theorem follows.

Lemma 18.4. *If* $\dim V \geq 1$, $A : V \to V$ *symmetric, then A has an eigenvalue on V.*

In order to motivate the proof, consider the *Rayleigh quotient*:

$$Q(x) = \frac{\langle Ax, x \rangle}{\|x\|^2}, \ x \in V, \ x \neq 0.$$

Notice that this is unchanged if one replaces x by a non-zero multiple:

$$Q(tx) = \frac{\langle A(tx), tx \rangle}{\|tx\|^2} = \frac{\langle Ax, x \rangle}{\|x\|^2} = Q(x), \ t \neq 0.$$

Let us now see how A behaves on a sum of eigenvectors

$$x = \sum_{j=1}^{m} v_j, \ v_j = N(A - \lambda_j I),$$

with λ_j distinct eigenvalues, so the v_j are automatically orthogonal to each other. Then (if not all v_j vanish)

$$Q(x) = \frac{\langle \sum_{j=1}^{m} Av_j, \sum_{i=1}^{m} v_i \rangle}{\langle \sum_{j=1}^{m} v_j, \sum_{i=1}^{m} v_i \rangle} = \frac{\langle \sum_{j=1}^{m} \lambda_j v_j, \sum_{i=1}^{m} v_i \rangle}{\sum_{j=1}^{m} \sum_{i=1}^{m} \langle v_j, v_i \rangle}$$

$$= \frac{\sum_{j=1}^{m} \lambda_j \sum_{i=1}^{m} \langle v_j, v_i \rangle}{\sum_{j=1}^{m} \langle v_j, v_j \rangle} = \frac{\sum_{j=1}^{m} \lambda_j \|v_j\|^2}{\sum_{j=1}^{m} \|v_j\|^2}.$$

In particular, if the λ_j are ordered, $\lambda_1 < \lambda_2 < \ldots < \lambda_m$, then

$$\lambda_1 = \frac{\sum_{j=1}^{m} \lambda_1 \|v_j\|^2}{\sum_{j=1}^{m} \|v_j\|^2} \leq \frac{\sum_{j=1}^{m} \lambda_j \|v_j\|^2}{\sum_{j=1}^{m} \|v_j\|^2} \leq \frac{\sum_{j=1}^{m} \lambda_m \|v_j\|^2}{\sum_{j=1}^{m} \|v_j\|^2} = \lambda_m.$$

Correspondingly, a reasonable idea is to look at the minimum of Q to find λ_1; similarly, one could look for the maximum of Q to find λ_m.

But this is easy. Since $Q(x) = Q(\frac{x}{\|x\|})$, with $\frac{x}{\|x\|}$ a unit vector, simply consider the constrained minimization problem: find the minimum of Q on the unit sphere $S = \{x : \|x\| = 1\}$. This is equivalent to finding the minimum of $f(x) = \langle Ax, x \rangle$ on S. But, assuming V is a real vector space, with the complex case requiring only minor modifications (cf. the proof of Lemma 17.15 and the arguments below),

$$f(x + h) = \langle A(x + h), x + h \rangle = \langle Ax, x \rangle + \langle Ah, x \rangle + \langle Ax, h \rangle + \langle Ah, h \rangle$$
$$= f(x) + 2\langle Ax, h \rangle + \langle Ah, h \rangle,$$

with $|\langle Ah, h \rangle| \leq C\|h\|^2$ for some C (for instance, $C = \|A\|$). Thus, directly from the definition of derivative, $\partial f(x)h = 2\langle Ax, h \rangle$, i.e. $\nabla f(x) = 2Ax$. Similarly, with $g(x) = \|x\|^2$, $\partial g(x)h = 2\langle x, h \rangle$, i.e. $\nabla g(x) = 2x$. Correspondingly, using the method of Lagrange multipliers (see below for the argument in the infinite dimensional case which does not use this), since critical points of f on S are exactly the points x at which $\nabla f(x) = \lambda \nabla g(x)$ for some $\lambda \in \mathbb{R}$, they are exactly the points x for which $Ax = \lambda x$, i.e. exactly the unit vectors which are eigenvectors of A. Since S is compact, f actually attains its maximum and minimum on S, and as we have just seen, these are necessarily attained at eigenvectors of A. Thus, A has an eigenvector and eigenvalue on V, proving the lemma, and thus the spectral theorem.

Note that this is actually more than just an existence proof: for any A, we actually found the smallest eigenvalue by minimizing the Rayleigh quotient (and the largest by maximizing it). One could proceed inductively to find all eigenvalues: once one has the smallest one, λ_1 with an eigenvector v_{11}, restrict A to span$\{v_{11}\}^\perp$, and find the smallest eigenvalue and an eigenvector there. The eigenvalue one finds is $\geq \lambda_1$; if it is $= \lambda_1$, one finds another, orthogonal to the first, eigenvector v_{12} in the λ_1-eigenspace; otherwise one finds the second smallest eigenvalue λ_2 and an eigenvector v_{21}. In the first case one now restricts A to span$\{v_{11}, v_{12}\}^\perp$ and in the second to span$\{v_{11}, v_{21}\}^\perp$, and proceeds inductively to find all eigenvalues and an orthonormal basis of eigenvectors.

2. The infinite dimensional minimization

We would now like to extend this to the situation where A is an operator on an infinite dimensional space. Of course, we cannot expect that the approach works in general; an example is $-\Delta_{\mathbb{R}^n}$, which is symmetric on $C_c^\infty(\mathbb{R}^n)$ or $\mathcal{S}(\mathbb{R}^n)$, but as the Fourier transform shows, has no eigenvalues. Another issue is that even when one has eigenvalues, they need not be bounded either above or below, as the example of $-i\partial_x$ on 2π-periodic functions shows: the eigenvalues are all the integers $n \in \mathbb{Z}$. The former problem arises because of the non-compactness of $\mathbb{R}^n$, while the latter because although $[0, 2\pi]$ is bounded, $-i\partial_x$ is not a positive operator (and more importantly not even bounded from below by a constant). Of course, even if we deal with a positive operator we expect infinitely many eigenvalues tending to infinity; indeed, this is the conclusion of Theorem 13.22. Thus, we cannot expect to be able to maximize the Rayleigh quotient, unlike in the finite dimensional case, but minimization could reasonably be expected to work.

So we consider bounded domains Ω as in Chapter 17, and the operator $A = -\Delta_c + q$ with Dirichlet boundary conditions, and $q \geq 0$ for now. Here we chose the negative sign for $-\Delta_c$ to make it into a positive operator (cf. Problem 13.3) so that the eigenvalues can be expected to tend to $+\infty$ rather than $-\infty$. Thus, as explained in Chapter 17, we will be working on the space $H_0^1(\Omega) \subset L^2(\Omega)$ with L^2 giving the inner product and will consider $A : H_0^1(\Omega) \to (H_0^1(\Omega))^*$, which means our Rayleigh quotient is

$$\mathcal{Q}(\phi) = \frac{\langle A\phi, \phi \rangle_{L^2}}{\|\phi\|_{L^2}^2}, \quad \phi \in H_0^1(\Omega).$$

Note that $\mathcal{Q}(t\phi) = t\mathcal{Q}(\phi)$ for $t \neq 0$ still. Then certainly $\mathcal{Q}(\phi) \geq 0$ for all ϕ. Correspondingly, one can minimize $\mathcal{Q}(\phi)$ in the sense that one may consider

$$\lambda_1 = \inf_{\phi \in H_0^1(\Omega) \setminus \{0\}} \mathcal{Q}(\phi) = \inf\{\mathcal{Q}(\phi) : \|\phi\|_{L^2} = 1, \ \phi \in H_0^1(\Omega)\},$$

where the second equality is from $\mathcal{Q}(t\phi) = t\mathcal{Q}(\phi)$ for $t \neq 0$, and the infimum exists since the set is non-empty and is bounded from below by 0. (The latter in particular implies that the greatest lower bound, λ_1, is ≥ 0.) The difference from the finite dimensional case is that now we do not know whether this minimum is attained — recall that in the finite dimensional space we used the compactness of the unit sphere to conclude this, but now we are dealing with a subset of the unit sphere in L^2! Nonetheless, we prove below:

Proposition 18.5. *Let* $\lambda_1 = \inf\{\mathcal{Q}(\phi) : \|\phi\|_{L^2} = 1, \ \phi \in H_0^1(\Omega)\}$. *Then there exists* $\phi_1 \in H_0^1(\Omega)$, $\|\phi_1\|_{L^2} = 1$, *such that* $A\phi_1 = \lambda_1 \phi_1$.

In order to get started with the proof of Proposition 18.5, we may take a sequence of functions $\psi_j \in H_0^1(\Omega)$, $\|\psi_j\|_{L^2} = 1$, with $\mathcal{Q}(\psi_j) \to \lambda_1$ by the very fact that λ_1 is the *greatest* lower bound, so $\lambda_1 + 1/j$ is *not* a lower bound, and thus there is ψ_j in the relevant set with $\mathcal{Q}(\psi_j) < \lambda_1 + 1/j$.

Note that so far this is very similar to the proof that the minimum is attained in the compact setting: there one would argue that ψ_j has a convergent subsequence by compactness and use the continuity of $\mathcal{Q}$ to get the result. Now, if we can show that at least a subsequence of ψ_j, say $\{\psi_{j_k}\}_{k=1}^\infty$, converges to some $\phi \in H_0^1(\Omega)$ in $H_0^1(\Omega)$, then we conclude by the continuity of $\mathcal{Q}$ that $\mathcal{Q}(\phi) = \lambda_1$. Further, as in the finite dimensional case, we compute $f(\phi + h)$, $g(\phi + h)$, where $f(\phi) = \langle A\phi, \phi \rangle$, $g(\phi) = \langle \phi, \phi \rangle$, $\phi \in H_0^1(\Omega)$, but as we are in an infinite dimensional setting, it is convenient to write $h = t\psi$, where $\psi \in H_0^1(\Omega)$ and $t \in \mathbb{C}$. Thus,

$$f(\phi + t\psi) = \langle A(\phi + t\psi), \phi + t\psi \rangle = f(\phi) + t\langle A\psi, \phi \rangle + \overline{t}\langle A\phi, \psi \rangle + |t|^2 \langle A\psi, \psi \rangle,$$

and similarly

$$g(\phi + t\psi) = \langle \phi + t\psi, \phi + t\psi \rangle = g(\phi) + t\langle \psi, \phi \rangle + \overline{t}\langle \phi, \psi \rangle + |t|^2 \|\psi\|^2.$$

Rather than considering the full derivative of $\mathcal{Q}$ in this infinite dimensional setting, we will just use the fact that the function $\mathbb{C} \ni t \mapsto \mathcal{Q}(\phi + t\psi) \in \mathbb{R}$ has a local minimum at $t = 0$, much as we did in the proof of Riesz' proposition, Proposition 17.14. For t real, using $\langle A\psi, \phi \rangle = \langle \psi, A\phi \rangle = \overline{\langle A\phi, \psi \rangle}$, so that

$$\frac{d}{dt} f(\phi + t\psi)|_{t=0} = 2\operatorname{Re}\langle A\phi, \psi \rangle, \quad \frac{d}{dt} g(\phi + t\psi)|_{t=0} = 2\operatorname{Re}\langle \phi, \psi \rangle,$$

by the quotient rule we have

$$\frac{d}{dt} \mathcal{Q}(\phi + t\psi)|_{t=0} = \frac{f'g - fg'}{g^2}\bigg|_{t=0} = \frac{2\operatorname{Re}\langle A\phi, \psi \rangle \|\phi\|^2 - 2\operatorname{Re}\langle \phi, \psi \rangle \langle A\phi, \phi \rangle}{\|\phi\|^4}.$$

Similarly, for $t = is$ pure imaginary,

$$\frac{d}{ds} f(\phi + is\psi)|_{s=0} = -2\operatorname{Im}\langle A\phi, \psi \rangle, \quad \frac{d}{ds} g(\phi + s\psi)|_{s=0} = -2\operatorname{Im}\langle \phi, \psi \rangle,$$

so

$$\frac{d}{ds} \mathcal{Q}(\phi + t\psi)|_{s=0} = \frac{f'g - fg'}{g^2}|_{s=0}$$

$$= -\frac{2\operatorname{Im}\langle A\phi, \psi \rangle \|\phi\|^2 - 2\operatorname{Im}\langle \phi, \psi \rangle \langle A\phi, \phi \rangle}{\|\phi\|^4}\bigg|_{s=0}.$$

Since $\mathbb{C} \ni t \mapsto \mathcal{Q}(\phi + t\psi) \in \mathbb{R}$ has a local minimum at $t = 0$, so do both $\mathbb{R} \ni t \mapsto \mathcal{Q}(\phi + t\psi) \in \mathbb{R}$ and $\mathbb{R} \ni s \mapsto \mathcal{Q}(\phi + is\psi) \in \mathbb{R}$ at $t = 0$, resp. $s = 0$, so their respective derivatives vanish there. Thus we conclude that

$$\operatorname{Re}\langle A\phi, \psi \rangle \|\phi\|^2 = \operatorname{Re}\langle \phi, \psi \rangle \langle A\phi, \phi \rangle, \quad \operatorname{Im}\langle A\phi, \psi \rangle \|\phi\|^2 = \operatorname{Im}\langle \phi, \psi \rangle \langle A\phi, \phi \rangle,$$

so using $\langle A\phi, \phi \rangle = \lambda_1 \|\phi\|^2$,

$$\langle A\phi, \psi \rangle = \lambda_1 \langle \phi, \psi \rangle,$$

and thus

$$\langle (A - \lambda_1 I)\phi, \psi \rangle = 0.$$

Since this is true for all $\psi \in H_0^1(\Omega)$, the element $(A - \lambda_1 I)\phi$ of $H_0^1(\Omega)^*$ vanishes, i.e. $A\phi = \lambda_1 \phi$ as desired.

Thus, we are reduced to showing

Lemma 18.6. *The sequence $\{\psi_j\}_{j=1}^{\infty}$ has an $H_0^1(\Omega)$-convergent subsequence.*

This in fact follows from the following lemma due to Rellich:

Lemma 18.7. *Suppose $\{\psi_j\}_{j=1}^{\infty}$ is bounded in $H_0^1(\Omega)$. Then $\{\psi_j\}_{j=1}^{\infty}$ has a subsequence which converges in $L^2(\Omega)$.*

We now prove Lemma 18.6 using Rellich's lemma, which we in turn prove later.

Proof. Since $\langle A\phi, \phi \rangle = \|\phi\|^2_{H_0^1(\Omega)}$, we have $\dfrac{\|\psi_j\|^2_{H_0^1}}{\|\psi_j\|^2_{L^2}} = \|\psi_j\|^2_{H_0^1} \to \lambda_1$ as $j \to \infty$, so in particular $\{\psi_j\}_{j=1}^{\infty}$ is bounded in $H_0^1(\Omega)$, and thus we have an L^2-convergent subsequence, $\{\psi_{j_k}\}_{k=1}^{\infty}$, converging to some $\phi \in L^2(\Omega)$; in particular $\|\phi\|_{L^2} = 1$ since $\|\psi_j\|_{L^2} = 1$. Let $\chi_k = \psi_{j_k}$ to simplify the notation. Now we use the parallelogram law as in the proof of Lemma 17.15 to show that $\{\chi_k\}_{k=1}^{\infty}$ is Cauchy in H_0^1. Namely

$$\|\chi_k - \chi_m\|_{H_0^1}^2 + \|\chi_k + \chi_m\|_{H_0^1}^2 = 2(\|\chi_k\|_{H_0^1}^2 + \|\chi_m\|_{H_0^1}^2),$$

so

$$\|\chi_k - \chi_m\|_{H_0^1}^2 = 2(\|\chi_k\|_{H_0^1}^2 + \|\chi_m\|_{H_0^1}^2) - 4\left\|\frac{\chi_k + \chi_m}{2}\right\|_{H_0^1}^2.$$

Note that $\chi_k \to \phi$ in L^2 and thus $\frac{\chi_k + \chi_m}{2} \to \phi$ as $k, m \to \infty$, so $\left\|\frac{\chi_k + \chi_m}{2}\right\|_{L^2} \to \|\phi\|_{L^2} = 1$. Further,

$$\left\langle A\frac{\chi_k + \chi_m}{2}, \frac{\chi_k + \chi_m}{2}\right\rangle = \mathcal{Q}\left(\frac{\chi_k + \chi_m}{2}\right)\left\|\frac{\chi_k + \chi_m}{2}\right\|_{L^2}^2 \geq \lambda_1\left\|\frac{\chi_k + \chi_m}{2}\right\|_{L^2}^2.$$

Thus,

$$\|\chi_k - \chi_m\|_{H_0^1}^2 \leq 2\|\chi_k\|_{H_0^1}^2 + 2\|\chi_m\|_{H_0^1}^2 - 4\lambda_1\left\|\frac{\chi_k + \chi_m}{2}\right\|_{L^2}^2,$$

with the first two terms each tending to $2\lambda_1$, while the last one is tending to $4\lambda_1$. Then $\|\chi_k - \chi_m\|_{H_0^1} \to 0$ as $k, m \to \infty$, which proves that $\{\chi_k\}_{k=1}^{\infty}$ is Cauchy in H_0^1, and thus converges to some $\chi \in H_0^1$. Thus in L^2. But $\chi_k \to \phi$ in L^2, so $\chi = \phi$, and thus $\chi_k \to \phi$ in $H_0^1(\Omega)$, completing the proof. $\qquad\square$

Taking into account Lemma 18.6 and the arguments preceding it, we have proved Proposition 18.5.

In fact, one can now proceed further, much as in the finite dimensional case. Thus, consider

$$\lambda_2 = \inf\{\mathcal{Q}(\phi): \ \|\phi\|_{L^2} = 1, \ \langle\phi, \phi_1\rangle = 0, \ \phi \in H_0^1(\Omega)\}.$$

Notice that $\lambda_2 \geq \lambda_1$, since the infimum is taken over a smaller set. Again, we may take a sequence, $\psi_j \in H_0^1(\Omega)$, $\|\psi_j\|_{L^2} = 1$, $\langle\psi_j, \phi_1\rangle = 0$, with $\mathcal{Q}(\psi_j) \to \lambda_2$. The analogue of Lemma 18.6 will still hold, and thus we obtain ϕ_2 with $\mathcal{Q}(\phi_2) = \lambda_2$, $\langle\phi_2, \phi_1\rangle = 0$. Then the above argument with $\mathcal{Q}(\phi_2 + t\psi)$, $\psi \in H_0^1(\Omega)$, $\langle\psi, \phi_1\rangle = 0$ still works and yields that

$$\langle(A - \lambda_2 I)\phi_2, \psi\rangle = 0, \ \psi \in H_0^1(\Omega), \ \langle\psi, \phi_1\rangle = 0.$$

This does not directly give $(A - \lambda_2 I)\phi_2 = 0$ because of the orthogonality condition $\langle\psi, \phi_1\rangle = 0$. However,

$$\begin{aligned}\langle(A - \lambda_2 I)\phi_2, \phi_1\rangle &= \langle\phi_2, (A - \lambda_2 I)\phi_1\rangle \\ &= \langle\phi_2, (\lambda_1 - \lambda_2)\phi_1\rangle = (\lambda_1 - \lambda_2)\langle\phi_2, \phi_1\rangle = 0.\end{aligned}$$

As for any $\psi \in H_0^1(\Omega)$ we may write $\psi = \psi_\| + \psi_\perp$ with the orthogonal decomposition relative to ϕ_1 using the L^2-inner product (recall that this does *not* require completeness since span$\{\phi_1\}$ is finite dimensional, cf. Chapter 13, which is important as $H_0^1(\Omega)$ is *not* complete when equipped with the L^2-norm). We conclude that in fact

$$\langle (A - \lambda_2 I)\phi_2, \psi \rangle = 0, \ \psi \in H_0^1(\Omega),$$

so

$$A\phi_2 = \lambda_2 \phi_2.$$

One can proceed in a similar fashion and construct a monotone increasing sequence $\{\lambda_k\}_{k=1}^\infty$ of reals and $\phi_k \in H_0^1(\Omega)$ with $\|\phi_k\|_{L^2} = 1$ and $\langle \phi_j, \phi_k \rangle = 0$ for $j \neq k$ and $A\phi_k = \lambda_k \phi_k$.

We still need to prove Rellich's lemma. One way to prove this is to start with the analogous result on 2π-periodic functions, i.e. functions on the circle.

Lemma 18.8. *If $\{\psi_j\}_{j=1}^\infty$ is an H^1-bounded sequence on $\mathbb{S}^1$, then it has an L^2-convergent subsequence.*

Proof. Suppose $\|\psi_j\|_{H^1}^2 \leq C^2$ for all j. It is easiest to work with the Fourier coefficients $\mathcal{F}\psi_j$ of ψ_j. Note that

$$\|\mathcal{F}\psi_j\|_{L^2}^2 = \sum_{n \in \mathbb{Z}} |(\mathcal{F}\psi_j)_n|^2$$

and

$$\|\mathcal{F}\psi_j\|_{H^1}^2 = \|\mathcal{F}\psi_j\|_{L^2}^2 + \|\mathcal{F}\psi_j'\|_{L^2}^2 = \|\mathcal{F}\psi_j\|_{L^2}^2 + \|\{in(\mathcal{F}\psi_j)_n\}_{n \in \mathbb{Z}}\|_{\ell^2}^2$$
$$= \sum_{n \in \mathbb{Z}} (1 + n^2)|(\mathcal{F}\psi_j)_n|^2.$$

Notice that as $\|\mathcal{F}\psi_j\|_{H^1}^2 \leq C^2$ for all j, the 'tail' of the series $\{(\mathcal{F}\psi_j)_n\}_{n \in \mathbb{Z}}$, i.e. the entries with $|n| > N$, where N is large, is small in L^2 for N sufficiently large. More precisely,

$$\sum_{n>N} (1 + n^2)|(\mathcal{F}\psi_j)_n|^2 \geq (1 + N^2) \sum_{n>N} |(\mathcal{F}\psi_j)_n|^2,$$

so

$$\sum_{n>N} |(\mathcal{F}\psi_j)_n|^2 \leq (1 + N^2)^{-1} C^2,$$

and thus they can be arranged to be $< \epsilon^2$ by choosing N sufficiently large (independent of j); i.e. given $\epsilon > 0$ there is N such that

$$(18.1) \qquad\qquad \sum_{n>N} |(\mathcal{F}\psi_j)_n|^2 < \epsilon^2$$

for all j. *This* is essentially the reason the result holds; while n takes infinitely many values, only finitely many are sizeable.

Now, for each n, $\{(\mathcal{F}\psi_j)_n\}_{j=1}^{\infty}$ is a bounded sequence of complex numbers since $|(\mathcal{F}\psi_j)_n|^2 \leq \|\mathcal{F}\psi_j\|_{H^1}^2 \leq C^2$, so by the compactness of closed bounded subsets of $\mathbb{C} = \mathbb{R}^2$, this has a convergent subsequence. Cantor's diagonal argument (see Problem 18.5) allows one to find a subsequence $\{\psi_{j_k}\}_{k=1}^{\infty}$ such that $\{(\mathcal{F}\psi_{j_k})_n\}_{k=1}^{\infty}$ converges for all n to some c_n. We claim that then in fact $\sum(1+n^2)|c_n|^2 \leq C^2$, and with $\phi = \mathcal{F}^{-1}\{c_n\}_{n\in\mathbb{Z}}$, $\psi_{j_k} \to \phi$ in L^2 as $k \to \infty$. To see

$$(18.2) \qquad \sum(1+n^2)|c_n|^2 \leq C^2,$$

note that it suffices to show that for each N, $\sum_{|n|\leq N}(1+n^2)|c_n|^2 \leq C^2$. But this follows from $\sum(1+n^2)|(\mathcal{F}\psi_{j_k})_n|^2 \leq C^2$, so $\sum_{|n|\leq N}(1+n^2)|(\mathcal{F}\psi_{j_k})_n|^2 \leq C^2$, and now with the sum being finite, we can simply take limits. Then $\psi_{j_k} \to \phi$ in L^2 as $k \to \infty$ follows from the convergence of $(\mathcal{F}\psi_{j_k})_n$ to c_n for each n, and fixing any $\epsilon > 0$, with N as in (18.1), $\sum_{|n|>N}|c_n|^2 < \epsilon^2$ holds (due to (18.2)) in addition to (18.1). Thus by the triangle inequality

$$\sqrt{\sum_{|n|>N}|(\mathcal{F}\psi_j)_n - c_n|^2} \leq \sqrt{\sum_{|n|>N}|(\mathcal{F}\psi_j)_n|^2} + \sqrt{\sum_{|n|>N}|c_n|^2} < 2\epsilon,$$

while

$$\sqrt{\sum_{|n|\leq N}|(\mathcal{F}\psi_j)_n - c_n|^2} \to 0$$

as $j \to \infty$ since for each n, $|(\mathcal{F}\psi_j)_n - c_n| \to 0$ and the sum is finite. So for sufficiently large j,

$$\sum_{n\in\mathbb{Z}}|(\mathcal{F}\psi_j)_n - c_n|^2 = \sum_{|n|\leq N}|(\mathcal{F}\psi_j)_n - c_n|^2 + \sum_{|n|>N}|(\mathcal{F}\psi_j)_n - c_n|^2 < 5\epsilon^2,$$

which is exactly the statement that $\psi_j \to \mathcal{F}^{-1}\{c_n\}_{n\in\mathbb{Z}}$ in L^2. $\square$

The same argument, using Fourier series in n-variables (see Problem 14.6 for the $n = 2$ case; the general case is similar), gives the higher dimensional analogue, i.e. that

Lemma 18.9. *If $\{\psi_j\}_{j=1}^{\infty}$ is an H_0^1-bounded sequence on $\mathbb{T}^n = \mathbb{S}^1 \times \ldots \times \mathbb{S}^1$ (an n-fold product), then it has an L^2-convergent subsequence.*

Now Rellich's lemma is easy to prove.

Proof of Lemma 18.7. Suppose ψ_j is a sequence in $H_0^1(\Omega)$. By the definition of $H_0^1(\Omega)$ as a completion, there exists a sequence ϕ_j with elements in $\mathcal{C}_c^{\infty}(\Omega)$ such that $\|\psi_j - \phi_j\|_{H_0^1} < 1/j$. Consider a large cube $[-R, R]^n$

with $\Omega \subset [-R/2, R/2]^n$. Then, via extension by 0, we can regard ϕ_j as a $2R$-periodic (in each variable) function on $[-R, R]^n$, i.e. on a large torus. Then by Lemma 18.9 $\{\phi_j\}_{j=1}^{\infty}$ has a subsequence $\{\phi_{j_k}\}_{k=1}^{\infty}$ converging to some ϕ in $L^2([-R, R]^n)$. But all the ϕ_j vanish outside Ω, and thus so does ϕ, and we conclude that $\phi_{j_k} \to \phi$ in $L^2(\Omega)$, and thus $\psi_{j_k} \to \phi$ in $L^2(\Omega)$ as well, completing the proof. $\square$

We can now in fact prove Theorem 13.22.

Theorem 18.10. *With λ_j, ϕ_j as above, $\lambda_j \to \infty$ as $j \to \infty$ and $\{\phi_j\}_{j=1}^{\infty}$ is a complete orthonormal set in L^2.*

Proof. Since the λ_j are monotone increasing, if they do not tend to infinity, they are bounded, say $\lambda_j \leq \Lambda$ for all j. But then $\|\phi_j\|_{H_0^1}^2 = \lambda_j \|\phi_j\|_{L^2}^2 = \lambda_j \leq \Lambda$, for all j, so by Rellich's theorem a subsequence of $\{\phi_j\}_{j=1}^{\infty}$ converges in L^2 to some $\phi \in L^2$. But then the subsequence is Cauchy, which is impossible since with the ϕ_j being orthogonal to each other, we have $\|\phi_j - \phi_k\|_{L^2}^2 = \|\phi_j\|_{L^2}^2 + \|\phi_k\|_{L^2}^2 = 2$ for all $j \neq k$. So the eigenvalues are not bounded, proving the first claim.

To see the second claim, notice that we only need to prove the L^2-convergence to ϕ when $\phi \in H_0^1(\Omega)$, or indeed simply $\phi \in \mathcal{C}_c^{\infty}(\Omega)$, much as in the case of the convergence of the Fourier series in Chapter 14. Indeed, for every element $\phi \in L^2$ there is a sequence $\phi_n \in \mathcal{C}_c^{\infty}(\Omega)$ converging to ϕ in L^2. If T is the generalized Fourier series operator, with

$$T\psi = \sum_{n=1}^{\infty} \langle \psi, \phi_n \rangle \phi_n$$

(recall that $\|\phi_n\|_{L^2} = 1$!), then $\|T\psi\| \leq \|\psi\|$ by Bessel's inequality. If we know the L^2-convergence of the generalized Fourier series to ψ when $\psi \in \mathcal{C}_c^{\infty}(\Omega)$, i.e. that $T\psi = \psi$ in this case, then $T\phi_n = \phi_n$, so taking the limit $n \to \infty$ we conclude that $T\phi = \phi$.

So it suffices to consider $\phi \in H_0^1(\Omega)$. First, note that by the very definition of λ_n, for $\psi \in (\text{span}\{\phi_1, \ldots, \phi_{n-1}\})^{\perp}$, $\psi \in H_0^1(\Omega)$, $\psi \neq 0$, we have $\mathcal{Q}(\psi) \geq \lambda_n$. But now the orthogonal projection of ϕ to $(\text{span}\{\phi_1, \ldots, \phi_{n-1}\})^{\perp}$ is $\psi = \phi - \sum_{j=1}^{n-1} \langle \phi, \phi_j \rangle_{L^2} \phi_j$, so $\mathcal{Q}(\psi) \geq \lambda_n$ gives

$$(18.3) \qquad \left\| \phi - \sum_{j=1}^{n-1} \langle \phi, \phi_j \rangle_{L^2} \phi_j \right\|_{H_0^1}^2 \geq \lambda_n \left\| \phi - \sum_{j=1}^{n-1} \langle \phi, \phi_j \rangle_{L^2} \phi_j \right\|_{L^2}^2.$$

But

$$\left\| \phi - \sum_{j=1}^{n-1} \langle \phi, \phi_j \rangle_{L^2} \phi_j \right\|_{H_0^1}^2$$

$$= \left\langle A\Big(\phi - \sum_{j=1}^{n-1} \langle \phi, \phi_j \rangle_{L^2} \phi_j \Big), \phi - \sum_{k=1}^{n-1} \langle \phi, \phi_k \rangle_{L^2} \phi_k \right\rangle_{L^2}$$

$$= \langle A\phi, \phi \rangle_{L^2} - \left\langle A\phi, \sum_{k=1}^{n-1} \langle \phi, \phi_k \rangle_{L^2} \phi_k \right\rangle_{L^2}$$

$$- \left\langle A \sum_{j=1}^{n-1} \langle \phi, \phi_j \rangle_{L^2} \phi_j, \phi \right\rangle_{L^2} + \sum_{j=1}^{n-1} \sum_{k=1}^{n-1} \left\langle A\langle \phi, \phi_j \rangle_{L^2} \phi_j, \langle \phi, \phi_k \rangle_{L^2} \phi_k \right\rangle$$

$$= \|\phi\|_{H_0^1}^2 - \sum_{k=1}^{n-1} \overline{\langle \phi, \phi_k \rangle_{L^2}} \langle \phi, A\phi_k \rangle_{L^2}$$

$$- \sum_{j=1}^{n-1} \langle \phi, \phi_j \rangle_{L^2} \langle A\phi_j, \phi \rangle_{L^2} + \sum_{j=1}^{n-1} \sum_{k=1}^{n-1} \langle \phi, \phi_j \rangle_{L^2} \overline{\langle \phi, \phi_k \rangle_{L^2}} \langle A\phi_j, \phi_k \rangle_{L^2}.$$

We now use $A\phi_j = \lambda_j \phi_j$, so this expression is equal to

$$\|\phi\|_{H_0^1}^2 - \sum_{k=1}^{n-1} \lambda_k \overline{\langle \phi, \phi_k \rangle_{L^2}} \langle \phi, \phi_k \rangle_{L^2} - \sum_{j=1}^{n-1} \lambda_j \langle \phi, \phi_j \rangle_{L^2} \langle \phi_j, \phi \rangle_{L^2}$$

$$+ \sum_{j=1}^{n-1} \sum_{k=1}^{n-1} \lambda_j \langle \phi, \phi_j \rangle_{L^2} \overline{\langle \phi, \phi_k \rangle_{L^2}} \langle \phi_j, \phi_k \rangle_{L^2}.$$

As the ϕ_j are L^2-orthonormal, the last term is $\sum_{j=1}^{n-1} \lambda_j \langle \phi, \phi_j \rangle_{L^2} \overline{\langle \phi, \phi_j \rangle_{L^2}}$, and the preceding two terms are also equal to the very same quantity. Thus, we conclude that

$$\left\| \phi - \sum_{j=1}^{n-1} \langle \phi, \phi_j \rangle_{L^2} \phi_j \right\|_{H_0^1}^2 = \|\phi\|_{H_0^1}^2 - \sum_{j=1}^{n-1} \lambda_j |\langle \phi, \phi_j \rangle_{L^2}|^2.$$

Then by (18.3),

$$\|\phi\|_{H_0^1}^2 \geq \|\phi\|_{H_0^1}^2 - \sum_{j=1}^{n-1} \lambda_j |\langle \phi, \phi_j \rangle_{L^2}|^2 \geq \lambda_n \Big\| \phi - \sum_{j=1}^{n-1} \langle \phi, \phi_j \rangle_{L^2} \phi_j \Big\|_{L^2}^2,$$

so as $\lambda_n \to \infty$ as $n \to \infty$, we conclude that $\|\phi - \sum_{j=1}^{n-1} \langle \phi, \phi_j \rangle_{L^2} \phi_j\|_{L^2}^2 \to 0$ as $n \to \infty$, completing the proof of the completeness. $\qquad\square$

Note that while this theorem implies the completeness of the Fourier basis, we used the completeness of the Fourier basis in its proof. Thus, it does *not* give a new proof of the completeness of the latter.

Problems

Problem 18.1. Show that the jth eigenvalue λ_j of the operator A can be calculated as follows: for any j linearly independent functions $\psi_1, \ldots, \psi_j$, consider

$$E(\psi_1, \ldots, \psi_j) = \sup_{c_1, \ldots, c_j} \mathcal{Q}\left(\sum_{k=1}^{j} c_k \psi_k\right),$$

where the supremum is taken over $c_1, \ldots, c_j \in \mathbb{C}$, not all 0. Then consider $\mu_j = \inf_{\psi_1, \ldots, \psi_j} E(\psi_1, \ldots, \psi_j)$, where the infimum is taken over all linearly independent choices $\psi_1, \ldots, \psi_j$. Show that $\lambda_j = \mu_j$ and the minimizing choice is $\psi_j = \phi_j$. *This is the min-max characterization of the eigenvalues.*

Hint: Since one possible choice is $\psi_k = \phi_k$ for all k, it should be easy to show that $\mu_j \leq \lambda_j$. So it suffices to show that for all choices ψ_k, $E(\psi_1, \ldots, \psi_j) \geq \lambda_j$, for that implies $\mu_j \geq \lambda_j$.

Problem 18.2. Suppose that Ω_k, $k = 1, 2$, are two domains in $\mathbb{R}^n$ and $\Omega_1 \subset \Omega_2$. Let $\lambda_{k,j}$, $j = 1, 2, \ldots$, be the eigenvalues of $-\Delta$ on Ω_j with Dirichlet boundary conditions. Show that $\lambda_{1,j} \geq \lambda_{2,j}$ for all j.

Hint: If $\psi_1 \in H_0^1(\Omega_1)$, consider its extension by 0 to Ω_2, ψ_2, and compare $\mathcal{Q}_1(\psi_1)$ and $\mathcal{Q}_2(\psi_2)$, where $\mathcal{Q}_k$ is the Rayleigh quotient for Ω_k. Now use Problem 18.1.

Problem 18.3. Let Ω be a domain in $\mathbb{R}^2$, and let λ_j be the eigenvalues of $-\Delta$ with Dirichlet boundary conditions. Show that there exist positive constants C_1, C_2 such that $C_1 j \leq \lambda_j \leq C_2 j$. (The importance is the behavior for j large.) This is a weak version of *Weyl's law.*

Hint: Give an estimate for the eigenvalues of the Laplacian on squares contained in Ω and containing Ω.

Problem 18.4. Consider the setting of Problem 17.9, namely the following operator on the 2-torus $\mathbb{T}^2 = \mathbb{S}_x^1 \times \mathbb{S}_y^1$:

$$A = -\nabla \cdot c(x, y)^2 \nabla + q(x, y),$$

$c, q > 0$. Recall that functions on $\mathbb{T}^2$ can be considered 2π-periodic on $\mathbb{R}^2$ in both the x and the y variables. Define $L^2(\mathbb{T}^2)$ to be the completion of $C^0(\mathbb{T}^2)$ in the inner product given in Problem 13.5 and $H^1(\mathbb{T}^2)$ to be the completion of $C^2(\mathbb{T}^2)$ with the norm

$$\|u\|_{H^1(\mathbb{T}^2)}^2 = \|u\|_{L^2(\mathbb{T}^2)}^2 + \|c D_x u\|_{L^2(\mathbb{T}^2)}^2 + \|c D_y u\|_{L^2(\mathbb{T}^2)}^2.$$

Show that the operator A is symmetric, its eigenvalues λ_j tend to $+\infty$, and its eigenfunctions form a complete orthonormal set in L^2.

Problem 18.5 (Cantor's diagonal argument). Show that if for each j, $a^{(j)}$ is a sequence, $\{a_n^{(j)}\}_{n \in \mathbb{Z}}$, and for each n, $\{a_n^{(j)}\}_{j=1}^{\infty}$ is a bounded sequence of complex numbers, then there is a subsequence $\{a^{(j_k)}\}_{k=1}^{\infty}$ such that for all n, $\{a_n^{(j_k)}\}_{k=1}^{\infty}$ converges to some $c_n \in \mathbb{C}$.

Hint: For each $n \in \mathbb{Z}$ one can certainly arrange the convergence of $\{a_n^{(j_k)}\}_{k=1}^{\infty}$ by the compactness of $\{z \in \mathbb{C} : |z| \leq C\}$, but one expects that the subsequence j_k depends on n, i.e. it is of the form $j_k^{(n)}$. If we had only finitely many values of n, say $1, \ldots, N$, we could arrange a subsequence $\{a^{(j_k^{(1)})}\}_{k=1}^{\infty}$ such that $\{a_n^{(j_k^{(1)})}\}_{k=1}^{\infty}$ converges for $n = 1$ and then take a subsequence of this, $\{a^{(j_k^{(2)})}\}_{k=1}^{\infty}$, such that $\{a_n^{(j_k^{(2)})}\}_{k=1}^{\infty}$ converges for $n = 2$; but as $\{a^{(j_k^{(2)})}\}_{k=1}^{\infty}$ is a subsequence of $\{a^{(j_k^{(1)})}\}_{k=1}^{\infty}$, $\{a_n^{(j_k^{(2)})}\}_{k=1}^{\infty}$ converges for $n = 1$ still. Proceed inductively; one is done in N steps. If one has countably many n, say $n \in \mathbb{N}$, then one acquires in the same way a sequence $\{a^{(j_k^{(n)})}\}_{k=1}^{\infty}$ for every n. Then $\{a^{(j_n^{(n)})}\}_{n=1}^{\infty}$ is the desired 'diagonal' (because $k = n$) subsequence.

Bibliography

[1] Lawrence C. Evans, *Partial differential equations*, Graduate Studies in Mathematics, Vol. 19, American Mathematical Society, Providence, RI, second edition, 2010.

[2] Gerald B. Folland, *Introduction to partial differential equations*, Princeton University Press, Princeton, NJ, second edition, 1995.

[3] Richard Johnsonbaugh and W. E. Pfaffenberger, *Foundations of mathematical analysis*, Dover Publications, Inc., Mineola, NY, 2002; Corrected reprint of the 1981 original [Dekker, New York; MR0599741 (82a:26001)].

[4] Leon Simon, *An introduction to multivariable mathematics*, Morgan & Claypool, 2008.

[5] Elias M. Stein and Rami Shakarchi, *Real analysis: Measure theory, integration, and Hilbert spaces*, Princeton Lectures in Analysis, III. Princeton University Press, Princeton, NJ, 2005.

[6] Walter A. Strauss, *Partial differential equations: An introduction*, John Wiley & Sons Ltd., Chichester, second edition, 2008.

[7] Michael E. Taylor. *Partial differential equations: Basic theory*, Texts in Applied Mathematics, Vol. 23, Springer-Verlag, New York, 1996.

Index

图字：01-2020-6628号

偏微分方程：理论和应用

Pianweifen Fangcheng: Lilun he
Yingyong

图书在版编目 (CIP) 数据

偏微分方程：理论和应用 = Partial Differential
Equations: An Accessible Route through Theory and
Applications : 英文 / (美) 安德拉斯·瓦西
(András Vasy) 著. -- 影印本. -- 北京 : 高等教育
出版社, 2021.2
ISBN 978-7-04-055651-3

Ⅰ.①偏… Ⅱ.①安… Ⅲ.①偏微分方程—英文
Ⅳ.①O175.2

中国版本图书馆CIP 数据核字(2021) 第020840 号

策划编辑　李华英　　　责任编辑　李华英
封面设计　张申申　　　责任印制　朱　琦

出版发行 高等教育出版社　　　开本　787mm×960mm　1/16
社址 北京市西城区德外大街4号　印张　18.75
邮政编码 100120　　　　　　　字数　500千字
购书热线 010-58581118　　　　版次　2021 年 2 月第 1 版
咨询电话 400-810-0598　　　　印次　2021 年 2 月第 1 次印刷
网址 http://www.hep.edu.cn　　定价　135.00 元
http://www.hep.com.cn
网上订购 http://www.hepmall.com.cn　本书如有缺页、倒页、脱页等质量问题，
http://www.hepmall.com　　　　请到所购图书销售部门联系调换
http://www.hepmall.cn　　　　　版权所有　侵权必究
印刷 保定市中画美凯印刷有限公司　[物 料 号 55651-00]